PUBLIC SCIENCE IN LIBERAL DEMOCRACY

Edited by Jene M. Porter and Peter W.B. Phillips

Regardless of whether science is practised in industry, the academy, or government, its conduct inescapably shapes and is shaped by democratic institutions. Moreover, the involvement of science with public policy formation and democracy has dramatically increased over the centuries and, by all accounts, will continue to do so. In order to understand the functioning of science and democracy, it is necessary to acknowledge the complex relationship between them. *Public Science in Liberal Democracy* aims to do this from an interdisciplinary perspective, presenting an array of substantively different positions on the issues that it explores.

The volume focuses on three major questions: Can science retain independence and objectivity in the face of demands to meet commercial and public policy objectives? In what ways is scientific discourse privileged in the formation of public policy? How can scientific knowledge and methodology be made compatible with the interdisciplinarity and integration required of public policy formation and discourse? Representing a wide range of viewpoints, the contributors to *Public Science in Liberal Democracy* come from Canada, Europe, the United States, and Australia, and include practising scientists as well as scholars working in the humanities and social sciences. This timely and thought-provoking collection makes an important contribution to the literature and will appeal to anyone interested in scientific research and its political and philosophical ramifications in democratic society.

JENE M. PORTER is a professor emeritus in the Department of Political Studies at the University of Saskatchewan.

PETER W.B. PHILLIPS is a professor in the Department of Political Studies at the University of Saskatchewan.

Edited by JENE M. PORTER and
PETER W.B. PHILLIPS

Public Science in Liberal Democracy

UNIVERSITY OF TORONTO PRESS
Toronto Buffalo London

Toronto Buffalo London
Printed in Canada

ISBN 978-0-8020-9359-2

Printed on acid-free paper

Library and Archives Canada Cataloguing in Publication

Public science in liberal democracy : the challenge to science and democracy / edited by Jene M. Porter and Peter W.B. Phillips.

Essays presented at a conference held in Saskatoon, Sask., Oct. 14–16, 2004.
ISBN 978-0-8020-9359-2

1. Democracy and science – Congresses. 2. Science and state – Congresses. 3. Research – Government policy – Congresses. I. Porter, J.M. (Jene M.), 1937– II. Phillips, Peter W.B.

Q175.4.P83 2007 303.48′3 C2007-903910-3

University of Toronto Press acknowledges the financial assistance to its publishing program of the Canada Council for the Arts and the Ontario Arts Council.

University of Toronto Press acknowledges the financial support for its publishing activities of the Government of Canada through the Book Publishing Industry Development Program (BPIDP).

Contents

Preface

In 2004 Canada's national facility for synchrotron light research (Canadian Light Source [CLS]) began operation at the University of Saskatchewan. The University-owned CLS is one of the largest scientific projects in Canadian history. It represents an extraordinary expenditure of government money; housed at a public university, it is sustained by research funded by industry, government, nongovernmental organizations, and universities. Its opening thus provided an opportunity to reflect upon the complicated and relatively unexplored relationship between public science and democracy.

Regardless of whether science is practised in industry, the academy, or government, its conduct inescapably shapes and is shaped by democratic institutions. Moreover, the involvement of science with public policy formation and democratic institutions has dramatically increased over the centuries. What does this complex relationship entail for the practice of science or of democracy? Reflection upon such a broad question is clearly outside the scope of any one discipline. Indeed, the very practice of science in different sectors of society is hard to uncover. The question cannot be limited to one nation-state but should be asked in all democratic systems. In short, people from a variety of academic disciplines and from all sectors in which science is practised are needed to address this question.

The essays in this volume were first presented at an international conference in October 2004 which was sponsored by the University of Saskatchewan and designed to begin the process of reflecting upon the complex relationship of science and democracy. Accordingly, the participants were from Canada, Europe, the United States, and Australia and included scientists from academia, industry, and governments, as well as academics from several disciplines in the humanities and social sciences.

The editors are grateful for the help provided by the Organizing Committee: Glen Caldwell, Ron Steer, Wilf Keller, Larry Stewart, Alan McHughen, Zaheer Baber, George Khatachatourians, Michael Atkinson, and Jeffrey Steeves. The Department of Political Studies provided necessary support.

The editors would especially like to thank our sponsors, starting with President Peter McKinnon of the University of Saskatchewan, whose support was foundational. We would also like to acknowledge financial support from the Colleges of Medicine and Arts and Science, Western Economic Diversification, Genome Prairie, Genome Canada, the Canadian Light Source, and the City of Saskatoon.

Jene Porter
Bella Vista, Arkansas

Peter W.B. Phillips
Saskatoon, Canada

Abbreviations

AGM	Annual General Meeting (CGIAR)
CGIAR	Consultative Group on International Agricultural Research
CIAT	Centro Internacional de Agricultura Tropical (CGIAR)
CIFOR	Center for International Forestry Research (CGIAR)
CIMMYT	Centro Internacional de Mejoramiento de Maïz y Trigo (CGIAR)
CIP	Centro Internacional de la Papa (CGIAR)
ESSD	Environmentally and Socially Sustainable Development Network (World Bank)
ExCo	Executive Council (CGIAR)
FAO	Food and Agriculture Organization of the United Nations
GFAR	Global Forum for Agricultural Research (CGIAR)
GM	Genetically modified
GMOs	Genetically modified organisms
ICARDA	International Center for Agricultural Research in the Dry Areas (CGIAR)
ICLARM	International Center for Living Aquatic Resources Management (CGIAR)
ICRAF	International Center for Research in Agroforestry (CGIAR)
ICRISAT	International Crops Research Institute for the Semi-Arid Tropics (CGIAR)
ICTs	Information and Communications Technologies
IFPRI	International Food Policy Research Institute (CGIAR)
IITA	International Institute of Tropical Agriculture (CGIAR)
ILCA	International Livestock Center for Africa (CGIAR)

ILRAD	International Laboratory for Research on Animal Diseases (CGIAR)
ILRI	International Livestock Research Institute (CGIAR)
IPGRI	International Plant Genetic Resources Institute (CGIAR)
IPR	Intellectual property right
IRRI	International Rice Research Institute (CGIAR)
ISNAR	International Service for National Agricultural Research (CGIAR)
ITPGR	International Treaty on Plant Genetic Resources for Food and Agriculture
IWMI	International Water Management Institute (CGIAR)
NARS	National agricultural research systems
NGO	Nongovernmental organization
NRM	Natural resource management
OED	Operations Evaluation Department (World Bank)
SGRP	System-wide Genetic Resources Programme
TAC	Technical Advisory Committee (CGIAR)
TRIPS	Trade-Related Aspects of Intellectual Property Rights (WTO)
TSR	Third System Review (CGIAR)
WTO	World Trade Organization
WARDA	West Africa Rice Development Association (CGIAR)

PUBLIC SCIENCE IN LIBERAL DEMOCRACY

Introduction
The History, Philosophy, and Practice of Public Science

JENE PORTER AND PETER W.B. PHILLIPS

Although the history of science, the sociology of science, and the philosophy of science are well-established fields of inquiry, the concept of 'public science' is relatively new. In the eighteenth century, public policy issues were dominated by questions of constitutional framing; in the nineteenth century, constitutional questions were joined by the issue of the proper role of government in the economy. It was not until the latter part of the twentieth century that science became a dominant player in the formation of public policy questions.

Science and its products have always had a public good character to them. Whether contributing to the conduct of war or the practice of medicine, scientific knowledge eventually becomes available to those who have had no hand in producing it. Controversies abound in the history of science, but toward the end of the twentieth century the generation, application, and diffusion of scientific knowledge became matters of intense public concern in Western democracies.

The nuclear debates of the 1950s illustrate the changing dialogue. Science, much of it produced by public funds, began forcing policymakers to undertake unprecedented calculations regarding risk and benefit. Regulatory challenges now taxed the knowledge of the state, and public interest groups began to form in response to perceived dangers. Science had become a public issue in the sense that its conduct and direction were subject to political pressures as never before. Also, as the century concluded, a host of new public science issues appeared on the policy agenda: global warming, stem cell research, reproductive technologies, and genetically modified foods. Moreover, a great many previously straightforward policy issues were now viewed as requiring a scientific analysis and defence.

Not only has bureaucratic and political debate been changed by the addition of a strong scientific component, but public discourse is now subject to the participation of scientists themselves. Public demand for more scientific research has been joined by an even more vociferous demand for justifications. These demands have drawn scientists into public discussions and have begun to sharpen differences in the ways in which scientists and the lay public evaluate scientific research agendas. Scientific education and the epistemology followed by scientists are not particularly compatible with public discourse. It makes little sense, for example, to call for negotiation and compromise in a chemistry or mathematics class. The personal or ethical stake of the individual in the discussion is irrelevant to the scientific knowledge claim.

Yet, public discourse requires compromise, negotiation, and sensitivity to the positions and stakes of the participants. Aristotle's dictum is apt: politics is the realm of the contingent and the probable, the realm of action rather than simply understanding or contemplation. The interconnectedness of public policy issues with human well-being means that scientific knowledge will remain a necessary element in modern public policy formation. But the tensions this produces cannot be glossed over. Scientists are recruited as privileged contributors to the substantive policy discussions, but the democratic nature of public policy formation requires scientists to take part in the discourse as only one of many participants.

Three Questions

The history of the complex relationship between science and democracy shows an increasing involvement of science in political life. This developing interrelationship puts in jeopardy the very features associated with the conduct of science that buttress its knowledge claims: autonomy, impartiality, independence, and objectivity. Public science clearly presents a severe challenge to science. Conversely, such features as impartiality and objectivity provide a status to scientists that separates them from others in the political process. Yet, democratic politics is hardly possible if some participants' status and knowledge claims are immune to negotiation and compromise, which are defining features of democratic politics. Once political issues are metamorphosed into scientific questions, the political process no longer is the locus of democratic politics. Public science thus presents a profound challenge to democracy. Lastly, these two challenges suggest a re-examination of the philosophic foundations of science, the practices of science, and the obligations of an educated citizenry. Is it possible, for example, for science to be so conceived that it contributes to the give and take of politics? Is it possible to have an educated citizenry sufficiently knowledgable about science that political discourse can be strengthened by scientific contributions? In sum, there is a pressing need to reflect philosophically on the compatibility of science and democracy.

The essays in this book address three key, cross-cutting questions. First, the chapters ask whether science can retain independence and objectivity in the face of demands to meet commercial and public policy objectives. In the twentieth century, it was possible to answer in the affirmative by taking refuge philosophically in the positivist position that facts and values are incommensurate. Scientists provided the facts; it was up to others to make the value judgments associated with using them. This position is no longer as comforting as it once was, for both philosophical and political reasons. Scientists – private, governmental, or university-based – cannot escape involvement. However, serious involvement in policy questions, because of the political character of the policy process, will give rise to doubts about the independence, objectivity, and the very competence of scientists.

Second, the chapters examine the ways in which scientific discourse is privileged in the formation of public policy. Can there be a genuine public discourse if one party is privileged? This may be an untenable arrangement. In the Middle Ages, when theologians were privileged by their special expertise and source of authority, other participants became intensely hostile. More importantly, the privileged position of theologians created obstacles to innovation. Theologians declined to participate in the give and take of discourse: they lectured down. This problem is particularly acute in polities organized around the concept of popular sovereignty. To the extent that contemporary scientists see their task as lecturing to the uninformed masses, the public policy results are likely to be suboptimal from both a scientific and a public welfare perspective.

Third, the papers investigate how scientific knowledge and scientific methodology can be made compatible with the interdisciplinarity and integration required of public policy discourse and formation. Creating closed communities of the competent has protected the reliability of objective scientific knowledge in the past. If public science requires those communities to become more open, will scientific knowledge itself become subjective and politicized?

The History of Public Science

While the concept of public science has been formally introduced only in recent years, there is ample evidence that science and the public have been connected for millennia. From the earliest days of social organization, those who control power in the system have acknowledged the potential importance of science. Tribal leaders, princes, emperors, and rulers of all types have sought to nurture, adapt, adopt, and control new science and technology to supplement or sustain their power. Sometimes this took the form of patronage and sponsorship (such as during the age of exploration and discovery, when princes funded expeditions for imperial expansion). At times it involved the expropriation and explicit ownership and control of new scientific and technological discovery (the state monopoly over nuclear technologies in much of the past seventy-five years is a good example). Other times it simply involved mercantilist support (through taxes, patents, or state monopolies) for private initiative. In this sense, many rulers saw science as an instrument of power to be exploited and not as an independent power system capable of challenging the legitimacy and power of the state.

While this concept of science as a handmaiden of the state remains in the minds of many, there has been a subtle shift in the relationship, such that the governors are now no longer in control. The advent of independent science, first in the guise of gentleman-discoverers and more recently in academia (with its academic freedoms) and in the corporate world, has loosened the hold of the rulers. The age of enlightenment and the age of reason loosened the grip of the church and state on science. Individuals and private enterprise filled that void. Academic scientists cast off their religious vestments and began in earnest to develop the 'scientific method,' which led scientists to investigate a wide range of new and potentially disruptive ideas. More recently, the market has usurped power and control over science. While much of the early for-profit science was sanctioned and supported through explicit and implicit government monopolies, increasingly private companies invested in and undertook science for their own interests. Ultimately, the internationalization of both science (through the international academies and increasingly cheap and accessible travel and communications) and industry (through foreign direct investment) further loosened the power of the state.

Now, one could argue that the state has become the handmaiden of science. It is becoming increasingly difficult for governments at any level to guide, influence, and control the science that is being done. Moreover, science in many cases challenges our understandings and relationships about the impacts of what we now do, or it develops and creates new technologies that have economic, social, cultural, and philosophical implications that demand public debate. While the state was generally happy with the evolution of the structural-materials science that underpinned the industrial revolution, the advancement of new information and communications technologies (ICTs) and biological technologies has been more contentious. While ICTs provide the analytic power that underpins the global scientific enterprise today, the resulting information and telecommunications revolution of the post-war period also undercut the power of the state and traditional power elites, precipitating a far less focused and structured public discourse. The state barely had time to realize and begin to consider the implications of this when the biological revolution occurred. The advent of new molecular technologies has challenged states to consider and decide on its appropriate use. This new science has raised serious questions about past practices (e.g., uses of chemicals), current opportunities (e.g., GM foods and reproductive technologies), and future implications (e.g., greenhouse gases and global warming). States and societies are grappling with how to address these new challenges.

Even this brief history of the relationship of science and society shows that the idea of public science will have various uses and meaning. It can mean science conducted within a government agency, within an institution or research project funded by public money, or within a business whose products and practices influence the public in some manner. Regardless of the usage, and readers will find various meanings among the essays, the term remains necessary to illuminate the evolution of science and its increasing impact on public policy formation and liberal democracy itself.

The four chapters in Section One examine the history of public science in theory and practice. Historian Larry Stewart opens the section with a historical examination of the role of the public as a witness in the scientific process. He argues that this role is so embedded in the scientific enterprise that it warrants a new 'element' entitled 'Publicum.' Stewart demonstrates through a brief look at the eighteenth and nineteenth centuries that the public provided from the earliest stages of modern science the essential trust upon which science could develop.

This early reliance on the public did clearly tie science with the public and the state. However, in Stewart's phrase, the 'modern priesthood of believers in experimental method' has evolved to where the original convergence of public and science has been increasingly strained by several factors. He concludes that while 'science increasingly scorns the hand that feeds it,' it remains true that in a liberal democracy public comprehension is possible and necessary for science.

The philosopher Carl Mitcham presents 'a historico-philosophical master narrative about relations between science, democracy, and philosophy.' By achieving some clarity about this history we can best address the crisis of public science. The classical world philosophy of Aristotle was the source of knowledge for both nature and the good and continued to be the source of authority for natural or scientific knowledge in the medieval age. In three successive stages the modern world of science and democracy was born. First, natural science replaced natural philosophy as the source of knowledge about nature. Second, revelation and theology as the source for moral knowledge was undermined by the application of science to its claims and to such texts as the Bible. Third, scientific naturalism undermined philosophy itself as the source of knowledge, in spite of efforts by Immanuel Kant and others. The result was that individuals themselves would 'openly state and adjudicate their values.' Thus, science replaces philosophy as the source for knowledge about nature, and liberal democracy replaces philosophy as the source of moral knowledge.

There are profound consequences, as Mitcham shows, for our own times, when philosophy is relegated to being the handmaid of science, of democracy, and of science–democracy relations. Mitcham, like Stewart, argues for the necessity of public involvement in the scientific enterprise. He uses the seminal works of Leon Kass and Albert Borgmann to illustrate that philosophy in the classical sense, with its concern for human well-being and the common good, can 'influence for the better both science and liberal democracy.'

The essay by Deborah Hutchinson and Richard Williams, Jr, of the U.S. Geological Survey delineates the long relationship between geology and the state. Indeed, geology has the longest record of all of the sciences of being an acknowledged public science. As the authors note, in 1787, two years prior to the adoption of the U.S. Constitution, Congress called for the mapping and classifying of public lands. The authors have provided a comprehensive historical survey of the evolution of public geoscience, beginning with the early mapping of the western land frontier and ending with the present focus on the ocean sea floor. This essay can be read as a case study illuminating the inescapable ties between the public and science and the inevitable difficulties for science and for public policy formation in a democracy. First, as the authors explain, there is the continual need for democratic support, and over the years various justifications have been used by the U.S. Geological Survey, such as economic development, national security, public safety from natural hazards, and environmental awareness. Second, the essay demonstrates that public policies and their implementation are interwoven with geoscience and with the scientists. And finally, the recognition of the consequent need to ensure the integrity and objectivity of public geoscience led Congress to implement the USGS Organic Act.

Peter Cook provides a fascinating overview of the evolution of the public–science interface from Pliny the Elder in the early years of the first millennium, up through the development of the academies in Europe in the 1600s, through the extension in the 1800s of the state into science through various operating agencies (such as the Geological Survey and the Meteorological Office in Britain), and up to today. He then offers a glimpse into the current challenges of the science–public interface with a detailed discussion of the role of new public–private–academic partnerships, such as the Cooperative Research Centres (CRC) in Australia. Cook, as chief executive of the CRC for greenhouse gas technologies, is ideally suited to examine how the system has changed and is now operating. He concludes that, in spite of increasing integration of science with other agencies, science can remain independent and objective despite commercial and policy demands, provided the interface between the actors is well defined and protected with transparent and accountable governance systems (as in the CRC system). He is less sanguine about the ability of society to control science. He acknowledges that scientists do enjoy privileged access to politicians and senior government officials, but he is not confident that science makes enough effort to assist policymakers. Most importantly, he believes that discourses of complex problems – such as greenhouse gases and climate change – include tensions between different areas of science, which tend to be portrayed in 'black and white when the reality is that there are many shades of grey.' He believes that institutions such as the Intergovernmental Panel for Climate Change (IPCC) offer a reasonable way to draw a huge community of scientists into a scientific assessment process and create a strong tie between science and policy.

Various themes, explicit and implicit, in these essays are scrutinized in later essays. Peter Cook, for example, is generally supportive of the institutional ties between scientists and policymakers as found in the IPCC. Both science and society are accommodated, he writes, although he is concerned that scientists may need to provide more leadership. Strikingly different assessments of the relation between science and political society are provided by the political scientists Grace Skogstad and Sarah Hartley and the economist Ross McKitrick. The philosophers too provide a different and less sanguine assessment of the effect of science on liberal democracy and vice versa.

Addressing the Issues: Philosophic Approaches

Our authors have tapped into a rousing philosophical debate about the interface between science and the public. This debate can be parsed into related and interconnected discussions about the foundations and conditions for liberal democracy, the nature and structure of science, and the need for links between the two solitudes.

Even though liberal democracy has been around for more than two centuries, debate continues regarding its stability, its limits, and its structure. Few, however, would question the general desirability of liberal democracy. Thus, for the authors in Section Two, a deleterious effect of science upon liberal democracy is both a theoretical and a practical issue.

The philosophy of science is rife with controversy. In the first instance, there is significant debate about the nature of the scientific method. The Mertonian norms – communism, universalism, disinterestedness, and organized scepticism – are generally accepted, but there is disagreement in and outside the science establishment about how to deal with new knowledge. Kuhn (1962) has offered a generalized picture of the process by which a science is born and undergoes change and development. According to Kuhn's model, the study of natural phenomena leads scientists to develop explanatory theories that, after an appropriate period of challenge and testing, lead to a 'mature' science that, if sustained through subsequent challenges, becomes paradigmatic. As a matter of course, the evolutionary process involves a period of 'normal science,' when efforts are made to articulate the paradigm, explore possibilities within the paradigm, and use the theory to predict facts, solve scientific puzzles, and develop new applications of theory. At times, however, further discoveries of natural phenomena may violate the paradigm-induced expectations that govern normal science.

This involves discovery of problems not previously known to have existed. In some cases, normal science cannot explain them fully. How this happens and how the public should respond to such a change are not clear.

The interface between science and society is a relatively new area of philosophical investigation. In the past the relationship was viewed generally as one-sided either way, with science being influenced by society or vice versa. Now we are confronted with a simultaneous problem, where science and society are both influenced by and influencing each other. These rapid, frequent, and heated exchanges are complicated by two main factors. First, the norms of science (developed over centuries but codified and normalized by Merton, Popper, and Kuhn) are significantly different from those embodied in liberal democracy. They are for the most part incommensurable, which makes accommodation between science and liberal democracy difficult if not impossible. Second, the language of discourse in the two realms exacerbates rather than facilitates communication and understanding. The subtle yet precise terminology and structure of scientific discourse may facilitate scientific progress but do not translate at all well into public debate. The concepts of paradigms and the premise of sustained scepticism that underlie the scientific establishment sound at times pedantic and often wishy-washy in the hurly-burly of political discourse, where the winners usually offer hyperbole and certitudes.

The seven chapters in Section Two investigate these issues from the philosophic perspective, presenting a range of assessments of the challenges of the co-determination of science and liberal democracy and a variety of options for improving the relationship.

As a philosopher, Robert Frodeman is acutely sensitive to the tension between science and democratic society. Indeed, he has argued in his writings for a philosophy of public science where the family of issues cited at the beginning of this introductory chapter could be addressed, if not solved. In this essay he notes that U.S. federal public agencies themselves have increasingly realized the need to engage the humanities and the social sciences. Just as Carl Mitcham argued for philosophy's salutary influence on science and democratic society, Frodeman presents the case that science and technology have 'transcended the boundaries of their disciplines' and are 'raising fundamental questions about what it means to be human and the nature of the good life,' and thus philosophy and the humanities in general must become engaged. In addition to illustrating the ethical component in such fields as biotechnology, nanotechnology, and climate science, Frodeman devotes the major part of his essay to indicating how the humanities can contribute to policymaking. Human well-being, the good life, and the common good provide perspectives for discussion within liberal democracy and for implementing public policies. We do not need philosopher-kings, he notes, but we do need philosophic bureaucrats.

Gordon McOuat, a historian of science, examines the changing relationship between the political, the social, and the public over the last two centuries. Karl Pearson (1892) has argued that the scientific frame of mind would produce good citizens with a healthy dose of scientific scepticism, with a lack of prejudice, and with passions under control. Thus, the scientific frame of mind would help to stabilize the society and state. Robert Merton, Thomas Kuhn, and Michael Polanyi are used to demonstrate the traditional norms of a pure science that in principle would be free from the instrumentalism of society and the state. McOuat approvingly discusses the more recent constructivist turn in the study of science and in particular recounts studies of research laboratories. The practice of science, it was found, was neither pure nor separate but involved 'long networks of negotiation, exchange and translation, involving social actors, political actors, aesthetic actors, rhetorical actors, technological actors, and so on.' The changing relationship between science and society includes changes in scientific practice itself. There is no doubt in McOuat's mind that scientific knowledge has become politicized. He believes that we 'can bring the formidable tools of recent close examination of science to examine the very ecology of that life.'

The philosopher of science Ian Jarvie powerfully defends the view that the inherent scepticism of science as traditionally understood is working relatively well. He finds the work of Karl Popper particularly useful in the defence of science. He contrasts his views with those found in the landmark work by Philip Kitcher, *Science, Truth, and Democracy* (2001). As Jarvie argues, a proper understanding of science and democracy stresses the sceptical frame of mind, contestation of ideas, and freedom of thought and discussion. A claim of absolute certainty, as Popper eloquently argued, is antithetical to both science and democracy. Where reform is needed is in the institutions and practices of science. Democratic responsibility and accountability, Jarvie writes, need to be injected into science. Once the democratic deficit is addressed in science, then the family of issues challenging science and democracy can be faced and handled.

The political philosopher Leon Craig takes direct issue with the assumption that democracy is able to address the issues coterminous with the evolution of science and technology. In particular, he does not believe that piecemeal innovations within the parameters of liberal democracy will be sufficient. We need, in short, a broader perspective. Craig describes the internal diseases within contemporary liberal democracy and the external dangers, 'the devil without,' that plague contemporary society. His verdict is that one cannot but be pessimistic about the possibility of a 'synthesis of scientific knowledge and political understanding.' Indeed, Craig views the challenge that science presents to liberal democracy as constituting the most 'urgent problem for political philosophy in our time.'

Steven Shapin writes in his essay that different kinds of scientific expertise bear different kinds of relationships to common sense and to lay concerns. In some cases, the technical nature of science is not a problem – people are willing to accept the conclusions of science and defer to experts. This may be regrettable, and one may well lament the public's failure to understand or the overreliance on experts. The public, or their democratically elected representatives, become involved only when they are directly impacted. Where people care deeply and personally about the issues, they may reject expertise and demand to make their own decisions. Shapin explains this behaviour as the science of the general and the science of the particular. He uses the debate about diet in the United States to illustrate how the science of the particular and the reliance on experts can lead to perverse but understandable results. The role of the experts, as he says, is now entrenched in our culture; moreover, their problematic role is enhanced by the very heterogeneity of expertise. As a historian of science, Shapin provides a case study of the popularity of the Atkins diet to illustrate the role of experts in our society, public support and popularity for different kinds of experts and science, and differences in academic and popular literature on diet. The credibility of science for the public is affected by the direct pertinence to the individual. The public understanding of science is accordingly limited.

Alan McHughen and Eric Sachs in their essays address the challenge of communicating about science (in their cases genetics and molecular science) and the roles of scientists and others in structuring discourse. McHughen is particularly concerned about the challenge to the Mertonian scientist. He believes that the demise of the 'academic scientist' in the twenty-first century is jeopardizing the public interest. As an expert on biotechnology, McHughen is well aware of the public lack of knowledge about genetics and the impact this has had on the discussions surrounding genetically engineered crops. The resulting discussions are so befuddled and misinformed that the public interest gets lost in the morass of confused views. Scientific illiteracy is so rampant that society desperately needs 'trusted public scientists to inform public debate involving scientific issues.' McHughen analyses the present economic and political factors that have led to an undermining of public trust. The decline of public support for science and the dependency on private support impair the perceived independence of science, thus resulting in the 'loss of an expert champion of the public good.'

Sachs takes a different approach to the same problem of public policy and scientific literacy. As a scientist who works at the nexus of industry, research, and products, he is concerned about the need for society to understand scientific advances and their actual effects on the public interest. He is less concerned about the provenance of the science and more concerned with what is communicated related to science. He identifies a science literacy gap that he believes impedes society's ability to critically evaluate new scientific developments, and he makes several suggestions for scientists, in universities and industry, and for the media as well, to respond to the science literacy needs of society. In the absence of true understanding, he believes, liberal democracy is impaired.

Although these essayists clearly delineate the tensions and dangers in the relationship between public science and liberal democracy, their suggestions provide for possible changes within our present political culture. In various degrees they all assume that, however difficult the tasks, our society does have recuperative powers, which are grounded in our historical openness and philosophic heritage.

Frodeman and Mitcham have persuasively argued that the humanities and philosophy would have a beneficial effect on the tensions within and between science and democracy. Jarvie calls for more democracy within science, that is, a democracy grounded on a sceptical frame of mind. McOuat shows that a politicized, if not democratic, scientific ethos already exists, and Shapin too notes the public's effect on the conduct of science. McHughen and Sachs strongly recommend an increased effort in improving the scientific literacy of the public precisely because of the inescapable interplay between science and liberal democracy.

Addressing the Issues: Institutional Approaches

Ultimately, the discourse about the interface between science and society leads to questions about how we can improve on what we have.

Given the diversity of both science and liberal democracy, one should not expect one single approach to managing this interface. Scholars and public commentators tend to look to one of three institutions to address the issues: the science establishment, industry, or government.

Philip Kitcher (2001) proposes 'well-ordered science' as the ideal to which the organization of the research enterprise should aspire. His optimum would involve a highly informed public engaging with a public-spirited research community to set overall research priorities. For the most part, he argues this can and should be done outside the 'vulgar democracy' of government. What is not clear, however, is how a disparate and increasingly specialized science establishment would be able to engage the public.

Others see industries as the source of the problem and necessarily part of the solution. While their rapid and significant engagement in science has destabilized the science policy system, their reach (both by discipline and by country) exceeds that of any other actor. Hence, some believe they must be part of the solution.

Most often, however, people look to government to take the lead in reconciling the goals and capacities of science and society. Here opportunities for reform abound but are ultimately constrained in many nations and in many circumstances by government's lack of focus, resources, authority, and resolve.

The six papers in Section Three examine a number of institutional responses to the challenge of managing specific complex scientific issues in a few key liberal democracies.

Two political scientists, Grace Skogstad and Sarah Hartley, argue that societies divide within and across one another in their views of the authority of science and their beliefs regarding what constitutes an appropriate role for science in public policy processes. Yet, in a world of integrated markets and with the rapid diffusion of norms and values across national borders, the willingness of other political communities to grant authority to science and to recognize it as a legitimate basis of policymaking bears on the status of science-based policymaking in domestic governance. The authors investigate the role and legitimacy of science in agrifood regulatory policies, more particularly those designed to regulate new technologies. They believe that the traditional authority that scientific knowledge has enjoyed as a basis of legitimate policymaking is eroding, so that nations (domestically and through international treaty) require a more democratic infrastructure such as standards of openness, accountability, and procedural fairness. The consequence would then be a substantive public deliberation which could provide the trust and legitimacy required for science to interface in a salutary fashion with liberal democracy.

Ross McKitrick, an economist, looks at three settings in which complex and uncertain information is used to support decisions that will have institutional implications and may thereby impinge on others in society: court trials, business prospectuses, and science-based policymaking. In the first two, he identifies formally coded rules that ensure balance, full disclosure, and due diligence, thereby putting the decisions on the most secure footings possible in terms of truthfulness and fairness. In the policymaking case, he finds little evidence that these principles apply in a systematic way and, with respect to the climate change issue, are conspicuously absent. He believes that in the context of the global warming debate, this lack of method has led to the tyranny of public or official science. McKitrick offers as an institutional response a variety of mechanisms for reforming the policy process that would bring public sector decision making up to standards approaching those in the courts and the private sector.

Zaheer Baber, whose area of research is the sociology of science, examines the constructivist perspective that seeks to question a 'scientism' where science is a value that transcends all other values and social concerns. He, like Philip Kitcher and Tom Gieryn, does not accept that science is simply the consequence of social, political, and moral struggles. Rather, the challenge is to deal with the fact that some of science is determined by nature and some by society. This creates, as Giddens has noted, a creative tension between the reality of constraints and the possibility of actions that contest those constraints. Baber believes the tension between constraints and possibilities is plainly evident in the republic of Singapore in its Biopolis, where he believes the most far-reaching scientific revolution in genomics and biotechnology is set to unfold. His analysis shows how the state can reassert its authority through a form of a neo-corporatist governance. While this leads to a reconciliation of science and society, it comes, as he notes in the case of Singapore, at some democratic costs.

Chris Gerrard writes from the perspective of an economist within the World Bank. He examines the public pressures from domestic constituencies in industrialized countries on the international research agenda embodied in the Consultative Group on International Agricultural Research (CGIAR). His assessment is that the CGIAR largely failed in the 1990s to retain scientific excellence in allocating its resources towards global public goods research that would deliver positive impacts on the poor in developing countries. He believes the directors of the fifteen Centers which comprise the CGIAR now find themselves in a chaotic marketplace for agricultural research in which the preferences of individual donors are not adding up to a global public good. He suggests a number of reforms that would make the system's governance and management relevant in today's environment.

David Guston, a political scientist, examines the challenge of the commercialized academy – which creates and exploits knowledge in the context of commercial incentives – and suggests public universities need to engage in greater scrutiny of the ethical, legal, and social implications of their role in knowledge creation. He makes the singular suggestion that universities create Centers for Responsible Innovation (CRIs), which would offer opportunities for reflexive pursuits in parallel to the universities' traditional missions in teaching, research, and service.

Rahul Dhanda brings the perspective of a scientific scholar and entrepreneur to the debate. He concludes that democracy, science, and industry form a precariously balanced triangle. While not necessarily by design, each institution represents individuals that are looking to provide some social benefit. He believes this reality is often easily lost in discussions of science and society. Resolving this tension will not be simple, nor is it likely to occur rapidly when effort is finally applied. Dhanda issues a clarion call for the state to become clearer about its expectations and actions related to science, especially big science initiatives such as the Human Genome Project.

Assessing the Three Questions

While none of our authors would claim to offer complete and definitive answers to the three interconnected questions posed in this volume, they individually offer thoughts that, combined, shed light on these issues.

First, we asked, 'Can science retain independence and objectivity in the face of demands to meet commercial and public policy objectives?' Although science was never wholly independent of society – as historians in this volume delight in illustrating – science now has far greater societal entanglements and is involved in a far greater number of public policies than ever before. The challenge to science cannot be reduced to idealism versus realism or epistemological purity versus social constructivism or privileged expert versus handmaiden of powerful institutions. Retaining the independence and objectivity of science cannot occur if public trust in science is lost. Here both philosophic and institutional approaches are required to nourish science and to develop public understanding of science. In short, it is possible for science to retain its independence and objectivity, but it will take much more effort than in the past.

Second, we asked, 'In what ways is scientific discourse privileged in the formation of public policy? Can there be a genuine public discourse if one party is privileged?' Although there is less consensus among our authors about whether science is truly privileged or whether it is simply one of many sources of power in the more diffuse global power system of the twenty-first century, there is no doubt that public science does constitute a challenge to liberal democracy. Our essays address these two questions from two directions: from the standpoint of modification of scientific discourse to make it more open and accountable and from the standpoint of improving scientific literacy in the public. Again, given the impact of public science on public policy, it will take much more effort than in the past to sustain a genuine public discourse.

Third, we asked, 'How can scientific knowledge and scientific methodology be made compatible with the interdisciplinarity and integration required of public policy discourse and formation?' Here our authors have provided us with a number of illustrative examples where interdisciplinary teams of scholars, policymakers, and entrepreneurs have engaged to manage the integration. If there is one lesson from the debates our authors generate, it is that these processes must be inclusive. While some disciplines may think they have all the answers, the debate about greenhouse gases offers a useful reminder of the scope of the challenge of being inclusive.

It is important to stress that our authors have not only addressed these three questions but have introduced other questions that inescapably occur. Several authors note that the compatibility of science and democracy is not possible without a quantum improvement in the scientific literacy of the public. Others suggest that there are potential institutional means for alleviating tensions between science and democracy. Some give examples of present institutions that were deemed successful; other authors argue that present institutions do not work and suggested radical alternatives. The philosphers see an urgent need for a philosophy of public science and for a rediscovery of the philosophic roots of our civilization. While some are confident that liberal democracy with a solid philosphic defence can meet the challenge of science to democracy, some believe that liberal democracy has become so weakened that it may not be able to rejuvenate itself. All of these questions would need to be examined if we are to have a fuller assessment of public science in liberal democracy.

Both science and liberal democracy are inescapably challenged by the evolution of public science and by its increasing impact on the public good. The challenges before us are fundamental and go to the heart of our civilization. Reflection is required by scientists and citizens, by philosophers, historians, and social scientists.

REFERENCES

Kitcher, P. 2001. *Science, Truth, and Democracy.* Oxford: Oxford University Press.

Kuhn, T. 1962. *The Structure of Scientific Revolutions.* Chicago: University of Chicago Press.

Pearson, K. 1892 (2004). *The Grammar of Science.* Dover Phoenix Edition. Mineola, NY: Dover Publications.

SECTION ONE

The History of Public Science in Theory and Practice

1 The Element Publicum

LARRY STEWART

Modern science relies on trust, even if that trust is often grossly ill informed. On that very trust rests foundations of credibility, status, moral authority, and claims to massive amounts of funding, both private and public. Yet, trust is not self-generated. It is, by definition, granted, an acknowledgment of expertise in a public world. The difficulty we have before us, however, is the dissonance between the apparent incoherence of much of modern scientific thinking – think of global warming, especially as it resonates among the broad public – and, at the same time, numerous obstacles of access to science in an age which surely could be characterized, in the West at least, as an age of access. One reason for the contradiction is the elevation of a modern priesthood of believers in experimental method who very often communicate among themselves in a language that few understand. What is peer review, if it is not this? This fact further represents a deep irony in the modern view of the natural world: more laboratories, vast research parks, more monuments to the march of science, even when measured by expenditure alone, actually mean less public comprehension and greater exclusion (Pollack 2005). Yet, by failing to seriously address the divide, science increasingly scorns the hand that feeds it.

The Discovery of the Public

I would like to examine not the current space of science in a liberal democracy so much as the emergence of the very public on which they both rest. If a historian might be indulged a little history, perhaps one might explore the dynamic whereby the discovery of the element PUBLICUM was the most important breakthrough in the early evolution of modern science. Without its public, science as we know it would not have been possible. The public has long mattered (Schaffer 2005: 301). This is a strong claim, but if it has any merit it raises very serious concerns about our current reliance on numerous government and regulatory agencies, especially those which purport to harness the intentions and authority of various co-called stakeholders, sometimes in an effort to sanitize suspicions of special interests and to sanctify the purity of market forces. This is more the case where efforts to influence scientific activity reside essentially in the local and the national rather than in the cosmopolitan. And yet regulatory agencies are at heart in contradiction with the authority which obviously resides in market forces and patent protection (cf. Shapin 2004: 58–9). To maximize markets and minimize the risk – say, in pharmaceuticals or even genomics – is a neat trick. It is self-evident that national governments, answerable to their own *res publica*, have, by definition, limited sovereignty, and, apart from enforcing patents, they never much mattered in the scientific world (Latour 2005). The scientific audience, by contrast, has always been international in its scope and will continue to be so. On the other hand, whether a modern public can now have much influence on the function of science is nevertheless deeply problematic (Marres 2005: 208). This is somewhat troubling, especially as, in my view, the public has historically proved essential to scientific change.

The natural world has long been a problem that has tested political will. Take, for example, the early modern world, where a strategy was conceived for collection and contemplation, in uniting the material world with a human comprehension, of bringing the mind to bear on a universe that, at least since the seventeenth century, seemed without boundary and contained variety seemingly without limit. Personal natural collections became museums, social boundaries were eased by widening access to public demonstrations, and private contemplation was transformed by broader insight. The hurdles to general access to knowledge were many. But how were the barriers broken?

Nature must have once seemed a riot of this thing and that, exotica without names, some mentioned perhaps in passing in ancient accounts, but lacking in arrangement or taxonomy. But what emerged in early modern Europe was a jumble of drawers and jars, no sense of vegetable, mineral, or even animal, making mad the senses, seducing the curious. Where was the use in such profusion? Was this just a tale of nature's plenty, a confusion of goods and curiosities, as though some visiting salesman forgot his samples? This was the world of Joseph

Priestley's laboratory and the confusing debate over vital air or oxygen, of Antoine Lavoisier's nomenclature, and the bundles and samples of James Watt's chemical workshop. Much promise would undoubtedly come of these, but promises were mostly what they would remain. Significantly, especially for the devout of the early modern world like Priestley, scripture also implied nature had its use (Stewart 2004). In the world of new philosophical instruments, like the microscope, the thermometer, or electrical machines, usefulness and observation distrusted the surface of things. The cultural consequences of such a proposition were profoundly appealing. No longer would trust reside in what was told. This meant that political and social elites in the early modern world could not capture the sources of knowledge as easily as they might buy paintings by the gross or libraries by the yard (Jordanova 1989). If elites could no longer appropriate natural knowledge as they once had galleries and cabinets, the reasons lay fundamentally in the rapid spread of sites where curiosities might be displayed and in the demands for utility (Latour 2005: 31). From the early seventeenth century, perhaps under the influence of Lord Chancellor Bacon, proposals were put for Repositories or Offices of Address, the purpose of which was the collection of useful knowledge and models of machines. Such a doctrine was revived, in the late seventeenth century, in the Royal Society's abortive History of Trades, but the notion continually resurfaced that natural and mechanical collections might disseminate useful applications. By the eighteenth century, the linking of utility with progress became essential to an enlightened vision of encyclopedias and to an industrial revolution in which so many came to believe (Jacob and Stewart 2004). Even the idiom of the American Revolution was the language of enlightened experiment (Delbourgo 2006: 152). An Enlightenment vision of the uses of creation had likewise no small effect on the politics of the American republic, which would engage in the greatest conquest of nature the world had then yet known. The Republic of Learning took a political form (Goldgar 1995; Goodman, 1994).

Science and the public were strikingly redefined in the North Atlantic Enlightenment. In the process, the promotion of natural utility became crucial to the growing democratic and republican movements of the eighteenth and nineteenth centuries. Science was soon for sale. By the eighteenth century, a rising clamour and curiosity elevated the distinction between the intellectually serious and those easily entertained by the optical displays and electrical effects revealed in Paris street theatre, in London's Covent Garden, and in fairs and gardens throughout urban Europe (Isherwood 1986). Science in the open meant a public engagement with new contrivances like hot air balloons and lightning rods (Lynn 2006:127). And in a North Atlantic world increasingly enamoured with the power of mechanical contrivances and on the verge of industrial upheaval, the participation of a public would have great consequence to those who would invest huge sums and to those especially who further argued that the security of states and empires depended on the conquest of nature's laws. Empire meant that exploitation followed conquest, as surely as colonists followed explorers. In this enlightened world, the aim of some philosophers was to enlist a public in the advancement of science and in the interests of the state. The state thus both needed nature and had to subdue it. In today's world of climate change and genetic modification, this is a relationship we are still evaluating. In the Enlightenment lay the roots of the battleground over what Andrew Ross has called the modern 'techno-body: the politics of reproductive freedom, immune deficiency, biogenetics engineering, worker safety, police surveillance, narcotraffic, clean air, homelessness, militarism, and so on.' Ross argues, from the Left, that 'there is little to be gained from holding on to the traditional humanist faith in the sanctity of the unalienated, "natural" self, nobly protected from the invasive reach of modern science and technology' (Ross 1996: 8). Our empire is not now simply geographic in its reach. It is increasingly personal. The individual body is the new frontier.

The nobility of nature was always a myth that sanctified simultaneously both bounty and resistance, exploitation and ecology. Even so, nations long enlisted nature in their empires. At the cusp of empire and industrial revolution, at a time when Joseph Banks trolled the Pacific on behalf of bread fruit and British conquest, the mad George III purchased the magnificent natural history drawings once owned by Cassiano dal Pozzo in the early seventeenth century (Freedberg 2002). On one level this appears simply as princely collecting. Yet, the King would also own a stunning cabinet of scientific apparatus impressively designed by the best British instrument-makers to reveal scientific principles to the curious of the court (Morton and Wess 1993). Enlightenment science and technology became increasingly interwoven and frequently confused, if largely in consequence of the many public lectures of the century.

In the escalating industrial revolution of the eighteenth century, mechanical achievement provided a powerful argument for the value of philosophers and their promotion of nature's use. By the middle of the century, new groups emerged from the rhetoric of improvement, such as the Society for the Encouragement of Learning and the Society for the Encouragement of Arts, Manufactures and Commerce. By 1778, within a few years of Captain Cook's *Endeavour* voyage, and when the botanist Joseph Banks became president of the Royal Society, it was asserted that a 'Harvest of Learning' in 'Arts, Sciences, Manufacturies, &c.' could be gathered from exhibitions at the British Museum, which was itself founded upon the grand cabinet of Hans Sloane (Goldgar 2000; Gascoigne 1998: 22–4). Thus, Eden's vast legacy, Captain Bligh's *Bounty*, and nature's laws served the interests of the imperial state (Drayton 2000). Imperial maps renamed and thus reclaimed all of nature as European. What role in early modern empires could a scientific public play? This situation has barely changed since the great days of Europe's colonial mission. During the period of the first British empire, amid an explosive industrialization, the 'shifting natures of public discourse' made science rhetorically resonant and fundamentally urgent in the search for political power (Taylor 1996). This was, and remains, the crucial issue. Let me provide one of many examples: experimental observation took on a new impetus, critically relying on audience for assurance and confirmation. By the eighteenth century, it was no longer sufficient to make claims to new natural knowledge and expect to be believed. This was precisely why Newton was not readily accepted for his apparent discoveries of optical refraction (Schaffer 1989). What Newton's experiments revealed more than a spectrum of colours was that evidence stood or failed on the accounts of witnesses who did not necessarily have any particular claims to expertise. To discover new facts or to confirm old ones was a process demanding not only replication by numerous practitioners but later confirmation through many, even untutored, eyes. Audience acutely mattered (Shapin 2004: 47; Schaffer 2005: 301). Witnessing also had to be replicated. Early modern science *depended* on its public.

Witnessing and affirming similarly lay behind the demonstration of the celebrated problem of the acceleration of falling bodies, which everyone daily saw but few might comprehend. Galileo, apparently, had once put paid to the idea of mass determining rate of fall by way of a crucial experiment, by dropping balls from the leaning *campanile* of the Cathedral of Pisa. Whether or not such an event is more than legend is not of concern here. It is certainly likely that Galileo's notion of the acceleration of bodies was confirmed by private experiments on balls rolling down inclined planes (Galilei 1945; Drake 1975; Harre 1984).

While his laboratory experiments might well have been solitary, the plunge from Pisa's tower could only have occurred in a public plaza. But the laboratory report which Galileo later provided of his experiments on falling bodies could best be believed only if reproducible before others' eyes. By the early eighteenth century, however, the Pisa experiment had been transported to Newton's London.

Newton's notion of universal gravitation, among the most profound discoveries of modern science, clearly had implications for this age-old problem of acceleration and free fall. When he later came to revise his great *Principia* (1687), Newton included in it experiments performed in 1710 by Francis Hauksbee, the Royal Society's curator of experiments, and subsequently those of 1719 by the Huguenot refugee John Theophilus Desaguliers (Hauksbee 1710; Desaguliers 1719; Newton 1971: 361–64; Heilbron 1983). The experiments performed by Hauksbee and Desaguliers, at Newton's behest, took place beneath the dome of St Paul's Cathedral, from a height greater than Pisa's 60 metres. Hauksbee dropped glass balls filled with mercury and one of cork from the top of the cupola at 220 feet. Nine years later Desaguliers's effort was somewhat more elaborate and at a distance of 52 feet higher yet, inside and sheltered from the wind, from the top of the lantern in the upper gallery (Royal Society 1719). Notably, efforts needed to be made to determine the effects of the lag of the experimental apparatus in releasing the weights and of the resistance of air to account for differences in the rate of fall from Galileo's reported results. Such experiments in the public space of Wren's great dome must have been seen by several, perhaps by many – we may never know. The distinction between an exclusive instrumental science in the Royal Society or in a philosopher's private laboratory, on the one hand, and a public moment, exploding on the deal boards beneath the cupola of St Paul's, on the other, is an important one. As instruments refined sensation, the places of experiment would also make a considerable difference.

The sites of science and trust collide most obviously when the human is directly involved, as witness most obviously, but sometimes as subject, voluntarily or not. One early example, among many, was the inoculation for smallpox on convicted felons in London's Newgate prison in 1721. Many intersections between the incarcerated and the innovative have since echoed across virtually three centuries. The early smallpox trials upon the felons were witnessed by some twenty-five physicians and surgeons. This thereby made the Newgate 'laboratory' subject to the same rules that had been evolving, especially among the gentlemen of the Royal Society, to ensure the veracity of experimental claims. No private, solitary philosopher or physician could present results without collaboration and corroboration by appropriate witnesses (Shapin 1995). It did not help that wags of *Applebee's Journal* declared of inoculation that 'any Person that expects to be hang'd may make Use of it' (Guerrini 2003: 53). Ultimately here, however, proof lay in the pustule.

Between the early modern experiment, published reports, and public speculation or alarm, there was no template or convention consistently adopted. In these circumstances, the human subject was essential for reading the effect of a natural event or experimental procedure. In the case of self-experimentation, the sentient being could reveal what the inanimate and even the animal could not (Scheibinger 2004: 389). As the promotion of experiment became the most essential characteristic of early modern science, the body became another instrument of revelation as much as had the telescope or microscope. The role of the human in experiment is epistemologically complex. Not only is the human an observer, but he or she is also potentially subject and, at the very least, interpreter of natural phenomena. The human, thus, is always instrumental and functional even in the most austere laboratory settings seemingly beyond the public gaze.

The Public Experiment

The promotion of experiment was significant for it relied entirely upon the growth of a public motivated to spend substantial amounts of money to attend lectures, to purchase the latest scientific books, and to invest in a plethora of apparatus emerging from the instrument-makers' shops. In no small measure the literate public was formed in the crucible of scientific discourse. The audience moreover was essential to the expansion of the experimental realm which we now accept as the foundation of modern science. One might argue, as did the chemist Joseph Priestley, that science was the hallmark of a true republic.

Empire and republic reflected each other in the early modern convergence between the aims of the nation-state and the passions of the scientific public. Scientific societies abounded, such as the American Philosophical Society, populated by gentlemanly amateurs and occasionally emerging from small groups of skilled tradesmen, alongside increasingly professional natural research at the Musée d'Histoire Naturelle in Paris and the Zoological Society in London's Regent's

Park. Empires in conflict and a nation's pride sought scientific pre-eminence. Even Kew, the Duchess of Somerset was embarrassed to say, did not measure up to the Jardin des Plantes, where 'they have a variety of lectures *gratis*' (Drayton 2000: 137–8; cf. Delbourgo 2006). Public interest and the encouragement of research were united in the imperial reach. Reflecting a plan of collecting that Hans Sloane had attempted to assert in the early eighteenth century, the naturalist Sir William Hooker in 1847 reminded the Royal Navy of the necessity of sending plants to the Museum of Economic Botany (Drayton 2000). Like the never-ending search for foods and dye stuffs, the rapid cultivation of museums, botanic gardens, and laboratories was a reflection of a widening scientific culture.

The nineteenth-century state increasingly took a role in the promotion of the uses of nature. In Britain, seemingly endless numbers of controls were legislated to bring science to bear on measurement and economic expansion – thus the Board of Trade (1832), the Factory Inspectorate (1833), the observatory at Kew (1842), and the Excise Laboratory (1842) were all precursors of our regulatory agencies (Schaffer 1997a). This undoubtedly owed much to the development of a cultural authority attached to science. As empires were 'more obviously commercial,' encouragement, enlightenment, and consumption merged. But empires were not merely commercial. They reflected as well as magnified the growing idiom of conquest of nature's expanding warehouse. The Crystal Palace in 1851 made transparent the crucial role of public understanding. The exhibition in Hyde Park revealed a bountiful ambition in the promotion of natural science and technical skill. Workers from provincial towns arrived on special trains to see the power that mechanics, mind, and nature might achieve (Jacob and Stewart 2004; Pickstone 2000). Crystal Palace and Mechanics' Institutes were not worlds apart. The public belief in the liberal mantra of material progress was thus rooted in the cabinets, lectures, gardens, expositions, zoos, and museums of the European empires (Stewart, 2004).

The Power of Science and Science in Politics

Since the Enlightenment, princes, republics, or the imperial state, and the broad public interest have repeatedly sought consilience. The political and the scientific were increasingly interwoven. This is particularly apparent over the last century, when significant scientific problems, like the on-going disputes over global warming, gained increasing public attention (Weart 2003: 69). Strategic, as well as economic, concerns meant applied science and technology significantly expanded the arsenal utilized in international conflict. Powerful spokesmen of preparedness charted the course of science and state in the interwar years. Senator Elihu Root, chairman of the Board of Trustees of the Carnegie Institute and winner of the Nobel Peace Prize in 1913, argued before the Armistice that 'competency for defence against military aggression requires highly developed organized scientific preparations. Without it the most civilized nation will be as helpless as the Aztecs were against Cortez' (Jenkins 2002: 16). The original expression of a 'Cortesian army' came from Robert Hooke's seventeenth-century vision of the conquest of nature, not of peoples (Hunter 1989: 233). The romance of the philosopher was best discovered at the frontier of knowledge. In the Cold War, this became a matter of national security.

A nascent democratic ideology, imperial reach, and national survival were quickly linked to the apparent utility of science. The task, as now, was to convince the public. Vannevar Bush, president of the Carnegie Institution, pronounced in 1949, in the aftermath of yet another horrific war, that 'the technological future is far less dreadful and frightening than many of us have been led to believe, and that the hopeful aspects of modern applied science outweigh by a heavy margin its threat to our civilization' (Bush 1949: 2–3). The public, the amateur, and the professional all had a place in this optimistic world. His analysis of the Cold War world placed science squarely within the arsenal of the West:

> Since the war ended, this elusive and powerful force, this mass concept, this public opinion ... most certainly rejected any idea that we should become a conquering nation, or strike early ourselves in the attempt to avert a later and a more desperate war. It insists that we control the traitors in our midst, and somewhat bewilderingly that we do not sacrifice our essential freedoms in the process. It has even begun to insist that special and selfish interests be regulated within a compass that will not wreck the national strength we need. It has placed us on the path we now pursue, yet does not know exactly where it leads. (Bush 1949: 3–4, 246; see also Greenberg 1967: 74–80, 99)

As amazing as this view now seems, in his earlier report to Roosevelt in 1945 Bush had been especially bullish. He stated bluntly, 'In this war it has become clear beyond all doubt that scientific research is absolutely essential to national security' (Bush 1949: 14). Among the mantras of the Cold War, national security and science were inextricably and immediately tied in the postwar strategy. He wrote to Roosevelt that 'the responsibility for the creation of new scientific knowledge rests on that small body of men and women who understand the fundamental laws of nature and are skilled in the techniques of scientific research.' Bush was not only a proponent of providing opportunities in basic research. He also adopted a view of the so-called Moe Committee: 'It would be folly to set up a program under which research in the natural sciences and medicine was expanded at the cost of the social sciences, humanities, and other studies so essential to national well-being' (Bush 1960: 23). In other words, from the view of democratic survival there resounded the affirmation of the cultural and the aesthetic as crucial to natural understanding (Kitcher 2001).

Educating the public about the value of science in the conflict with communism was a natural sell but baseless in fact. As Daniel Greenberg has demonstrated, the lamentable state of public understanding of science in the postwar years had no effect on the rising expenditure on research. On the contrary, spending on science significantly increased despite the end of the Cold War (Greenberg 2001: 210–11). The 'Bush card,' no matter how often it was played, was obviously self-serving; yet it also played to those who believed that public comprehension was a significant obstacle in the funding of science.

During the Cold War, there was something much more subtle and long-standing at work in the transformation of Western culture. The new race was not in arms alone, but in expanding the public support for science and industry. But efforts to expand understanding, through presidential commissions, public museums, science centres or, latterly, Discovery channels, were not simple statism at work. In the American circumstance, this was not simply a response to the much-vaunted Soviet challenge on yet another frontier. If the media had any role in the debate over funding, it was invariably drawn to the dramatic. Even debates on global warming were muddled by rapid iceberg calving and endangered polar bears, or even by theories of the relation of planetary alignment to earthquake probabilities. Repeated demands for enhancing public understanding were the appropriate response to a media unable to distinguish between scientific facts and fantasy (Pollack 2005: 30, 36). And in the midst of the twentieth century, the promotion of science relied on dubious historical comparisons. Thus Glen Seaborg, winner of the Nobel Prize in chemistry and chairman of the Atomic Energy Commission, would mobilize the past: 'The educated man of today and tomorrow can no more ignore science than his predecessors of the Middle Ages could ignore the Christian church or the feudal system' (Greenberg 1967: 37). Seaborg at least understood that *Sputnik* and the later *Apollo* program had a long trajectory. Such interplay between the public and the practitioners of science had become inescapable.

In serious current problems, observers and even participants have not always been well informed. Let us take one of many compelling examples. The critical struggles with so-called 'mad cow disease' have caused much economic misery and have confounded the scientific and political communities in Western democracies. Competing interests, both protectionist and health related, cloud decision making. But the long history of BSE (bovine spongiform encephalopathy) or TSE (transmissible spongiform encephalopathy), and its relationship to scrapie and to Creutzfeldt-Jakob disease in humans, is remarkable for revealing the way in which the public sometimes plays a crucial role. Like the spectators of physics experiments in the eighteenth century, observers often could have added much to the study of this cluster of diseases. It is remarkable, for example, that a degenerative distemper was reported in sheep from the early eighteenth century and was closely followed through flocks. It reappeared throughout the nineteenth century and was repeatedly noted by farmers and veterinarians into the early part of the twentieth century (Schwartz 2003). Notably, this was not the kind of evidence that would find its way into the medical literature relating to humans. Even the creation of concepts of infectious micro-organisms, by Pasteur or Koch in the nineteenth century, did not lead to any clear understanding of the problem of transmissibility of scrapie. One key breakthrough did come in the 1930s, when the experiments by French veterinarians showed that scrapie could be transmitted across species, in this case from sheep to goats, rabbits, and guinea pigs and, moreover, that the incubation period varied by species and even breed (Schwartz 2003).

The issue here was not discovery but dissemination of this information. There was little resonance until the 1960s, when Richard Chandler managed to transmit scrapie to mice, and, most importantly, his research was published in *The Lancet*, a medical journal. The audience suddenly widened even if the variability of infection and incubation period appeared confused and to differ widely. It was not until the 1970s that the similarities with Creutzfelt-Jacob disease were noticed in autopsies, notably showing that human-to-human transmission was possible (Schwartz 2003). My point here is not that persistent research and experiment have helped solve much of the puzzle, especially on the role of proteins in the disease, but rather that the disease was long known. What was of concern to a small number of veterinarians was now a matter of public health. But as Schwartz points out in her study of the various varieties of the disease, one of the crucial problems is 'the compartmentalization that exists between fields of study' (Shapin 2004: 47–8). Undoubtedly this remains a serious problem. It has led to a fundamental confusion which pays lip service to public input without any real means of attesting to the value of public opinion in determining research priorities. But there is a potential solution. This resides in broad awareness of how the scientific process operates. Yet the public is blithely unaware that disciplinary boundaries obstruct.

One critical difficulty remains the convergence of pure and applied research, especially when the public, and politicians who purport to speak for it, demands tangible results. This is the 'value for money' argument. Thus, as Philip Kitcher has shown, Leon Rosenburg of Princeton in his 1998 report on the National Institutes of Health (NIH), *Scientific Opportunities and Public Needs*, tried to respond to criticism that the NIH had failed to encourage public consultation. The apparent answer was in reference to a democratic process whereby those 'affected by disease and disability' would provide their views on the research direction of the NIH. The real decisions, of course, had to remain with the scientific professionals. But expertise, while it provides authority, also necessarily narrows discourse and limits audience. As Kitcher has put it, the real issue remains how to 'connect the abstract ideal of well-ordered science with the kinds of decisions policymakers face' (Kitcher 2001). What was once obvious from the fascination and interest of audience and public in the European Enlightenment has long since been overtaken by the prohibitive cost of much research, the institutionalization of big science, and notions of accountability that are so readily manipulated for ideological or political gain. Disputes, for example, over the ownership of DNA demonstrate clearly that science cannot avoid political questions (Shapin, 2000).

Science, therefore, has come to have a transparently ideological function. This enables some groups to make claims to profound objectivity within an apparently homogeneous professional enterprise called science. This is highly suspect but rhetorically powerful. Science has become self-conscious in its promotion not only of experimental method but of the stunningly massive apparatus that this seems occasionally to require. Hence the modern limitations on a public are no longer philosophical or rhetorical as they once were; they are economic, market-driven, and expressly political (Taylor 1996). This may account, ironically, for the fact that despite all of the assertions of the political activity of scientists, politicians and policymakers remain indifferent to their media hype (Greenberg 2001: 247, 251, 281). No amount of public consultation, Congressional hearings, or royal commissions can address scientific concerns in a meaningful manner. Even as thorough an analysis as the recent Stern report on climate change, despite being commissioned by the British government, is unlikely to convince sceptics on the Right.

The unification of science and state over the past three centuries has politicized the public understanding of research. Funding of science museums, university research, and even swarms of ethicists increasingly rely on governments, industry, and private foundations for which scientific literacy has become one means to an end. As Simon Schaffer recently commented about museums, 'the problem for curators, is that authority and the facts are contemporaries of each other: one gets defined as the other does.' This was the problem of portraits of nature that had confronted Frederico Cesi and the Accademia dei Lincei in the early seventeenth century. The debate was not really about natural objects but about the authority to determine the facts of the natural world in a situation no less 'conflicted and debatable' than issues like cold fusion or BSE where so much might be at stake (Schaffer 1997b). Science and the state were joined by purse and national survival. At the height of the Cold War in 1950, the National Science Foundation in the United States was explicitly established by Act of Congress to 'promote the progress of science.' Indeed, scientific hegemony was ultimately a major issue in the new American empire. But where public access to science was desired, accountability and comprehension failed to follow (Greenberg 2001).

There are obviously numerous examples whereby public fascination with and, indeed, belief in the methods of science need to be recognized as more than a historical curiosity to be politely acknowledged. When an audience watched demonstrations of the effects of new airs in Enlightenment Britain, they learned much that would expand the experimental enterprise and ultimately, if I may be so bold, to lay the foundation for the vast industrialism which would come later. That is an argument which has engaged historians for generations and need not detain us here. However, my assertion is that replication of results required an audience, both as witnesses and participants, and not merely the limited affirmation of self-defined experts. Likewise, if we are to acknowledge that there might be something to be learned from the experience of the uninitiated, this would require *from science* a degree of sophistication, and perhaps simply the taking of some trouble, that would expand the support from a clearly fascinated public. Go now into any decent bookstore and you will find there shelves upon shelves of scientific books, from diet books to the fantastic, popular as well as starkly professional. No rational argument can possibly be mounted that the literate public is not interested. The public has always been drawn to the flame. The history of scientific change will show numerous examples whereby the fundamental complexities of science have learned from the broader world. In a liberal democracy, likewise, science is far too important to be left to the scientists and the politicians.

As had long been the case, public understanding was essential. So the politics of scientific research has invariably proved Janus-faced – looking for immediate useful advantage and simultaneously anticipating the dangers of unforeseen consequences. Both potential and risks appeared to increase, notably in the gap between research and application. The experience of the two World Wars and the Cold War made it clear that scientists must seek to manage public scientific knowledge (Kevles, 1978). In so doing, magazines, newspapers, museums, and science centres all played an important part. Likewise, it was inevitable that museums and collections would feel pressure when exhibits engaged controversial issues and engendered public or political heat, especially where scientific problems were unresolved, as in the hot topics of genetic engineering, race and intelligence, or global warming. Political liberals might feel drawn to pollution controversies, while conservatives prefer sometimes to save money and let markets make their choices. Both, of course, operate in the name of responsive politics (Schaffer 1997b; Oppenheimer 1987). Barely a day goes by without the press reporting some unfortunate controversy befalling those in drug trials or embryo testing for genetic disease. We cannot easily disentangle the motives from the ideology of the practitioners. Yet it is abundantly clear that debates over global warming, DNA, stem cells, or genetic engineering have returned us to a place where the public will inevitably have a compelling impact on the practice of science (Weart 2003).

REFERENCES

Bush, V. 1949. *Modern Arms and Free Men. A Discussion of the Role of Science in Preserving Democracy.* New York: Simon and Schuster.

– 1960 (reprinted). *Science. The Endless Frontier. A Report to the President on a Pro-*

gram for Postwar Scientific Research. Washington, DC: National Science Foundation.

Delbourgo, J. 2006. *A Most Amazing Scene of Wonders. Electricity and Enlightenment in Early America.* Cambridge, MA, and London: Harvard University Press.

Desaguliers, J.T. 1719. An Account of Some Experiments Made on the 27 Day of April, 1719, to Find How Much the Resistance of the Air Retards Falling Bodies. *Philosophical Transactions* 30 (362): 1071–8.

Drake, S. 1975. Free Fall from Albert of Saxony to Honore Fabri. *Studies in the History and Philosophy of Science* 5: 347–66.

Drayton, R. 2000. *Nature's Government. Science, Imperial Britain, and the 'Improvement' of the World.* New Haven and London: Yale University Press.

Freedberg, D. 2002. *The Eye of the Lynx. Galileo, His Friends, and the Beginnings of Modern Natural History.* Chicago and London: University of Chicago Press.

Galilei, G. 1945. *Dialogues Concerning Two New Sciences.* Translated by H. Crew and A. de Salvio. New York: Dover.

Gascoigne, J. 1998. *Science in the Service of Empire. Joseph Banks, the British State and the Uses of Science in the Age of Revolution.* Cambridge and New York: Cambridge University Press.

Goldgar, A. 1995. *Impolite Learning. Conduct and Community in the Republic of Letters, 1680–1750.* New Haven and London: Yale University Press.

– 2000. The British Museum and the Virtual Representation of Culture in the Eighteenth Century. *Albion* 32 (Summer): 195–211.

Goodman, D. 1994. *The Republic of Letters. A Cultural History of the French Enlightenment.* Ithaca and London: Cornell University Press.

Greenberg, D.S. 1967. *The Politics of Pure Science. An Inquiry into the Relationship between Science and Government in the United States.* New York: Plume Books.

– 2001. *Science, Money, and Politics. Political Triumph and Ethical Erosion.* Chicago and London: University of Chicago Press.

Guerrini, A. 2003. *Experimenting with Humans and Animals. From Galen to Animal Rights.* Baltimore and London: Johns Hopkins University Press.

Harre, R. 1984. *Great Scientific Experiments.* Oxford and New York: Oxford University Press.

Hauksbee, F. 1710. Experiments Concerning the Time Required in the Descent of Different Bodies, of Different Magnitudes and Weights, in Common Air, from a Certain Height. *Philosophical Transactions* 27 (328): 196–8.

Heilbron, J.L. 1983. *Physics at the Royal Society during Newton's Presidency.* Los Angeles: William Andrews Clark Memorial Library.

Hunter, M. 1989. *Establishing the New Science. The Experience of the Early Royal Society.* Woodbridge: Boydell Press.

Isherwood, R.M. 1986. *Popular Entertainment in Eighteenth-Century Paris*. Oxford and New York: Oxford University Press.
Jacob, M.C., and L. Stewart. 2004. *Practical Matter. From Newton to the Crystal Palace Exhibition of 1851*. Cambridge, MA: Harvard University Press.
Jenkins, D. 2002. *The Final Frontier. America, Science, and Terror*. London and New York: Verso Press.
Jordanova, L. 1989. Objects of Knowledge: A Historical Perspective on Museums. In *The New Museology*, edited by P. Vergo, 22–40. London: Reaktion Books.
Kevles, D.K. 1978. *The Physicists. The History of a Scientific Community in Modern America*. New York: Knopf.
Kitcher, P. 2001. *Science, Truth, and Democracy*. Oxford: Oxford University Press.
Latour, B. 2005. From Realpolitik to Dingpolitik or How to Make Things Public. In *Making Things Public. Atmospheres of Democracy*, edited by Bruno Latour and Peter Weibel, 14–41. Cambridge, MA, and London: MIT Press and ZKM Center for Art and Media Karlsruhe.
Lynn, M.R. 2006. *Popular Science and Public Opinion in Eighteenth-Century France*. Manchester and New York: Manchester University Press.
Marres, N. 2005. Issues Spark a Public into Being. A Key but Often Forgotten Point of the Lippmann-Dewey Debate. In *Making Things Public. Atmospheres of Democracy*, edited by Bruno Latour and Peter Weibel, 208–17. Cambridge, MA, and London: MIT Press and ZKM Center for Art and Media Karlsruhe.
Morton, A.Q., and J. Wess. 1993. *Public and Private Science. The King George III Collection*. Oxford and London: Oxford University Press, in association with the Science Museum.
Newton. 1971. *Principia*, vol. 1, edited by F. Cajori. Berkeley and Los Angeles: University of California Press.
Oppenheimer, F. 1987. Exploratorium. In *Interactive Science and Technology Centres*, edited by S. Pizzey, 5–20. London: Science Projects Publishing, 1987.
Pickstone, J.V. 2000. *Ways of Knowing. A New History of Science, Technology and Medicine*. Chicago and New York: University of Chicago Press.
Pollack, Henry N. 2005 *Uncertain Science ... Uncertain World*. Cambridge and New York: Cambridge University Press.
Ross, A. 1996. *Strange Weather. Culture, Science, and Technology in the Age of Limits*. London: Verso.
Royal Society. 1719. *Journal Book XI* (1714–1720), 30 April 1719 and 22 October 1719 reports to the Society.
Schaffer, S. 1989. Glass Works: Newton's Prisms and the Uses of Experiment. In *The Uses of Experiment. Studies in the Natural Sciences*, edited by D. Gooding, T.

Pinch, and S. Schaffer, 67–104. Cambridge and New York: Cambridge University Press.
- 1997a. Metrology, Metrication, and Victorian Values. In *Victorian Science in Context*, edited by B. Lightman, 438–74. Chicago: University of Chicago Press.
- 1997b. Temporary Contemporary: Some Puzzles of Science in Action. In *Here and Now. Contemporary Science and Technology in Museums and Science Centres*, edited by G. Farmelo and J. Carding, 31–9. London: Science Museum.
- 2005. Public Experiments. In *Making Things Public. Atmospheres of Democracy*, edited by Bruno Latour and Peter Weibel, 298–307. Cambridge, MA, and London: MIT Press and ZKM Center for Art and Media Karlsruhe.

Scheibinger, L. 2004. Human Experimentation in the Eighteenth Century: Natural Bodies and Valid Testing. In *The Moral Authority of Nature*, edited by Lorraine Daston and Fernando Vidal, 384–408. Chicago and London: University of Chicago Press.

Schwartz, M. 2003. *How the Cows Turned Mad*. Translated by E. Schneider. Berkeley and Los Angeles: University of California Press.

Shapin, S. 1995. *A Social History of Truth. Civility and Science in Seventeenth-Century England.*. Chicago and London: University of Chicago Press.
- 2000. Trust Me. A Review of Paul Rabinow, *French DNA: Trouble in Purgatory. The London Review of Books* 22 (27 April): 2–8.
- 2004. The Way We Trust Now: The Authority of Science and the Character of the Scientists. In *Trust Me, I'm a Scientist*, edited by Pervez Hoodbhoy, Daniel Glaser, and Steven Shapin, 42–62. London: British Council.

Stewart, L. 2004. *La buona e non macherata filosofia*, or the Exhbitionists. In *Creating Connections. Museums and the Public Understanding of Current Research*, edited by David Chittenden, Graham Farmelo, and Bruce V. Lewenstein, 29–46. Walnut Creek and Oxford: AltaMira Press.

Taylor, C.A. 1996. *Defining Science. A Rhetoric of Demarcation*. Madison: University of Wisconsin Press.

Weart, S.A. 2003. *The Discovery of Global Warming*. Cambridge, MA, and London: Harvard University Press.

2 Science, Democracy, and Philosophy: From Marginal Achievements to Impossible Opportunities

CARL MITCHAM

Taking relations between science, democracy, and philosophy as a theme, let me begin by erecting a scaffold for reflection on these three key elements in European history. Consider a generalized historico-philosophical narrative that goes something like this: In the Greek and Roman worlds philosophy was the source of knowledge (as opposed to opinion) about both the good and nature. In the medieval period philosophical knowledge or reason about the good was displaced by theological knowledge or revelation, but philosophy remained as the basic source of knowledge about nature. This is why Aristotle, the archetypical natural philosopher of antiquity, after his recovery in the West, was so often invoked as an authority for natural scientific (that is, systematically organized) knowledge.

The modern period witnessed, indeed was constituted by, a major plot shift in this narrative. The shift took place in three (overlapping) phases. In phase one, modern natural science arguably superseded natural philosophy as the basic form of knowledge about nature. But natural science established its new primacy with a paradoxical appeal to revealed theology. Because revelation offers more certainty than philosophy concerning morals, Francis Bacon argued that human beings could forgo further discussion about ethical issues – a discussion which, in the Socratic tradition, produced a kind of uneasiness about worldly activity – and simply undertake the task assigned by God: that is, worldly dominion. As Bacon concludes the first book of his *Novum Organum* (I, 129), the inventions of printing, gunpowder, and the compass have done more to change 'the whole face and state of things throughout the world' than politics, sectarian religion, or philosophy. Human beings should therefore turn their attention and energies from politics and speculation, whether supernatural or natural, to scientific investigation and technological transformation of the world – and philosophers should promote and assist in this turning, as Bacon himself sought to do.

In phase two, supernatural theology or revelation was gradually marginalized as a source of moral knowledge. Despite its original appeal to and dependency on theology, science exercised a corrosive effect on the idea of revelation. Being difficult to restrain within its original bounds, scientific naturalism was applied by Thomas Hobbes, Baruch Spinoza, and others to historical texts such as the Bible. The naturalization of Scripture led to its reinterpretation as a human, all too human, construction. The former presentation of scriptural texts as revealed truth was replaced by their presentation as disguised statements of human interest or aspiration, as metaphorical proclamations of value, which could be criticized by philosophy. Thus did philosophy reclaim its role as a source of moral knowledge, a position historically manifested in the Enlightenment promotion of liberal democracy as that social order which is at once rational and good.

Then in phase three, philosophy itself became subject to the corrosive influence of scientific naturalism. In premodern philosophy, the claim to moral knowledge rested on the existence of a reality to which philosophy claimed privileged access. In the famous allegory of the cave and the myth of the divided line at the centre of Plato's *Republic*, the ascent from illusion to reality leads ultimately to a recognition of the good itself as 'beyond being,' the experience of which is available through a unique form of activity known as *noesis* or rational intuition. Even in the more earthy Aristotelian science, natures were understood to be imbued with *teloi* or ends that were philosophically discernible by those who exercised the intellectual virtues.

In the modern period the existence of such a supersensual or noumenal reality beyond or within phenomenal appearances was defended as well by Immanuel Kant. For Kant, it was moral metaphysics that best connected with this level of reality and thereby disclosed the categorical imperative. But for all the effort in Kant's critical idealism to delimit the power of science 'in order to make room for faith' (*Kritik der reinen Vernunft*, Bxxx) in the philosophical as well as religious senses, the best that such special knowledge provided was a regulative ideal of personal autonomy as the metaphysical foundation of ethics. Any more full-bodied disclosure of the good was to be found neither in theology nor in philosophy – but in democracy. Let human beings more openly state and adjudicate their values and interests.

In the modern period, then, science has replaced philosophy as the source of knowledge about nature, and liberal democracy has replaced philosophy as the source of knowledge about the good. The historical claim of philosophy to have been the matrix out of which arose both science and democracy is no more than homage to the past. Where does this leave philosophy in the future? That is, what can philosophy contribute to the pursuit of science in the context of liberal democracy? The three standard answers are that philosophy can help clarify logical and conceptual difficulties in science, defend and explicate procedural issues in democracy, and provide guidelines for the best relations between scientific expertise and democratic decision making. Is there nothing more? Before we consider the last question, it is appropriate to acknowledge the real if limited achievements in each of the three basic responses, which may be termed the projects of philosophy as handmaid to science, to liberal democracy, and to science–democracy relations.

Handmaid to Science

Insofar as philosophy became the handmaid of science, the branches of philosophy – metaphysics, epistemology, and logic – were transformed into philosophy of science. The pioneering achievements of this transformation were, on the one side, the empiricist methodologies of Bacon and his followers and, on the other, the rationalist methodologies of René Descartes and his followers. The bridge between these philosophical founders of modern science and the more modest contributions of such twentieth-century philosophers as Rudolf Carnap, Karl Popper, and Carl Hemple was the nineteenth-century transitional figure of William Whewell.

Unlike Bacon or Decartes, Whewell no longer proposed to argue a foundational, synoptic metaphysics and ethics, but he had not yet become a specialist in the analysis of scientific method like Carnap et al. Instead, he was what could be called an interdisciplinary scholar who wrote on mechanics, mineralogy, geology, astronomy, politics, theology, ethics, education, international law, architecture, as well as the history and philosophy of science. Inspired by the poet Samuel Taylor Coleridge, he famously coined the term 'scientist' to denominate those previously identified as 'natural philosophers' or 'men of science.' Instead of a philosophical school, he founded or promoted social institutions such as the British Association for the Advancement of Science, the Royal Society, and the Geological Society (serving at different times as president of each). In his effort to advance science he further contributed to the conceptualizing and naming of 'anode,' 'cathode,' and 'ion' for William Faraday. More generally, he promoted the self-understanding of science as a culturally distinct form of human thought and action.

It was his notion of philosophy as an assistant to the advancement of science that has become the leading vision in the philosophy of science. Although there have been disagreements about how it can best serve this role – from proposals for positivism, rational reconstruction, conceptual clarification, and more – the idea that philosophy should be dedicated to helping demarcate science from non-science and to enhancing the internal logic of science has since Whewell remained central to the philosophy of science. Philosophy of science in turn has come to be understood as the strongest form of philosophy. Its contributions to science are not unworthy of this reputation, but they are thinner than those that characterized pre- and early modern practices of philosophy. There has been a shift, one might say, from philosophical reflection on science to philosophical reflection in science. Within the philosophy of science it is true that there has existed a persistent debate between realism and anti-realism (or instrumentalism) as alternative views of the cognitive character of scientific knowledge. Yet since neither pole in this debate questions the view that science is the most reliable form of knowledge of nature, it is sufficient here simply to note rather than to review such discussions.

Handmaid to Democracy

Insofar as philosophy became the handmaid of liberal democracy, political philosophy became democratic theory. Like the philosophy of science, classic modern democratic theory has advanced democratic self-understanding, especially in relation to issues of equality (in the work, for instance, of John Locke, Jeremy Bentham, and John Stuart Mill) and solidarity (Jean Jacques Rousseau, Immanuel Kant, and G.W.F. Hegel). The tensions between utilitarian individualism, in the British and North American traditions of democratic theory, and the more organic or communitarian emphases of the continental European tradition, continue to play out in the visions and practices of globalization.

For a time, during the mid-twentieth century, mainstream political theory seemed to have exhausted itself in conceptual examinations of 'the vocabulary of politics' (Weldon, 1953) and/or the 'end of ideology' (Bell, 1960) – a version of the latter being revived in post–Cold War 'end of history' speculations (Fukuyama, 1992). In the last third of the century, however, democratic theory was re-energized in North America by John Rawls (1971) and Robert Nozick (1974). The ensuing debate between the proponents of equality (Rawls) and liberty (Nozick) is nevertheless no more than a debate within democratic theory rather than a fundamental reflection on democracy itself. Just as the philosophy of science, insofar as it assumes science as the primary source of valid knowledge of nature, narrows the scope of epistemological reflection, so too does the assumption of democracy as the sole legitimate social order establish boundaries for political philosophy.

In modern democratic theory the expression of individual interests is taken as the primary source for knowledge of the good. There are two interlocking arguments related to such a belief. Ethically, since nothing is morally superior to the person, no one is more qualified to speak about persons' interests than persons themselves. All citizens are presumed to have equal access to knowledge of the good precisely because it is their good that is ultimately at issue. Equality among free citizens is the necessary and perhaps sufficient condition for the manifestation of this knowledge. Epistemologically, as in science, the collective is further argued to be wiser than any individual. Collaborative knowledge production is superior to that of individuals insofar as collaboration can correct for individual biases. A challenge for democratic theory working within such a framework is to develop the most adequate versions of such arguments and to determine in detail their implications and applications under conditions increasingly influenced by science and technology.

Reflecting the tensions between ethics and epistemology, democratic theory has been riven by questions concerning the ability of democratic practices to yield knowledge of the good. Precisely because such questions affect the ways philosophy can function as a handmaid to democracy, it is appropriate to consider in modest detail the example of responses to a widely cited article by political scientist Philip Converse (1964) that challenged the rationality of democratic voting behaviours. According to Converse, only 10 per cent of the public in mass democracies, those he terms ideologues, exercise voting choices rationally as part of some 'political belief system.' The majority are non-ideologues whose votes reflect simply perceived self-interest (42 per cent), whether times are good or bad (25 per cent voting for incumbents in good times and against in bad), or no 'issue content' factor at all (22 per cent). Moreover, in many cases neither self-interest nor the character of the times is accurately assessed. Mass democracy appears to be largely irrational. As social critic Louis Menand (2004) observed forty years later, empirical studies imply that democratic practice no more expresses the true interests of the majority than does the divine right of kings.

Short of accepting the failure of democracy, there have arisen three responses. The first is that policy elites act as rational leaders. The second is that people use heuristics that do in fact reflect their true interests. The third is an argument for the distinctive wisdom of collective opinion.

The policy elites theory can be traced back to late nineteenth- and early twentieth-century ideas of engineer-sociologist Vilfredo Pareto and were given democratic interpretation even before Converse in the work of Gabriel Almond (1950). In an analysis of democratic foreign policy formation, Almond imagined a pyramid with a small group of leaders at the top, supported by a stratum of policy elites, themselves above an attentive public. At the base of the pyramid remained the inattentive general public. Although little rational communication might take place between the leaders and the general public, rational interactions did occur between the policy elites and the leaders (going up) and policy elites and attentive publics (going down). Further quasi-rational interactions were allowed to take place between the attentive and the non-attentive publics, with individual citizens altering their attention levels depending on particular issue interests. Rationality and rational disclosure of true interests, and therefore knowledge of the good, were mediated by a series of interconnected linkages.

As if to confirm the policy elites theory, the rise of both the progressive-liberal movement in the 1930s and the neoconservative movement in the 1980s grew out of elite policy work: the first among pragmatist progressives such as John Dewey, Walter Lippman, Reinhold Niebuhr, and George Kennan; the second among conservative theorists such as Irving Kristol and his son, William Kristol, Allan Bloom, and William F. Buckley, Jr. In each case, issue-focused elites arose from and then attracted attentive publics, who in turn mediated between the policy elites and the inattentives.

The heuristics theory is that citizens vote by means of legitimate short cuts unrespected by intellectuals. The theory of heuristics in decision theory was initially developed by mathematician George Pólya (1945) but then took root among computer scientists and cognitive psychologists who recognized the existence of types of reasoning in between strict logic and purely arbitrary decision making. Pólya called this 'plausible inference.' Unlike algorithms or procedures that always crank out the correct answer, while being unable to address complex problems, heuristics not only produce generally correct answers but can deal with massively complex problems. One example of heuristics is political parties. People vote for candidates nominated by parties to which they are loyal because of the long-term policies of the parties. Given the fact that politicians do not always follow party discipline, and that parties themselves sometimes paper over divergent interests, any one vote may be less than fully rational. In aggregate, however, party votes will tend toward rationally representing an individual member's interests. But even voting for candidates who are tall or photogenic may on occasion be rational short cuts: tall people do in fact tend to be more self-confident and decisive, and many photogenic people bring with them powers of persuasion and influence. When for whatever reason state leadership calls for decisiveness or influence, using such cues can serve rational ends. Political scientist Samuel Popkin (1991) defends heuristic-based democracy as what he terms 'low information rationality.'

Collective opinion theory, although advanced initially by political scientists during the early 1990s, has more recently been articulated by James Surowiecki (2004). According to Surowiecki, the expert distrust of mass opinion is not always justified. 'Under the right circumstances, groups are remarkably intelligent, and are often smarter than the smartest people in them' (xiii). One illustration is the search engine Google, which increases intelligence on the basis of collective search responses by users who are mostly non-experts. For Surowiecki the 'collective intelligence' of a crowd emerges from opinion diversity (which introduces different perspectives), member independence (which keeps others from being swayed by a dominant member), decentralization (which allows diverse errors to balance out), and an appropriate aggregating method (which guarantees full inclusion). Under such conditions the results are often 'smarter' than any individual expert – and in many instances such conditions are approximated in voting behaviours.

Each of these three efforts to defend democracy against the Converse challenge – that is, the theories of elite leadership, heuristics, and collective wisdom – constitutes a political theoretical effort to serve the interests of liberal democracy and in this sense may be described as placing boundaries on political philosophy. Democracy is defended as the primary source for knowledge of the good. At the same time – as the previous references to sociology, mathematics, cognitive science, and economics indicate – democratic theory work often breaches boundaries through the practice of interdisciplinarity. Moreover, because the work at issue has a bearing on democratic practice, this broad interdisciplinarity on occasion even makes an anti-disciplinary foray into the public realm to address problems of how to make democracy more functional. Today, however, the functioning of liberal democracy is more than an issue of voting behaviour. It involves as well multiple interdisciplinary challenges related to science and technology, especially that merger of science and technology known as technoscience.

Handmaid to Science–Democracy Relations

Natural knowledge comes from science, moral knowledge from democracy. What then is the relation between science and democracy? Insofar as philosophy becomes the handmaid of science–democracy relations it takes on a role beyond philosophy of science and democratic theory by arguing for the democratic guidance of technoscience – that is, for the establishment of relations between modern knowledge of nature and knowledge of the good.

The argument for integrating democracy and technoscience takes place in relation to both aspects of the complex amalgam of technological science and scientific technology: that is, in relation to science and to technology. With regard to modern science, the scientific community has conceived of itself as necessarily a self-governing, representative democracy. Although scientists understand themselves as an elite in relation to the rest of society, elites are to be strictly limited within science. Only by means of the internal rejection of any aristocratic appeals to hierarchies of inheritance or other unearned forms of authority can scientific knowledge be validated through democratic reproducibility and peer review. This is an argument that has been advanced in various forms by both social scientists (see, e.g., Robert Merton 1942) and physical scientists (e.g., Michael Polanyi 1962). Concepts of the aristocratic autonomy of science from any external democratic control go hand in hand with arguments for internal democratic governance. Indeed, for some scientists the functioning of democracy within science provides a model for larger democratic social orders.

With regard to modern technology, however, there arises an argument for some kind of subjection to external democratic control. The argument here is closely associated with the interdisciplinary field of science, technology, and society (STS) studies, and its articulation was pioneered by and often associated with the STS scholar Langdon Winner (1980). Indeed, Winner's argument for the political character of technological artifacts has undergirded much subsequent discussion of this issue. Moreover, insofar as science becomes technoscience, the idea of some external democratic participation concerning technoscientific decision making tends to trump arguments for scientific autonomy.

Further important statements of this argument have come from Andrew Feenberg (1995) and Richard Sclove (1995), both of whom make extended briefs for public participation in technoscientific decision making. Feenberg proceeds from a perspective influenced by Frankfurt School–based critical social theory. Sclove situates himself within a framework provided by Benjamin Barber (1984). Both argue that, in the words of Feenberg,

> as technology becomes central to more and more aspects of our lives, its legislative authority increases. But [when] technology is so powerful, then surely it should be measured by the same democratic standards as other political institutions. (Feenberg 1995: 5)

Mark Brown, a student of Barber's, has developed the argument in further institutional detail by means of an extended case study on a California electric vehicle program from the late 1990s. According to Brown,

> the assertion of political sovereignty over technical activities and artefacts begins with the recognition that it is human agents [who conduct] inquiries and [build] machines ... Because science and technology are human creations, they are subject to governance by human communities ... [Moreover,] because science and technology continuously transform political life, democratic citizens require opportunities to participate in the construction and regulation of science and technology. (Brown 2001: 8–9)

As Brown himself reveals, the ideal of democratic participation is not something simple. There exist a variety of kinds and degrees of participation. For many turbo-capitalists the trial-and-error responsiveness of the market economy is the best form of participation. For journalists, media exposure and governmental response to public opinion serve the same end. Since liberal democracy is also representative democracy, participation will further entrain some theory of representation. But prefatory to sorting out the kinds and degrees of participation with which to confront the dangers of scientism and the challenges of technocracy – not to mention their adequacies and inadequacies – it would be useful to unpack more fully the reasons for such confrontation. In the present instance, one can distinguish deontological, consequentialist, and virtue ethics justifications for some form of democratic participation in technoscientific decision making (but see Mitcham 1997 for a related analysis).

First, perhaps the Feenberg, Sclove, and Brown reasoning can most easily be interpreted in deontological terms. Those who are affected by technical decisions – whether the effects be good or bad – should have some say over what affects them. To some degree this simply updates the Kantian concept of moral autonomy for a technoscientific context. Human beings are moral agents whose agency is abridged when decisions that affect their lives are made heteronomously by others, including scientists and engineers or the products, processes, and systems they design, construct, or operate. In the special area of biomedical research, the well-established principle of requiring free and informed consent from human subjects is a regional application of Kantian deontology that also exemplifies democratic concerns. But the 'say' that citizens in general should be able to exercise across technoscientific affairs need not be veto power, after the manner of informed consent in biomedicine. It is better encapsulated in something like Steven Goldman's motto 'No innovation without representation' (Goldman 1992).

Second, the argument here can also be framed in diverse consequentialist terms. The most common consequentialist argument is simply that of utilitarianism: public participation increases happiness. Of course, some technocrats would contest this idea with the counterargument that the experts are actually better at producing whatever goods or services promote happiness. But practically, in many instances the public simply demands some kind of participation before a particular technoscientific project is going to be allowed to move forward. As the NIMBY (Not in My Back Yard) and BANA (Build Absolutely Nothing Anywhere) syndromes show, there are many instances in which without some form of public participation little will get done – or will get done only at the expense of excessive cost and with the generation of considerable ill will.

In a more positive consequentialist vein, public participation has been argued to produce better outcomes. Like the public demand argument, this stresses the need to face the reality of citizen veto power in order to get something done. But here the idea is not just getting something done faster and cheaper, as it were, but better. Involving an affected public, and drawing on local knowledge, can often make for better products – and thus more happiness.

A further consequentialist-related argument could also be called an argument from technosocial realism. Experts simply cannot escape non-expert influence. There will either be corporate influence or influence of some other kind. Influence is pervasive, whether technoscientific experts like it or not. Indeed, psychological studies have confirmed that experts inevitably tend to promote their own self-interests, often at the expense of those of the public. Technoscientific decision making is never neutral or objective. Of course, the technoscientific establishment can try to minimize such influence. Another and complementary strategy would be to make whatever in fact exists at least clear and open: to turn influence into well-articulated participation. Transparency is closely allied to participation.

Another consequentialist argument concerns postmodern culture. The over-riding ethical feature of the postmodern social order is pluralism or a lack of any thick political consensus about many issues. Tolerance, diversity, public ethical minimalism are the marks of the technoscientific world. The best to be had in such a situation is participatory, democratic, give-and-take consensus. Otherwise technoscience will create its own incentives and be unopposed, undirected, more or less autonomous.

The third and final justification for democratic participation in technoscientific decision making involves a virtue ethics argument. Only from participation will citizens be able to learn and become more intelligent about their political and economic support of technology as well as about the real risk-and-benefit complexities of technical action. With roots in the educational theory and social philosophy of Dewey, a related and perhaps richer version of this argument stresses the processes of societal learning and emerging intelligence for dealing with the complexities of technoscientific life.

By means of such arguments and their relatives – including many of those associated with the more specialized fields of applied ethics such as biomedical ethics, environmental ethics, engineering ethics, and computer ethics – activists and scholars in the STS studies field have promoted the building of bridges between the scientific pursuit of knowledge of nature and democratic efforts to articulate the good.

Impossibilities

Under the conditions associated with modernity, philosophy has found itself restricted to clarifying logical or conceptual difficulties in science, defending democracy, and providing guidelines for relations between technoscientific expertise and democratic decision making. These are not trivial activities, and the achievements – from conceptual clarity in science, increased appreciation of the complexities of democracy, and contributions to democratic science and technology policy – are not without significance. Nevertheless, is it unreasonable to long for something more? The argument here is that there exist what may be termed impossible opportunities that come into focus especially with regard to the issue of public science in liberal democracy.

To begin, consider what may be meant by 'public science.' In the background is the idea of a socially constructed public within and for experimental science (see, e.g., Shapin and Schaffer 1985; Latour, 1993). But in the foreground are three other interrelated meanings. First, public science can be publicly understood science. The International Public Science Day program sponsored by the American Association for the Advancement of Science (AAAS) seeks, for instance, to promote the more public understanding of science or what is also known as scientific literacy. The adequacy of the top-down education of the public by scientists has nevertheless been challenged, especially in Europe, with arguments for a parallel learning from the public by scientists (Stilgoe, Wilsdon, and Wynne 2005).

Second, 'public science' can refer to publicly funded science. Such funding is sometimes argued to depend on a measure of scientific literacy and associated appreciation of the significance of science. It may also be argued that while public funding is usually identified with government science agencies such as the U.S. Geological Survey and the National Institutes of Health or extramural funding by agencies such as the U.S. National Science Foundation or even the military, funding by private corporations likewise depends to some degree on the public, since it is consumers who ultimately pay by purchasing the results. Marketplace funding is another kind of public funding of science.

Third, 'public science' can be another name for public activism by scientists and engineers. Here examples would be the Federation of American Scientists, the Union of Concerned Scientists, and International Pugwash. Idealistic scientific activism, as illustrated by the movement for science in the public interest, is an often neglected aspect of public science (see Mitcham 2003).

But it is possible to imagine another meaning for public science, one not so much opposed to the three previous conceptions as complementary and of more immediate significance: public science as science (and technology) that is integrated with public flourishing, with human well-being – well-being not in the sense of simple material welfare, but in Aristotle's sense of *eudaimonia*, what may also be called, from a slightly different perspective, the common good. There is, for instance, an emerging body of research on public policy and happiness that has direct bearing on science policy (see, e.g., Lane 2000; Layard 2005). Public science in this sense is coordinate with public art and public architecture. The possibility is that, contrary to modern arguments and assumptions, philosophy might have the potential to contribute a knowledge of both nature and the good that goes beyond what is available through philosophy of science and democratic theory, and that this philosophical knowledge can influence for the better both science and liberal democracy.

This is a difficult, not to say impossible, idea. It is certainly not one that can be fully developed in the present instance. It is nevertheless worth provisional exploration, if only to make us more fully aware of its impossible character. In this spirit, then, consider two emerging philosophical discussions: those centred in the new humanism debate and those in technological design criticism. For present purposes, the work of Leon Kass and of Albert Borgmann will be used as illustrative.

Kass, as chair of President George W. Bush's Council on Bioethics from 2001 to 2005, is a controversial figure. But he is relevant to the argument at issue precisely because of the way he has promoted disciplined reflection on what he has termed the dangers of 'soft dehumanism' associated with the popular faith in biomedical science and technology. As he commented on the establishment of the Bioethics Commission in early 2002:

> The moral challenges the Council faces are very different from the ones confronting the President and the nation as a result of September 11. In the case of terrorism, as with slavery or despotism, it is easy to identify evil as evil, and the challenge is rather to figure out how best to combat it. But in the realm of bioethics, the evils we face (if indeed they are evils) are intertwined with the goods we so keenly seek: cures for disease, relief of suffering, preservation of life. Distinguishing good and bad thus intermixed is

often extremely difficult. (President's Council on Bioethics, 2002, session 1, paragraph 3)

In the middle of the twentieth century, the hard threats of Nazism and Communism were characterized by top-down, centralized control in the name of an ideal that went against the human grain – threats with which terrorist fanaticism is arguably continuous. By contrast, the soft threats of the present are bottom-up, decentralized appeals to desires that are all too in harmony with democratic, market-driven consumerism. Indeed, Kass's new humanism is opposed by democratic transhumanism and arguments for personal human enhancement as both a right and a responsibility (Hughes 2004).

In an extended reflection that points in the opposite direction, *Beyond Therapy: Biotechnology and the Pursuit of Happiness*, Kass and colleagues (2003) identify four areas in which democratic desires are in the process of problematically exploiting biotechnological possibilities: aspirations for better children, superior performance, ageless bodies, and happy souls. In bioethics it is not uncommon to examine critically a number of issues associated with such pursuits: health, safety, and risk; unfairness; equality of access; and personal liberty. But for Kass such familiar sources of concern remain somewhat superficial, and he proposes four others that cut deeper and are much less well examined: the dialectic of mastery, the possible unnaturalness of technoscientific means, challenges to identity and individuality, and the dubious character of the very ends in view – that is, especially, better children, ageless bodies, and happy souls.

In short, Kass invites philosophically inclined citizens to reflect more deeply than is common on what it means to be human and on the nature of the human condition. In this he asks people to draw on the wisdom of the humanities as found in epic poetry, classical tragedy, and the great novels of modernity. In such works Kass suggests there exists a caution about the limits of human aspiration and action that people in advanced techoscientific societies would do well to revisit as scientific knowledge and technological powers reach outward to the stars and inward to the nanoscale reconstruction of the material matrix of both non-living and living forms.

Borgmann's work is less overtly controversial (in part, no doubt, because it is less close to political power). It also contributes to an emerging philosophical reflection on engineering and technological design – the nature and good of artifacts. The central concept in Borgmann's philosophy of technology is the device paradigm as contrasted with focal things and practices. According to Borgmann (1984), contemporary life has been and is being structured in a new way as technological devices permeate the culture.

One telling example concerns heat. Household heating was once centred in the hearth, which served both comfort and cooking but required care. Wood had to be harvested, split, and brought into the house; the fire had to be tended. Such activities called forth skills and created familial divisions of labour. The hearth itself was also a particular artifact, a thing of stone, eventually cast iron, around which the family gathered for warmth, for the preparation and sharing of a meal, or perhaps just to talk over the experiences of the day. This romantic description must no doubt be complemented by the acknowledgment of difficulties arising from shortages of wood and quarrels concerning the distribution of chores.

Today few houses in the developed world have hearths or wood stoves. People in the advanced technoscientific world have been disburdened from the tasks of heating through the gradual creation of the central heating system, a device which recedes into the quotidian background. People simply become consumers of a commodity inconspicuously provided by such a device: warmth. There is no hassle, no aroma of burning wood, no conspicuous artifact occupying a place in the room – and no central gathering point either.

Borgmann's homey example aims to disclose how the device paradigm alters the character of a culture in its ambit. The disburdenment it brings into the human world tends to undermine both care and well-being in the premodern senses. Work transformed by devices often loses the pleasures of skilled making or finds those pleasures taken away from the body and resituated in some virtual space. The job becomes merely a means for earning money by sitting at a computer screen. Leisure declines in character as well, insofar as it is filled with excessive consumption and entertainment.

One may summarize Borgmann's analysis of the character of contemporary life by describing it as having become through the paradigm of device technology 'unfocused' or 'disintegrated.' His response is thus to propose the reintroduction of anti-device focal things and practices. What technoscientific culture lacks in its material richness is physically or practically centred depth. Technoscientific disburdenment alone is unable to yield a deep or meaningful life. At most, technology can free people to enjoy things and practices in an old-fashioned sense. Indeed, for Borgmann it is 'when we allow ourselves to be guided by focal things, matters of ultimate concern that are other and greater than ourselves' that the good life comes to the fore in poetic discourse (Borgmann 1984: 169).

In such discourse, speech becomes testimonial and poetic in character. One cannot force others to recognize focal things. Yet once people have experienced the orienting power of focal phenomena, they commonly want to bear witness. Focal things thus not only concentrate life but act as touchstones for technology assessment. Focal reality is a philosophically articulated good to be offered democratic citizens, a good they are to be both philosophically and poetically encouraged to choose.

Borgmann's views are often dismissed as unrealistic if not idealistic. But in an essay that is already echoed in the preceding review of Borgmann's thought, Topi Heikkerö (2005) suggests that focal things and practices such as Kendo, a martial art, and Chado, the tea ceremony, have actually functioned in Japan to help maintain some balance in the technoscientific world. Japan's long and philosophically sophisticated tradition of focal practices, Heikkerö argues, can be used to reveal how Borgmann's argument is not so unrealistic after all:

> Acknowledging this parallel provides support for Borgmann's stance that consciously cherished focal things and practices are the ground for a deep and meaningful life within the technological world. Without them, life lacks gravity and loses reality. Focal things provide a metatechnological viewpoint that enables us to weigh technological devices. (Heikkerö 2005: 258)

Taking his argument further, Heikkerö inquires why it is that Japan has developed a vigorous culture of consciously cherished focal things and practices such as Kendo and Chado. As a speculative response he references a provocative comment found in a neo-Hegelian analysis of the end of history by Alexandre Kojève. According to Kojève (1969) the end of history is defined by the appearance of the universal and homogeneous state in which nature is mastered, there are no wars, and people are equal. But this is a condition that has been experienced in partial forms well before the dreams of globalization. For Kojève, Japan during the Tokugawa or Edo period (1603–1867) had in important respects arrived at such an 'end of history.'

What happens with the coming of such an end time to a humanity previously defined by its worldly struggles? Will human beings revert to an animal life without ideals once there is nothing 'real' to do? Will they become Friedrich Nietzsche's 'last humans'? Using Tokugawa Japan as a hint, Kojève argued that there is another possibility. In their end state the Japanese became cultivators of such refined and 'snobbish' practices as the martial arts and the tea ceremony – practices that fully transcend animal behaviour. In fact, in the 1950s Kojève forecast that in the future the world, including Europe, might well become more Japanized than Americanized.

As the reference to Kojève suggests, there is also an important qualifier to any idea that philosophy can bring knowledge of the good and nature to bear on science and democracy. This impossible opportunity depends on a philosophy that has more in common with religion, especially religion as exemplified by Japanese Buddhism, than with contemporary professional philosophy. Indeed, premodern philosophy, in both its Greek and medieval forms, also had more in common with Buddhism than with analytic professionalism. Platonism especially was more a way of life than is contemporary philosophy, one that involved conceptual analysis but also moral discipline. What became the intellectual methodology of science, as a means to knowledge of nature, was in the Platonic tradition a broader experiential *askesis*, designed to yield special insight into both human aspirations and nature (see Hadot 2002). That recent work on the social construction of science has revealed the practical dimensions of scientific methodologies, including their dependencies on technological instrumentation, provides further support to this contrast.

In Indian Buddhism one of the words translated as 'philosophy' was *darshana*, which means 'to see,' 'to hear,' or 'to understand.' As Malcolm David Eckel (2001) has argued in a historical introduction to Buddhism, Buddhist philosophy is meant as a tool to help students see reality free of the illusions that cause suffering, especially the illusions manifested in the most common desires. The same word names the emotionally charged disclosure of an image in early Buddhist rituals of worship. It is likewise etymologically related to a form of meditation known as *vipassana*, meaning discriminating vision or insight. Philosophy in the Buddhist tradition is not simply a theoretical activity, but a form of practice that aims to torpedo those desires that Kass sees as motivating our scientific and technological quests for better children, ageless bodies, and happy souls.

In a provocative Buddhist history of ideas in the West, David Loy (2002) interprets the dynamics of European culture as a series of case studies confirming the truth of the doctrine of no-self. According to Buddhism, the self is an illusion that is continuously trying to avoid facing up to its illusory nature, about which it is nevertheless vaguely aware and anxious. Anxiety is heightened by the combination of humanism and Christianity and their attempts to defend or redeem the self. But there remains an 'inescapable trace of nothingness in my "empty" (because not really self-existing) sense-of-self that is experienced as a sense-of-lack' that people in the West have repeatedly tried to overcome by various means (Loy 2002: 6). Today these means are primarily science and technology, and thus in thrall to illusion.

In the early medieval period in the West, St Anselm defined theology as faith seeking understanding. In a like manner one may describe Buddhist – and Platonic – philosophy as a seeing through illusion in search of understanding. The understanding at issue requires a clarification of the mind that is more than intellectual. Only through such more than intellectual exercise is it possible to become aware of otherwise obscured dimensions of the good and of nature.

By way of conclusion, consider again the faux-master narrative with which this exposition began: the historical trajectory that runs from premodern philosophy as a source of knowledge of both the good and nature, through medieval philosophy as reduced to natural knowledge alone, and finally to modern philosophy as the handmaid of science (true knowledge of nature) and democracy (the process for determining or projecting the good). Although no doubt a caricature of European intellectual history, it nevertheless throws into relief the challenges facing relations between public science and liberal democracy. Can science be public in the sense of public art, architecture, and policy? Is its democratic governance sufficient to build a bridge between scientific knowledge, technological power, and the good? The thesis presented for consideration here is that only philosophy in a quite traditional – that is, non-modern – sense offers the impossible opportunity to realize such an end. Of course, the revival of philosophy in this sense is not without its own impossibilities. But if Buddhism can remain a viable practice, even in the West, why not Platonism?

REFERENCES

Almond, Gabriel. 1950. *The American People and Foreign Policy.* New York: Harcourt Brace.

Barber, Benjamin. 1984. *Strong Democracy: Participatory Politics for a New Age.* Berkeley: University of California Press.

Bell, Daniel. 1960. *The End of Ideology: On the Exhaustion of Political Ideas in the Fifties.* Glencoe, IL: Free Press.

Borgmann, Albert. 1984. *Technology and the Character of Contemporary Life: A Philosophical Inquiry.* Chicago: University of Chicago Press.

Brown, Mark B. 2001. *Civilizing Science: The Joint Construction of Science and Citizenship in Modern Democracy.* Doctoral dissertation. New Brunswick, NJ: Rutgers, The State University of New Jersey.

Converse, Philip. 1964. The Nature of Belief Systems in Mass Publics. In *Ideology and Discontent*, edited by David E. Apter, 206-61. New York: Free Press.

Eckel, Malcolm David. 2001. *Buddhism.* Chantilly, VA: Teaching Company.

Feenberg, Andrew. 1995. *Alternative Modernity: The Technical Turn in Philosophy and Social Theory.* Berkeley: University of California Press.

Fukuyama, Francis. 1992. *The End of History and the Last Man.* New York: Free Press.

Goldman, Steven L. 1992. No Innovation without Representation: Technological Action in a Democratic Society. In *New Worlds, New Technologies, New Issues,* edited by Stephen H. Cutcliffe, Steven L. Goldman, Manuel Medina, and José Sanmartín, 148–60. Bethlehem, PA: Lehigh University Press.

Hadot, Pierre. 2002. *What Is Ancient Philosophy?* Translated by Michael Chace. Cambridge, MA: Harvard University Press.

Heikkerö, Topi. 2005. The Good Life in a Technological World: Focal Things and Practices in the West and in Japan. *Technology in Society* 27, no. 4: 251–9.

Hughes, James H. 2004. *Citizen Cyborg: Why Democratic Societies Must Respond to the Redesigned Human of the Future.* Cambridge, MA: Westview Press.

Kass, L. 2003. *Beyond Therapy: Biotechnology and the Pursuit of Human Improvement.* Paper presented to the President's Council on Bioethics, 16 January. Washington, DC.

Kojève, Alexandre. 1969. *Introduction to the Reading of Hegel: Lectures on the Phenomenology of Spirit.* Assembled by Raymond Queneau, edited by Allan Bloom, translated by James H. Nichols Jr. Ithaca, NY: Cornell University Press. (Original French publication, 1947.)

Lane, Robert E. 2000. *The Loss of Happiness in Market Democracies.* New Haven, CT: Yale University Press.

Latour, Bruno. 1993. *We Have Never Been Modern.* Translated by Catherine Porter. Cambridge, MA: Harvard University Press.

Layard, Richard. 2005. *Happiness: Lessons from a New Science.* New York: Penguin.

Loy, David R. 2002. *A Buddhist History of the West: Studies in Lack.* Albany: State University of New York Press.

Menand, Louis. 2004. The Unpolitical Animal. *The New Yorker,* 30 August.

Merton, Robert K. 1942. A Note on Science and Democracy. *Journal of Legal and Political Sociology* 1: 115–26. Reprinted as The Normative Structure of Science, in Merton, *The Sociology of Science: Theoretical and Empirical Investigations,* ed. Norman E. Storer, 267–78. Chicago: University of Chicago Press, 1973.

Mitcham, Carl. 1997. Justifying Public Participation in Technical Decision Making. *IEEE Technology and Society Magazine* 16, no. 1 (Spring): 40–6.

– 2003. Professional Idealism among Scientists and Engineers: A Neglected Tradition in STS Studies. *Technology in Society* 25, no. 2 (April): 249–62.

Nozick, Robert. 1974. *Anarchy, State, and Utopia.* New York: Basic Books.

Polanyi, Michael. 1962. The Republic of Science: Its Political and Economic Theory. *Minerva* 1, no. 1 (Autumn): 54–73.

Pólya, George. 1945. *How to Solve It: A New Aspect of Mathematical Method.* Princeton, NJ: Princeton University Press.

Popkin, Samuel L. 1991. *The Reasoning Voter: Communication and Persuasion in Presidential Campaigns.* Chicago: University of Chicago Press.

President's Council on Bioethics. 2002. Transcript of 17 January meeting. http://www.bioethics.gov.

Rawls, John. 1971. *A Theory of Justice.* Cambridge, MA: Harvard University Press.

Sclove, Richard. 1995. *Democracy and Technology.* New York: Guilford Press.

Shapin, Steven, and Simon Schaffer. 1985. *Leviathan and the Air Pump: Hobbes, Boyle, and the Experimental Life.* Princeton, NJ: Princeton University Press.

Stilgoe, Jack, James Wilsdon, and Brian Wynne. 2005. *The Public Value of Science: Or How to Ensure that Science Really Matters.* London: Demos.

Surowiecki, James. 2004. *The Wisdom of Crowds: Why the Many Are Smarter than the Few and How Collective Wisdom Shapes Business, Economies, Societies, and Nations.* New York: Doubleday.

Weldon, T.D. 1953. *The Vocabulary of Politics.* Baltimore: Penguin.

Winner, Langdon. 1980. Do Artifacts Have Politics? *Daedalus* 109, no. 1 (Winter): 121–36.

3 Public Geoscience at the Frontiers of Democracy

DEBORAH R. HUTCHINSON AND
RICHARD S. WILLIAMS, JR

Introduction

In common with science, the American form of democratic government functions best in an atmosphere of open, tolerant exchanges of ideas by an educated citizenry. From the earliest days of the nation, 'geoscience' has been a part of the U.S. democratic system. In the broadest sense, geoscience means the study of the Earth, the distribution of its materials, and the processes that modify these materials. The geoscience subdisciplines of geography (cartography), mining geology, petroleum and coal geology, and hydrogeology were essential in the early settlement and economic progress of the citizenry and contributed to the later manufacturing economy. Geoscience continues to be important in the current 'information age' to both the domestic and international security of the nation.

The U.S. Geological Survey (USGS) celebrated its 125th anniversary in 2004. This public, national geoscience agency provides a model for tracing how democracy and science evolved together and interacted since its founding in 1879 as an agency within the U.S. Department of the Interior (DOI). The first part of this essay addresses how the federal need for geoscience evolved and expanded with the discovery and then settlement of the western land frontier, including Alaska (figure 3.1). The second part examines the modern ocean sea-floor frontier (figure 3.2) and compares the different ways geoscience was utilized in governance of these two frontiers.

Justification for Spending Public Monies on Science

The foundation of American democracy is the U.S. Constitution. Sci-

Figure 3.1 Land acquisitions of the United States (excluding the territories). The white boundaries outline the nine land additions to the United States subsequent to the original thirteen states of 1776: (1) the Louisiana Purchase from France in 1803; (2) a small northern area ceded by Great Britain in 1818; (3) the Florida region ceded by Spain in 1819; (4) the annexation of Texas in 1845; (5) the Oregon Territory in 1846; (6) the Mexico lands acquired by treaty in 1848; (7) the Gadsden Purchase of 1853; (8) Alaska purchased from Russia in 1867; and (9) Hawaii annexed in 1898. Shading indicates relative elevations, with white representing high (positive) topography and black indicating low (negative) bathymetry. Map is modified from USGS, *US Public Lands Surveys Map*, 1:3,000,000, U.S. Geological Survey, 1965, plotted on a basemap utilizing elevation data from ISCIENCES, *TerraViva! SRTM30 Enhanced Global Map – Elevation/Slope/Aspect* (Ann Arbor, MI: ISCIENCES, 2003), 2 CD-ROMs.

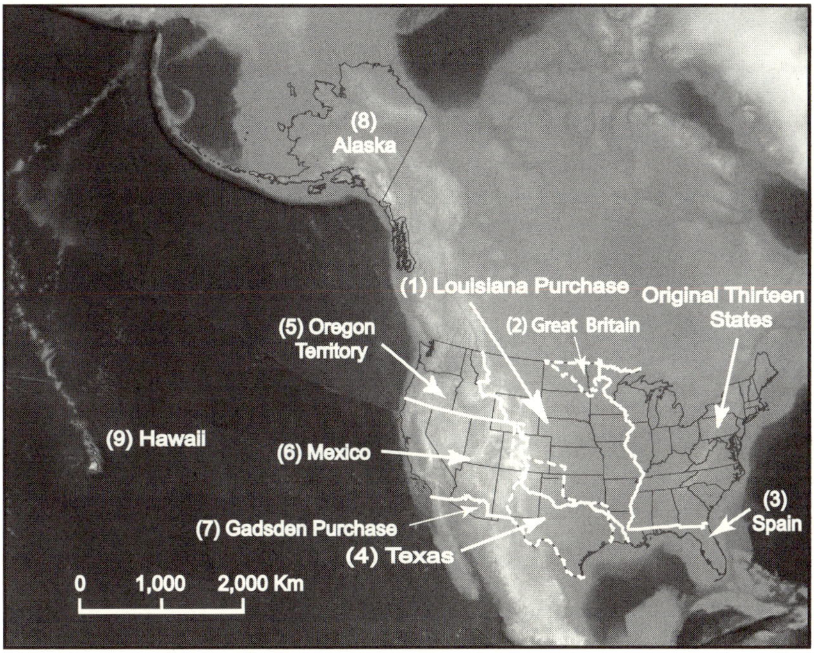

ence is mentioned in Article I, Section 8, in reference to the powers of Congress 'to promote the Progress of Science and useful Arts, by securing for limited Times to Authors and Inventors the exclusive Right to their respective Writings and Discoveries.' Nowhere else in the original document, however, is there mention of an obligation for the federal government to support science. It is from the preamble of the U.S. Constitution, in which the goals of government are listed, that Congress

Figure 3.2 Exclusive Economic Zone (EEZ) of the United States (excluding territories and possessions). The white boundaries outline the EEZ proclaimed by President Ronald Reagan on 10 March 1983, ensuring the United States sovereign rights to its living and non-living resources within 200 nautical miles of its coastline. Basemap is the same as figure 3.1.

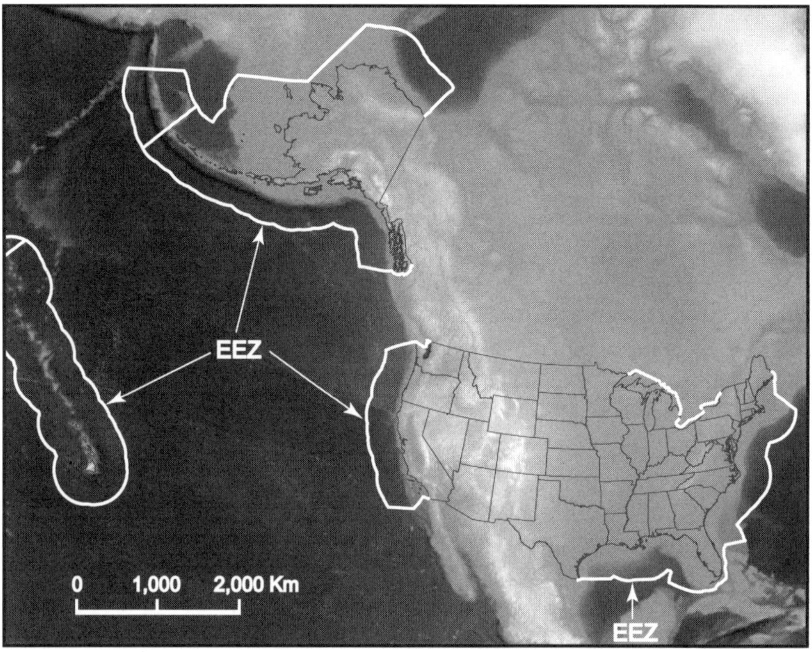

derives its authority for spending money on science: 'We the people of the United States in Order to form a more perfect Union, establish Justice, insure domestic Tranquility, provide for the common defence, promote the general Welfare, and secure the Blessings of Liberty to ourselves and our Posterity, do ordain and establish this constitution for the United States of America.'

The general wording of the preamble leaves justification for spending public monies up to the priorities and wisdom of Congress. As the United States evolved from an agricultural economy through a manufacturing economy into a service economy and attained the status of a global military and economic superpower, societal priorities about what constitutes 'domestic Tranquility ... common defence ... general

Welfare' have changed. The success of the Second World War Manhattan Project, which developed the first atomic bomb, showed how important science (especially physics) was to the national security of the United States (Drell 2000) from the perspective of the 'military-industrial complex' (Eisenhower 1961). Accordingly, the vast majority of science advisers (and directors of the Office of Science and Technology Policy) to the president of the United States have been physicists.[1]

One of the most eloquent visions of how science is relevant to postwar societal priorities was written by electrical engineer and scientist Vannevar Bush in his seminal document 'Science – The Endless Frontier' sent to President Truman in 1945 (Bush, 1945). Bush linked democratic support of science to improvements in national health, standards of living, cultural progress, national security, new enterprises, and new jobs. The 1945 Bush Report, the 1947 Steelman Report (Steelman, 1947; see also Blanpied, 1999), and the First Hoover Commission Report[2] were catalysts in establishing many governmental research institutions during the postwar era, both inside and outside the U.S. Department of Defence (DOD)[3] and further strengthened scientific-research programs within existing federal agencies, especially programs deemed to be 'defence-related.'[4] Since the 1940s, the growth in public funding of science has created several multi-billion-dollar-per-year scientific agencies (e.g., National Science Foundation [NSF], National Aeronautics and Space Administration [NASA], and National Oceanic and Atmospheric Administration [NOAA]) that support scientific programs that monitor and investigate various components and processes of the Earth and planetary systems. Between fiscal years 1976 and 2005, approximately 55 per cent of the federal monies used to support scientific research and development has gone to various components of DOD.[5] If DOD programs 'embedded' in other agencies are included, the annual percentages are even larger. Whether it is DOD or non-DOD funds, the federal support of research on the Earth system consumes tens of billions of dollars each year.

The Western Land Frontier – Early Public Geoscience

When the American Revolution took place, the study of rocks was in its infancy, and the terms 'geology' and 'geoscience' were not commonly used (Mather and Mason 1967). The term 'geology' became common in the nineteenth century, particularly as manufacturing replaced agriculture as the primary economic base of the United States. In the latter part of the twentieth century, the term 'geoscience' has been used to more broadly summarize both classical geology and its related disciplines, including, for example, geophysics, hydrology, and geographic information systems.

The history of 'public geoscience' in the young U.S. democracy began with the passage of the Land Ordinance of 1785, which provided a means for the government to survey and allocate land located mostly west of Pennsylvania. The act called for 'one-third of all gold, silver, lead, and copper mines to be sold or otherwise disposed of, as Congress shall thereafter direct,' implying that the lands also needed to be mapped and classified. This act was passed two years before the signing of the U.S. Constitution (in 1787), at a time when holding public land was not considered a federal responsibility and mapping was generally considered a state responsibility (Rabbitt 1989). In the agricultural economy of the late eighteenth century, this act encouraged settlement by the citizenry and generated revenue for the government. Even though this act was passed more than one hundred years before the creation of the USGS, its purpose of mapping and deriving public benefit from mineral lands provided part of the rationale for the establishment of the USGS in 1879.

The goals of westward expansion, discovery, mapping, and classification are common themes in congressional funding of 'geoscience' in the first half of the nineteenth century. Funding of science was initially considered a state responsibility or a private concern. The famous Lewis and Clark expedition was a notable exception. President Thomas Jefferson sought and received both authorization and funding from Congress in 1803 for exploration of the western lands. The Louisiana Purchase, also completed in 1803, approximately doubled the territory of the United States (figure 1); the exploration of this vast wilderness expanse by the 'Corps of Discovery,' led by Captain Meriwether Lewis, included initial observations about the nation's geologic and mineral wealth. In 1807, Thomas Jefferson established the first public science agency with formation of the Survey of the Coast (the U.S. Coast and Geodetic Survey from 1878) to provide better hydrographic and nautical charts for navigation in coastal waters.

Congress was basically slow to realize how mapping aided commerce (Rabbitt 1989). The creation of new agencies and continued authorization to explore the west proceeded cautiously and without coordination. Congress directed the Topographical Bureau of the U.S. Army in 1834 to create a geological map of the United States (Nelson 1999), though work on the map was discontinued after only two years (Goetzmann 1959). In 1836, a U.S. Exploring Expedition was authorized as an aid to commerce and navigation (Stanton 1975). This was followed in 1838 by the creation of the Corps of Topographical Engineers to explore and map the continent. In 1839, Congress authorized geological studies to classify public lands and identify mineral lands in the Upper Mississippi Valley. Eight years later, in 1847, Congress directed that geological classification of mineral lands in parts of Michigan and Wisconsin be completed prior to their sale. As noted by Rabbitt (1989: 4), these actions show the 'growing realization that certain economic purposes could be aided by science, or ... that the scientific activities of the Government should serve the great economic interests of the country.'

By the middle of the nineteenth century, the United States was transitioning from primarily an agricultural to a manufacturing economy (Rabbitt 1989). Coal and iron, mined from locations such as Pennsylvania, helped fuel the industrial revolution. Gold was discovered in California in 1848; silver in Nevada in 1859. Also in 1859, the first oil well was drilled in northwestern Pennsylvania. Recognizing the importance of the sale of mineral lands to the Treasury of the United States, Congress established the U.S. Department of the Interior (DOI) in 1849,[6] placing the General Land Office (GLO) and several offices from other departments within it.

In the 1860s federal support of 'geoscience' blossomed. In an act that further recognized how science contributed to the economy and general well-being of the nation, the 37th Congress established the Department of Agriculture in 1862 and 'for the first time authorized "practical and scientific experiments" to obtain "useful information on subjects connected with agriculture."'[7] A year later, during the Civil War, President Lincoln signed the Act of Incorporation[8] to create the National Academy of Sciences (NAS) to provide scientific advice on technological advances important to the war-time effort. Creation of the NAS formalized a type of discourse that is still used today – for example, bringing together groups of experts to deliberate issues and make recommendations to Congress and the administration (president).[9] In 1862, the Morrill Act established the land grant system of donated universities; the sale of land to states and territories by the federal government resulted in the eventual establishment of seventy-two universities and colleges and twenty-two Native American tribal colleges 'for the benefit of agriculture and mechanic arts.'[10] In the fall of 1866, the Commissioner of the GLO of DOI stated 'that the proper development of the geological characteristics and mineral wealth of the country was a matter of the highest concern to the American people.' (Rabbitt 1989: 5). In 1867, Congress subsequently funded exploration of the western frontier for which geology was the primary objective: the United States Geological Exploration of the Fortieth Parallel, led by Clarence King (the King survey), to survey the lands and resources flanking the route of the transcontinental railway, and the survey that later became the U.S. Geological and Geographical Survey of the Territories, led by Ferdinand Hayden (the Hayden survey) to survey natural resources of Nebraska and later Wyoming, Idaho, Colorado, and Montana (Bartlett 1997). This congressional recognition and funding led King, as first director of the USGS, to later remark: '1867 marks ... a turning point, when science ceased to be dragged in the dust of rapid exploration and took a commanding position in the professional work of the country' (King 1880).

The 1860s also brought to a head a funding issue that was simmering beneath the surface: whether mapping and surveying should be a military or civilian responsibility.[11] The King survey was under the U.S. Army Corps of Engineers; the Hayden survey was within the GLO and then under the Secretary of Interior. Two other surveys were also funded: the U.S. Geographical and Geological Survey of the Rocky Mountain Region, led by John Wesley Powell, within DOI, and the U.S. Geographical Surveys West of the One Hundredth Meridian (1872–9), led by Lieutenant George Wheeler, also of the Army Corps of Engineers, to map lands for locating military posts and transportation routes. Each of these surveys boasted significant scientific accomplishments and developed separate plans for geological mapping of the western frontier. Inevitable conflicts among the latter three surveys (Hayden, Powell, and Wheeler) arose in the competition for funds from Congress. This led Congress, in 1878, to ask the National Academy of Sciences to make recommendations on efficiencies of mapping the west. In addition to the National Academy report, several other factors contributed to congressional action a year later to establish the U.S. Geological Survey within the Department of the Interior. Two such factors were congressional intent to improve the civil service and the need for a federal organization that could address immediate concerns of the mineral industry, primarily for materials for construction and currency

(Rabbitt 1979; Nelson and Rabbitt 1991). The purpose of the USGS was for 'classification of the public lands and examination of the geological structure, mineral resources, and products of the national domain' (USGS Organic Act: 20 Stat. 394).

By the time the USGS was established, the land and inland water area of the United States had increased 400 per cent from 2.3 million square kilometres (0.6 billion acres) in 1790 to 9.3 million square kilometres (2.3 billion acres) in 1890 (Farmighetti 1997). The population increased from 1.4 million in 1800 to 50 million in 1880. These large changes in both land area and population increased national diversity and posed challenges for maintaining unity and governance (the most extreme example being the Civil War). 'Geoscience' during this period played a growing role in government in exploration and economic development. The establishment of the USGS recognized the importance of federal sponsorship of 'public geoscience' in the survival of the democracy. In effect, the USGS became to the mineral industry what the Department of Agriculture was to the farming base, a source of science and information that was important to the management and economy of the country.

When the 1879 Organic Act establishing USGS was written, a key clause showed how well the legislators interested in reform of the civil service understood ethics in public science. The USGS Organic Act (20 Stat. 394) also stated: 'The Director and members of the Geological Survey shall have no personal or private interests in the lands or mineral wealth of the region under survey, and shall execute no surveys or examinations for private parties or corporations.' This statement prevents USGS scientists from having or giving the appearance of having a conflict of interest in conducting their scientific work and most importantly preserves objectivity and integrity in agency data (Rabbitt and Nelson 1990). While all U.S. federal employees are subject to ethical guidelines, the financial restrictions on USGS employees are among the most extensive of any federal agency, allowing no ownership by the employee or immediate family members of any stock, bonds, or mutual funds trading in energy and mineral resources, and no involvement in any non-USGS surveys of any kind. This ensures that USGS employees are free of real or perceived bias in conducting their science, thereby protecting the objectivity entrusted to them by the government and the American people. Historical context may also make this ethical restriction understandable. In 1872, seven years prior to the creation of USGS, Clarence King, who helped write and pass the USGS founding legislation, had unravelled one of the biggest hoaxes of U.S. mineral exploration by showing how a supposed discovery of the mother lode of diamonds in extreme northwestern Colorado was in fact the work of unscrupulous speculators (Hausel and Stahl 1995; Wilson 2004; Wilkins and Hinkley 1988).

The notion of defining the scope and purpose of research conducted by USGS taps a theme central to the role of public geoscience. When the USGS was created, the federal government controlled more than 1.2 billion acres of land, mostly west of the Mississippi River; only about 15 per cent had been surveyed (Rabbitt 1989). The 1879 Organic Act directed the newly formed USGS to work in the 'national domain' (all lands) as opposed to the 'public domain' (only lands still held in title by the federal government). An attorney for DOI interpreted 'national domain' to mean 'public domain,' exacerbating the uncertainty about the limits on USGS studies (Rabbitt 1989; Nelson 1990). Clarence King, as first director of the USGS (1879–81), sought a clarification from Congress about the scope of the work so that USGS could legally undertake work nationwide. Congress failed to act on this before recessing, so King focused the earliest USGS studies on mining geology in the west and extended the work nationwide through cooperative work with the Tenth Census. When John Wesley Powell became second director of the USGS (1881–94), research by USGS geologists expanded to include topographic mapping, general geology, and paleontology as the agency's overriding objectives. Powell is credited with recognizing the aridity of much of the west and the importance of irrigation. However, a number of his failed policies and operations led Congress to enact drastic staff and budget cuts (1890, 1892), and Powell resigned in 1894 before his salary cut went into effect (Rabbitt 1980). Throughout its history, USGS has grappled with how its science can best serve the needs of government.

During the first fifty years of the USGS, until about the start of the Great Depression, the 'geoscience' needs of the western land frontier evolved: the land was becoming extensively settled, and nine states and two territories (Alaska and Hawaii) were added to the United States. The study of water supply became a continuing responsibility of the USGS after Powell's departure in 1894. A USGS geologist and geophysicist, G.F. Becker, participated in studies of the proposed route in Panama for a canal connecting the Atlantic and Pacific oceans (Becker 1917). In the early 1900s, President Theodore Roosevelt fully endorsed the concept of utilizing science to manage the nation's resources, and the Reclamation Service became an adjunct of the USGS in 1902. The need for raw materials, particularly iron and coal, to fuel the manufacturing economy grew. To this was added oil, as the oil industry began to expand in the early twentieth century. Conservation lands required an inventory of their natural resources. The First World War introduced a new concept of 'strategic minerals' to support the military effort, expanding the scope of USGS studies to the international arena, primarily in the Caribbean and in Central and South America. Postwar oil shortages brought greater focus on energy resources and policy.

An under-appreciated aspect of how public geoscience has influenced government structure and function can be seen in the creation of federal land management agencies from entities within USGS.[12] The Bureau of Reclamation began as the Reclamation Service in the USGS in 1902 and was made an independent agency within DOI in 1907. The Bureau of Land Management was established in 1946 through the combination of several federal programs, including the General Land Office and Grazing Service in DOI. USGS had responsibility for specialized mapping of the Forest Reserves (later the National Forests). The Minerals Management Service (established in 1982; formerly called the Conservation Branch from 1925 and the Conservation Division of USGS from 1949) supervises operations and collects royalties from extraction of petroleum, gas, and minerals from onshore federal lands and offshore lands within the Exclusive Economic Zone (EEZ), including continental and insular shelves, slopes, and adjacent submarine areas of the United States and its territories. Finally, the Bureau of Mines was established from the USGS in 1910; it was abolished in 1996, but its Minerals Information Program came full circle back to the USGS.

The Ocean Sea-Floor Frontier – Mature Public Geoscience

The Truman Proclamation of 1945[13] set the international legal precedent for a nation to adjudicate and manage the sea floor on its contiguous continental shelf. Prior to this, nations had mostly concerned themselves with navigation and other water-column (e.g., fishery) issues in the oceans. The date of Truman's proclamation marks the start of U.S. governance in the ocean sea-floor frontier.

The ocean sea floor is a modern frontier, where policy, science, and management have taken a wholly different path than what occurred on the western land frontier. By the time of the Truman Proclamation in 1945, the United States had the benefit of more than one hundred years of experience utilizing 'geoscience' to explore, map, settle, develop, and preserve land. The concerns and priorities of the United States in 1946 differed from those of the earlier century and a half when western land was the frontier. Two world wars had demonstrated the importance of strategic mineral and energy resources and geographic knowledge. The need to mitigate the impact of submarine warfare fuelled an intense program to understand the water column and better describe the sea floor to detect submarines. Petroleum was emerging as the primary energy resource supporting the U.S. economy. And, as the United States grew to be a global military and economic superpower, its quest for knowledge to support and enhance this leadership role increased.

Two additional factors set apart the government's attitude toward supporting science in the ocean sea-floor frontier compared with the western land frontier. First, in 1962, the publication of *Silent Spring* by Rachel Carson (Carson, 1987) started an environmental awareness movement that affected public opinion and eventually resulted in substantial federal funding for environmental research. Second, mitigating natural hazards became a national priority after the 8.6 magnitude Good Friday earthquake destroyed much of Anchorage, Alaska, on 27 March 1964 and triggered a powerful tsunami that damaged the port cities and marine oil terminals of Valdez and Seward, Alaska, killing forty-four people. The Good Friday earthquake remains the most violent natural cataclysm in the history of North America. It destroyed about 75 percent of Alaska's commerce and industry and left thousands homeless (Ward 1989).

Perhaps the biggest difference between the ocean sea-floor and western land frontiers lay in the fact that individual ownership of the sea floor was never contemplated. Ownership was by the government, placed under the jurisdiction of the secretary of DOI by executive order (no. 9633) in 1945 until enactment legislation was implemented. Legislative authority was resolved eight years later when the Submerged Lands Act and the Outer Continental Shelf Act was passed, giving states jurisdiction of the sea floor up to three nautical miles (nm) beyond the coastline and asserting federal authority for the sea floor beyond three nm.[14] Neither act affected science per se, but they established an early precedent on an issue that had also complicated the opening of the western land frontier: deciding individual vs. state vs. federal control of this large public resource. The decision that the federal government had superior rights, unless specifically granted to the states by legislation such as the Submerged Lands Act, laid the groundwork for federal investment in the ocean sea-floor frontier.

As was the case in the western land frontier, science in the ocean seafloor frontier was initially supported by a mix of private and federal, mostly military, sources. The Scripps Institute of Oceanography was founded in 1903 with a gift from interested San Diego citizens (Revelle 1980). Following the First World War, the U.S. Naval Research Laboratory opened in 1923 to investigate ocean acoustics and submarine detection (Gaffney and Petrik 2000). Seven years later, the Rockefeller Foundation responded to a 1927 National Academy of Sciences Committee on Oceanography report by awarding a grant that led to the creation of the Woods Hole Oceanographic Institution (Revelle 1980). The Office of Naval Research was created in 1946 (Gaffney and Petrik 2000). Congress responded to the 1945 Vannevar Bush report and the 1947 Steelman report with the National Science Foundation Act (64 Stat. 152) of 1950, which established the National Science Foundation 'to promote the progress of science; to advance the national health, prosperity, and welfare; to secure the national defence; and for other purposes.' As the Cold War escalated in the 1950s, the stage was set for a major expansion in public funding of scientific studies of the ocean and its sea floor.

Another pivotal event integrating democracy and science in the ocean frontier was the 1958 Geneva Conventions on the Law of the Sea, in which nations of the world came together to develop an international understanding of and agreement on jurisdiction and adjudication of the world's oceans, including resources in the water column (for example, fisheries), on the sea floor (manganese nodules), and beneath the sea floor (oil and gas). Partly in response to this potential new national need, and partly because the Conservation Division[15] of the USGS had been given leasing authority for energy resources beneath the continental margin, the USGS marine program was created in 1962 for the purpose of evaluating 'potential mineral resources on or beneath the sea floor and [aiding] in solving the problems caused by rapid population growth, urbanization, and industrial expansion in the coastal areas' (Rabbitt 1989: 40).

The final United Nations Convention on the Law of the Sea (UNCLOS), approved 10 December 1982 in Montego Bay, Jamaica, and officially implemented on 16 November 1994, heavily utilizes science in a legal context. A key provision that was hotly debated was the definition of a juridical continental shelf (Hosler 1988). Wording of the treaty with respect to the sea floor was negotiated with extensive scientific and legal discussion (Hedberg 1983; McKelvey 1984; Duff 2004). By logical extension, understanding a nation's resources on and beneath the sea-floor frontier was an integral part of negotiating what constituted a juridical continental shelf. Although USGS monies spent on ocean research are small compared with DOD, NSF, NOAA, and NASA outlays (Merrell and Boledovich 2000), USGS studies form the bulk of the federal effort to acquire regional data for understanding of the sea-floor and sub-sea-floor geology and resources of the submerged lands of the United States (Hutchinson et al. 2004).

Less than twenty years after the Truman Proclamation, the U.S. Congress called for a major review of national marine science and ocean policy by creating two complementary bodies: the Council on Marine Resources and Engineering Development and the Commission on Marine Science, Engineering, and Resources.[16] In creating these bodies, Congress sought to bring attention to the marine environment and potential resources of the oceans, as well as to promote marine exploration, science, technology, and investment. The Council on Marine Resources and Engineering Development brought together representatives of the federal agencies involved in ocean matters for coordination. The Commission on Marine Science, Engineering, and Resources, ultimately known as the Stratton Commission, was charged to 'examine the Nation's stake in the development, utilization, and preservation of our marine environment ... review all current and contemplated marine activities to assess their adequacy to achieve the national goals set forth in the Act,[17] ... formulate a comprehensive, long-term, national program for marine affairs designed to meet present and future national needs in the most effective possible manner, ... [and] to recommend a plan of Government organization best adapted to the support of the program and to indicate the expected costs' (Commission on Marine Science, Engineering, and Resources 1969: vi). This effort to unify national ocean policy grew from the government's need to integrate marine sea-floor research with federal policy priorities.[18] Commission members were a multidisciplinary group that included representatives from the disciplines of science, law, economics, fisheries, government, education/academia, private industry, non-governmental organizations (NGOs), and philanthropy, as well as four congressional advisers.

The Stratton Commission report, released in 1969, had an immediate and lasting impact on U.S. ocean policy, ocean science research, and the organization of the executive branch of the U.S. government. A new federal department, the National Oceanic and Atmospheric Administration (NOAA), was established in 1970 from recommendations of the Stratton Commission, with one of its core missions being to develop a national oceanic and atmospheric research and development program. The National Marine Sanctuary Program promoting seabed conservation was established in 1972.[19] The National Environmental Policy Act (NEPA) was passed in 1970 for the purpose of protecting the environment especially as a result of federal activities (Alm 1988). The Coastal Zone Management Act of 1972[20] promoted sustainable development of the land and offshore waters and established the National Estuarine Research Reserve System. Each of these acts demonstrates how the ocean sea-floor frontier has integrated science into policy from the beginning, in contrast to the much slower and sometimes intermittent incorporation of science into policies for managing and exploring the western land frontier.

Perhaps the best example of geoscience being promoted and utilized in policy applications is Article 76 of the United Nations Convention on the Law of the Sea (UNCLOS). This international treaty[21] was founded on the concept that the oceans are for the 'common heritage of mankind' (UNCLOS 2001). The treaty codifies the concept of an Exclusive Economic Zone (figure 2) in which states have rights for both living and non-living resources in the water and subsoil of the continental margin out to approximately 200 nm. Article 76 further defines how a state defines its juridical continental shelf beyond 200 nautical miles from its territorial baseline and can therefore exercise legal jurisdiction over its more distant mineral resources (article 76, sections 1–10). When a state makes a submission to the United Nations quantifying the boundaries of its Article 76 juridical continental shelf, the representative body that examines the appropriateness of the boundaries and makes recommendations to the UN Secretary-General is the Commission on the Limits of the Continental Shelf, consisting of twenty-one members from signatory nations who are experts in the field of geology, geophysics, or hydrography (annex II, article 2). Clear conflict-of-interest clauses are intended to protect the impartiality of the commissioners (annex II, article 5). As of summer 2007, there have been seven submissions to UNCLOS for an extended continental shelf based on Article 76; two (Russia, Brazil) have received final commission recommendations. Therefore, the actions of geoscience in UNCLOS remain untested, despite extensive, and sometimes speculative, literature about the interpretations of Article 76 (Nordquist, Moore, and Heidar 2004).

On 10 March 1983, President Ronald Reagan proclaimed an EEZ for the United States,[22] thereby asserting a minimum 200-nautical-mile zone of jurisdiction consistent with international law (figure 2). Shortly thereafter, the USGS undertook a major mapping effort of the U.S. continental margins (Gardner, Field, and Twichell 1996). Many USGS legacy data will be utilized when the United States ratifies the UNCLOS treaty and undertakes its own submission.

One area where public geoscience in the ocean sea-floor frontier serendipitously anticipated national need is the study of methane hydrates, a peculiar ice-like form of concentrated methane found in near-sea-floor continental margin sediments in water depths greater than about 500 m. Since the early 1980s, USGS scientists provided some of the most comprehensive analyses of naturally occurring gas hydrate, which led to the international recognition of hydrate as a likely future global energy resource, a possible trigger for global climate change, and a potential hazard to deep-water drilling (Max 2000). Much of this USGS research laid the foundation for the multiple field and drilling programs that have occurred nationally and internationally in the last decade and a half[23] and raised awareness of hydrate issues to the extent that Congress passed (2000) and reauthorized (2005) the Methane Hydrate Research and Development Act[24] to promote coordination in research and technological development of hydrates. Hydrates remain a frontier research discipline with the potential to provide alternative fuel and alter the geopolitical global energy balance (Max 2000).

Thirty-two years after establishing the Stratton Commission, the U.S. Congress, in 2000, established another blue-ribbon commission to conduct a major overhaul and review of national ocean policy to address the numerous changes that made the Stratton Commission report outdated:[25] the rapid growth of population centres and associated pressures along coastal areas; the increasingly complex bureaucratic and legal framework pertaining to the oceans; the realization that ocean resources may be limited; the inadequate understanding of the fragility of ocean resources and the threats to them; and new technological advances that provided opportunities for both increased understanding and utilization of the ocean environment, including reducing the costs of natural hazards. The Oceans Act of 2000 (PL 106-256) established the Commission on Ocean Policy, a sixteen-member group that, like the Stratton Commission before it, reached across disciplines and stakeholders in its makeup: education, research, marine transportation, government, law, business, policy, philanthropy, and military. The commission was charged with making recommendations to the president and Congress 'for coordinated and comprehensive national ocean policy.'

The report of the Commission on Ocean Policy (2004), entitled *An Ocean Blueprint for the 21st Century*, embraces a holistic vision of national policy founded on 'ecosystem-based management ... that reflects the relationships among all ecosystem components, including human and nonhuman species and the environments in which they live ... rather than political boundaries' (xii). The report makes 212 recommendations covering comprehensive coastal and ocean policy. One of the three crosscutting themes that the commission highlights is strengthening science to meet information needs; this theme is embedded throughout the the report's thirty-one chapters and is the primary subject of the four chapters in Part IV, 'Science-Based Decisions: Advancing Our Understanding of the Oceans.'

It is still too soon to know the impact of the Commission on Ocean Policy's report on science, geoscience, or democratic governance. The final report was submitted on 20 September 2004. In accordance with the Oceans Act of 2000, the president responded to the report within the required ninety days after its submission by signing an executive order establishing a secretarial-level Committee on Ocean Policy within the Executive Office of the President and releasing the U.S. Ocean Action Plan (Council on Environmental Quality 2004) outlining both short- and long-term actions about changes in ocean governance and funding. Two of the president's six actions supported science: advancing our understanding of the oceans, coasts, and Great Lakes; and advancing international ocean policy and science. However, none of the actions specified budget goals or timelines. Congressional action since the delivery of the Ocean Commission report has included the introduction of legislation for an NOAA Organic Act (H.R. 50) and a national ocean exploration program (S.39). The president's budgets (e.g., White House 2004) have emphasized austerity in most areas except national security, which poses challenges to funding and implementation of its ocean policy recommendations.

In an interesting twist of public advocacy, the Pew Charitable Trusts released their own Oceans Commission report (Pew Oceans Commission 2003) to Congress and the nation in June 2003. This report posed a more sobering reality for America's oceans than the U.S. Commission on Ocean Policy, stating that 'America's oceans are in crisis and the stakes could not be higher ... Without reform, our daily actions will increasingly jeopardize a valuable natural resource and an invaluable aspect of our national heritage' (v). The focus of the Pew Oceans Commission was narrower: 'Our mission is to identify policies and practices necessary to restore and protect living marine resources in U.S. waters and ocean and coastal habitats on which they depend' (ix). They called for a new perspective in which the oceans are treated as a public trust and for which government needs to exercise its authority with an ethic of stewardship. As with the U.S. Commission on Ocean Policy, the Pew Commission report also emphasized science, education, and funding in fulfilling its stated objectives. Each of the Stratton, Pew, and U.S. Ocean Commission reports represents a holistic, comprehensive, and authoritative attempt to shape ocean policy in a framework that integrates science into the priorities of government.

Conclusions

The western land frontier and the modern ocean sea-floor frontier provide two examples of the role of 'public geoscience' in a democratic context. In the land frontier, the overriding goals of the young democracy were for land to provide revenue and settlement. Geoscience provided a means to explore and classify the lands for these broader goals. This geoscience was initially funded in a somewhat intermittent way, with Congress reacting to specific national needs with sympathetic presidents or secretaries of the Interior. With the challenges posed by both land and population increases during the nineteenth century, 'public geoscience' was formally incorporated into governance with the creation of the USGS in 1879. Manufacturing and industry, which required energy and natural resources, were the basis of the nation's economy; USGS provided a federal focus on minerals issues. Federal needs for geoscience also evolved as the nation embraced other goals, such as land conservation and public safety from natural hazards, as legitimate roles for government. Congress imposed strict ethical guidelines in the USGS Organic Act to ensure integrity and objectivity in the conduct of this public geoscience.[26]

The ocean sea-floor frontier is a relatively young frontier that, arguably, has not yet been fully explored. The federal role in governing its submerged sea floor shares some goals that characterized the land frontier, such as mapping, exploration, classification, and conservation. Governance of the ocean sea-floor frontier also includes broader goals of protecting coastal communities from natural hazards, maintaining healthful marine environments, and undertaking ecosystem-based management. While many of these goals are now applied to traditional land management, they were not recognized to be a federal responsibility during development of the western land frontier. Perhaps most obviously different, the federal government has recognized the need to incorporate science into all aspects of ocean policy and governance, as illustrated by the overarching recommendations of the Stratton, Pew, and U.S. Ocean Commissions reports. Public geoscience plays only a small part of overall ocean governance, but the USGS has been integral to regional understanding of the sea-floor and sub-sea-floor resources and processes.

As the United States moves into the twenty-first century, with a population that reached 300 million in late 2006,[27] the linkages between public geoscience and democracy remain strong. Article 76 of the UNCLOS places geoscience at the centre of this international treaty with regard to jurisdiction over the sea-floor resources of a nation's continental shelf. President Bush has pledged to make ratification of this treaty a priority of his second administration (CEQ 2004). Geoscience plays an integral role in the U.S. Commission on Ocean Policy report. It is still too soon to know whether the recommendations in this report will make a lasting change in governance, but the report makes bold recommendations that are intended to strengthen the U.S. democracy and allow it to most effectively meet the environmental, economic, and security challenges of the new century.

NOTES

Constructive reviews by Clifford M. Nelson and S. Jeffress Williams significantly improved earlier versions of this article.

1 Frank Press, science adviser under President Jimmy Carter, is a geophysicist who worked on the 'Vela Uniform' project, which, among other aspects, successfully used the principles of seismology to discriminate between natural earthquakes and those caused by explosion of underground nuclear devices (Barth 2003). This important achievement in geoscience provided the scientific basis for a Comprehensive Test Ban Treaty completed in 1999 but still to be ratified by the U.S. Senate (similar to the status of the United Nations Convention on the Law of the Sea, the Kyoto Protocol, and the United Nations Framework Convention on Climate Change).

2 The Commission on the Organization of the Executive Branch (more familiarly known as the First Hoover Commission) produced twenty-one reports. The first of these reports was The Commission on the Organization of the Executive Branch of the Government, *General Management of*

the Executive Branch (Washington, DC: Government Printing Office, 1949).

3 The National Defence Establishment, created in 1947, became DOD in 1949; it had been called the War Department from 1789 to 1947 (Farmighetti 1997).

4 New scientific agencies that were established within the DOD were the Office of Naval Research (ONR, 1946–present), Army Research Office (ARO, 1951–present), and the Air Force Office of Scientific Research (AFOSR, 1951–present). Other federal scientific agencies established were the National Science Foundation (NSF, 1950–present), National Advisory Committee for Aeronautics (NACA, 1915–58; later National Aeronautics and Space Administration, NASA, 1958–present), Environmental Science Services Administration (1965–70, a predecessor agency to NOAA), National Institute of Standards and Technology (NIST, 1988–present, originally the Office of Weights and Measures [1832–1901] under U.S. Coast Survey), National Bureau of Standards (1901–88), Atomic Energy Commission (AEC, 1946–75, replaced by the Energy Research and Development Administration [1975–7], now in the U.S. Department of Energy, 1977–present), and others.

5 American Association for the Advancement of Science, R&D Budget and Policy Program (graph of Trends in Federal R&D, FY1976–2006 at http://www.aaas.org/spp/rd/trtot06p.pdf).

6 An Act to Establish the Home Department, 3 March 1849 (9 Stat. 395).

7 An Act to Establish a Department of Agriculture (12 Stat. 387). The Soil Erosion Service (1933–5) was established in 1933, later called the Soil Conservation Service (1935–94), now the Natural Resources Conservation Service (1994–).

8 An Act to Incorporate the National Academy of Sciences, 3 March 1863 (12 Stat. 806).

9 A listing of reports can be found at http://www.nap.edu/info/browse.htm.

10 See University of Florida website on 'Acts, History, and Institutions' for information about the 1st Morrill Act (1862), 2nd Morrill Act (1890), and related federal legislation and information about the Sea Grant Act (1966) that established Sea Grant colleges and programs administered by NOAA. 'Sea Grant is a partnership of academia, government, and industry focusing on coastal and marine resources' (http://ifas.ufl.edu/ls_grant).

11 This policy is being revisited as part of the domestic response to international terrorism. It is not clear whether the National Geospatial-Intelligence Agency (NGA, a military agency) or the National Geospatial Programs Office (NGPO) of USGS (a civilian agency) should have the primary responsibility for large-scale maps of urban areas and other 'security-sensitive'

areas or themes of the United States (showing, for example, dams, reservoirs, and other potential targets for terrorism).

12 Department of Interior, *People, Land, and Water* (November 2004): 32.

13 'Having concern for the urgency of conserving and prudently utilizing its natural resources, the Government of the United States regards the natural resources of the subsoil and sea bed of the continental shelf beneath the high seas but contiguous to the coasts of the United States as appertaining to the United States, subject to its jurisdiction and control.' Truman Proclamation of 1945, 28 September 1945 (Presidential Proclamation no. 2667).

14 The exceptions to this 3-nautical-mile boundary are in the Gulf of Mexico, where Texas and Florida have rights to 'three marine leagues,' or approximately 9 nautical miles from their Gulf of Mexico coastlines. Submerged Lands Act of 22 May 1953, 67 Stat. 29; Outer Continental Shelf Act of 7 August 1953, 67 Stat. 469.

15 The Conservation Division became part of a new independent agency within the Department of the Interior, the Minerals Management Service (MMS) in 1982.

16 Both bodies are products of the Marine Resources and Engineering Development Act of June 1966 (80 Stat. 203).

17 Marine Resources and Engineering Development Act of June, 1966 (80 Stat. 203).

18 For example, two recommendations from the report are: (a) 'The Commission recommends that the United States establish as a goal the achievement of the capability to occupy the bed and subsoil of the U.S. Territorial sea. The Commission also recommends that the United States learn to conduct surface and undersea operations to utilize fully the continental shelf and slope to a depth of 2,000 feet' and (b) 'The Commission recommends that the United States establish as a goal the achievement of the capability to explore the ocean depths to 20,000 feet within a decade and to utilize the ocean depths to 20,000 feet by the year 2000' (32).

19 The Marine Protection, Research, and Sanctuaries Act of 1972. Public Law (PL) 106-580.

20 PL 92-583.

21 Ratifications of, accessions to, and successions to UNCLOS as of 1 February 2005 number 148 states parties. http://www.un.org/Depts/los/convention_agreements/convention_agreements.htm.

22 Presidential Proclamation no. 5030.

23 For example, Ocean Drilling Program (ODP) Leg 164 on the Blake Ridge off the southeastern U.S. (see http://www-odp.tamu.edu/publications/164_SR/164sr.htm).

24 The Methane Hydrate Research and Development Act of 2000, PL 106-193. See also http://www.netl.doe.gov/scngo/Natural%20Gas/Hydrates/rd-program/legislation.htm.
25 Senate Report 106-301.
26 A similar restriction has recently been proposed for scientists in the National Institutes of Health (NIH). On 7 December 2003, the *Los Angeles Times* reported that scientists at the NIH received millions of dollars in fees and stock options from biomedical companies for consulting on potential drugs and therapies. This sparked congressional concern about conflict-of-interest violations, which resulted in congressional hearings in 2004. On 3 February 2005, the NIH published revised conflict-of-interest and ethical guidelines in the Federal Register. The most notable change is restrictions on participating in and being remunerated for outside consulting contracts with companies who are substantially affected by federal drug and biomedical regulation.
27 In 2005, the population of the United States was about 295 million, about five times (245 million) greater than the population at the time USGS was created. See Farmighetti (1997).

REFERENCES

Alm, A.L. 1988. NEPA: Past, Present, and Future. *EPA Journal* (January/February). http://www.epa.gov/history/topics/nepa/01.htm.

Barth, K.H. 2003. The Politics of Seismology; Nuclear Testing, Arms Control, and Transformation of a Discipline. *Social Studies of Science* 33: 743–81.

Bartlett, R.A. 1997. Scientific Exploration of the American West 1865–1900. In *North American Exploration*, vol. 3: *A Continent Comprehended*, edited by J.L. Allen, 461–520. Lincoln and London: University of Nebraska Press.

Becker, G.F. 1917. Mechanics of the Panama Canal Slides. *U.S. Geological Survey Professional Paper* 98–N: 253–261. Washington, DC: U.S. Government Printing Office.

Blanpied, W.A. 1999. Science and Public Policy: The Steelman Report and the Politics of Post–World War II Science Policy. In *Science and Technology Policy Yearbook*, edited by A.H. Teich and others. Washington, DC: American Academy for the Advancement of Science. http://www.aaas.org/spp/yearbook/chap29.htm.

Bush, V. 1945. *Science – The Endless Frontier.* A report to the President by V. Bush, Director of the Office of Scientific Research and Development. Washington, DC: U.S. Government Printing Office. http://www.nsf.gov/about/history/vbush1945.htm.

Carson, R.L. 1987. *Silent Spring*. 25th anniversary edition. Boston: Houghton Mifflin.

Commission on Marine Science, Engineering, and Resources. 1969. *Our Nation and the Sea, A Plan for National Action*. Washington, DC: U.S. Government Printing Office. http://www.lib.noaa.gov/edocs/stratton/title.html.

Commission on Ocean Policy. 2004. *An Ocean Blueprint for the 21st Century*. http://www.oceancommission.gov/documents/prepub_report/welcome.html.

Council on Environmental Quality (CEQ). 2004. Committee on Ocean Policy. http://ocean.ceq.gov/actionplan.pdf.

Drell, S.D. 2000. Special Issue on Physics and National Security. *Physics Today* 53.

Duff, J.A. 2004. A Note on the United States and the Law of the Sea: Looking Back and Moving Forward. *Ocean Development and International Law Journal* 35: 195–219.

Eisenhower, D.D. 1961. Military-Industrial Complex. Speech. In Public Papers of the Presidents, D.D. Eisenhower, 1960. New Haven, CT: Yale Law School, the Avalon Project: 1035–40.

Farmighetti, R. 1997. *The World Almanac Book of Facts 1998*. Mahway, NJ: World Almanac Books.

Gaffney, P.G. II, and J. Petrik. 2000. *A Way of Life Enriched and Secured: The Nation's Debt to Oceanography*. Washington, DC: The H. John Heinz III Center for Science, Economics, and the Environment. 33–44. http://www.heinzctr.org/meeting_reports.htm.

Gardner, J.V., M.E. Field, and D.C. Twichell. 1996. *Geology of the United States Sea Floor; the View from GLORIA*. Cambridge: Cambridge University Press.

Goetzmann, W.H. 1959. *Army Exploration in the American West 1803–1863*. New Haven, CT: Yale University Press.

Hausel, W.D., and S. Stahl. 1995. The Great Diamond Hoax of 1872. In *Resources of Southwestern Wyoming*, edited by R.W. Jones, 13–27. Laramie: Wyoming Geological Association 46th Conference Guide Book.

Hedberg, H.D. 1983. Letters and Comments; a Critique of Boundary Provisions in the Law of the Sea Treaty. *Ocean Development and International Law Journal* 12: 337–48.

Hosler, A.F. 1988. Offshore Lands of the USA: The Exclusive Economic Zone, Continental Shelf and Outer Continental Shelf. *Marine Policy* 12: 2–8.

Hutchinson, D.R., J. Childs, E. Hammar-Klose, S. Dadisman, N.T. Edgar, and G. Barth. 2004. A Preliminary Assessment of Geologic Framework and Sediment Thickness Studies Relevant to Prospective U.S. Submission for Extended Continental Shelf. *U.S. Geological Survey Open-File Report* 04-1447, CD-ROM. Abstract at http://pubs.usgs.gov/of/2004/1447/.

King, C. 1880. *United States Geological Survey.* First Annual Report. Washington, DC: U.S. Government Printing Office.

Mather, K.F., and S.L Mason. 1967. *A Source Book in Geology.* Cambridge, MA: Harvard University Press.

Max, M. 2000. Hydrate Resource, Methane Fuel, and a Gas-Based Economy? In *Natural Gas Hydrate in Oceanic and Permafrost Environments,* edited by M. Max, 361–70. Boston: Kluwer Academic Publishers.

McKelvey, V.E. 1984. Interpretation of the UNCLOS Definition of the Continental Shelf. In *Proceedings of the Annual Conference of the Law of the Sea Institute, University of Rhode Island* 16: 465–72.

Merrell, W.J., and G. Boledovich. 2000. *Oceanography: The Making of a Science – People, Institutions, and Discovery.* Washington, DC: The H. John Heinz III Center for Science, Economics, and the Environment. http://www.heinzctr.org/meeting_reports.htm.

Nelson, C.M. 1990. Clarence King as the Interior Department's Principal Geologist. *Geological Society of America Abstracts with Programs* 23: A322.

– 1999. Toward a Reliable Geologic Map of the United States, 1803–1893. In *Surveying the Record: North American Scientific Exploration to 1930,* edited by E.C. Carter II. Philadelphia: Memoirs of the American Philosophical Society 231: 51–74.

Nelson, C.M., and M.C. Rabbitt. 1991. Why President Hayes Chose Clarence King. *Geological Society of America Abstracts with Program* 23: 108.

Nordquist, M.H., J.H. Moore, and T.H. Heidar. 2004. *Legal and Scientific Aspects of Continental Shelf Limits.* Boston: Martinus Nijhoff Publishers.

Pew Oceans Commission. 2003. Final Report Issued. http://pewoceans.org.

Rabbitt, M.C. 1979. *Minerals, Lands, and Geology for the Common Defence and General Welfare,* vol. 1: *Before 1879.* Washington, DC: U.S. Government Printing Office.

– 1980. *Minerals, Lands, and Geology for the Common Defence and General Welfare,* vol. 2: *1879–1904.* Washington, DC: U.S. Government Printing Office.

– 1989. The United States Geological Survey: 1879–1989. *U.S. Geological Survey Circular* C-1050.

Rabbitt M.C., and C.M. Nelson. 1990. Origins of the U.S. Geological Survey. *Geological Society of America Abstracts with Program* 22.

Revelle, R. 1980. The Oceanographic and How It Grew. In *Oceanography the Past,* edited by M. Sears and D. Merriman, 10–24. New York: Springer-Verlag.

Stanton, W. 1975. *The Great United States Exploring Expedition of 1838–1842.* Berkeley, Los Angeles, and London: University of California Press.

Steelman, J.R. 1947. *Science and Public Policy: A Program for the Nation.* Washington, DC: Government Printing Office.

United Nations Convention on the Law of the Sea (UNCLOS). 2001. Agreement Relating to the Implementation of Part XI of the Convention. http://www.un.org/Depts/los/convention_agreements/texts/unclos/closindx.htm.

Ward, K. 1989. *Great Disasters*. Pleasantville, NY: The Reader's Digest Association Inc.

White House. 2004. Budget of the United States Government Fiscal Year 2006. http://whitehouse.gov/omb/budget/fy2006/.

Wilkins, T., and C.L. Hinkley. 1988. *Clarence King: A Biography.* Albuquerque: University of New Mexico Press. (Initially published in 1958.)

Wilson, R. 2004. The Great Diamond Hoax of 1872. *Smithsonian Magazine* 35: 71–9.

4 Public Science, Society, and the Greenhouse Gas Debate

PETER J. COOK

One place to start the debate about public science in liberal democracy is with definitions. The *Concise Oxford Dictionary* defines *democracy* as '(a state having) government by all the people, direct or representative; form of society ignoring hereditary class distinctions and tolerating minority views' and *liberal* as 'favourable to democratic reform and individual freedom.' One of the key defining features of a liberal democracy is a vigorous and open media, one willing to question the range of political, social, and scientific issues confronting society. We cannot address this theme without also considering the role of the media in communicating complex scientific issues, even though the media seldom devote much time, or many words, to science.

Similarly, political parties seldom make science a centrepiece of an election campaign, although components of an election campaign or a policy are frequently underpinned by science. On occasions a national science policy may have an iconic component to it such as 'putting man on the moon' or, more generically, creating 'a smart nation.' More commonly, it is the environment that provides the headlines or the political focus that is underpinned by science. In such circumstances, should the scientist be seeking to express a 'view' in order to influence the outcome or dispassionately present the 'facts' and let others make the decisions? Does the scientist have any role in the development of public policy? Or should the scientist seek no role beyond that of any other voter? Scientists must of course present information, facts, and evidence and take care to maintain their reputation for integrity and impartiality. But they must be part of the debate not least because many critical public issues are scientifically complex and some for the moment cannot be solved by the use of a 'scientific fix.'

Public science underpins key parts of the machinery of government in most liberal democracies, and the public expects that governments will both support and use science. As a result, scientists are important, even privileged players in major issues affecting liberal democracies. The potential of science and scientists to influence society depends in large part on the quality of the scientific evidence and the compelling logic of the scientific arguments, but the influence of science can also be affected by public uncertainty and fear arising from a lack of understanding, poor communication between the scientist and the community, and on occasions a distrust of science and scientists. The community has in the past had grounds for caution in accepting the opinions of the scientific community, as evidenced by such examples as BSE (mad cow disease), thalidomide, nuclear weapons, Chernobyl, and Creutzfeldt-Jakob disease. At the same time scientists and their evidence have been ignored, and as a consequence, the public and politicians have been unduly alarmed or have been denied the benefits of scientific knowledge. Examples might include genetically modified organisms (GMOs), stem cell research, nuclear waste disposal, and fluoridation of drinking water.

Misunderstanding and mistrust of science are important issues that must be considered in any debate on public science. There is a particular concern relating to the source of funding for research or the institutional affiliations of the researcher, for on occasions these can be perceived as representing a conflict of interest that in turn threatens the credibility of the science.

Many of these scientific issues have been discussed in a national context, but increasingly, the most profound issues can only be considered (or effectively addressed) from a global perspective. For example, the greenhouse gas issue is national and international in scope, science is critical to its resolution, and there are complex social and economic issues that cannot be ignored. Therefore, in considering public science in a liberal democracy, we have to transcend national and disciplinary boundaries if we are to fully address this theme. I will provide focus by drawing on my own experience as director of major earth science organizations and describing their work on environmental issues.

The History of Public Science

The first 'modern' liberal democracies were established in the eighteenth century, but science impacted on government for many centuries before that time. The philosopher-scientist Aristotle (384–22 B.C.) and his disciples had a major influence on Greek culture. The scientific observations of Pliny the Elder (24–79 A.D.) and his followers influenced Roman government and society. The Arab philosophers from about the eighth or ninth century A.D. until the Renaissance carried the light for science, set up great universities and libraries, and influenced their society.

Leonardo da Vinci (1452–1519) is the quintessential example of the Renaissance artist-scientist who profoundly influenced society and responded to the military aspirations of rulers. Georgius Agricola (1492–1555) was a much lesser-known but nonetheless highly influential earth scientist who observed and documented a range of geological topics relating to mining that in turn affected central European society in particular. Galileo (1564–1642) was the outstanding scientist of his age, whose observations were supported by the state despite the opposition of the Church.

Little if any of this science was institutionalized; it was focused on support for one person. However, the development of the great universities of the Islamic world from the eighth century and those of the Christian world from about the twelfth century onwards did provide a broad institutional framework for science.

The start of a more formalized and ongoing relationship between science and a liberal society (or government) was perhaps marked by the setting up and formal endorsement of the scientific academies of Europe, such as the Royal Society in Britain in 1663. This was to give a significant boost to the science–government interface in Britain and in other countries of Europe. Perhaps the best early example is provided by Isaac Newton, who was not only the outstanding scientist of his day but also president of the Royal Society, a member of Parliament, and then head of the Royal Mint (Westfall 1983). In the United States Ben Franklin provides a similar but later example of mobility (and cross-fertilization) between science and government. By the eighteenth century, voyages of discovery such as those of Cook began to routinely include scientists. One of the outstanding success stories of an eighteenth-century government defining a scientific problem is that of John Harrison and the development of the technology to establish longitude at sea to a high degree of accuracy (Sobell 1995).

As a geologist, let me make some observations on the importance of geology in the early history of public science. The industrial revolution was born in Britain in the eighteenth century, bringing with it a need to identify the nature and extent of the raw materials required to support industrial activity – coal, iron ore, limestone, clay – and the need to transport the goods produced, initially by canals and later by railways. Those material and transport needs required a systematic understanding of Britain's geology. James Hutton (1726–97) developed the principles of geology including the concepts of geological processes and the nature of geological time (Playfair 1802). In the late eighteenth century, William Smith (1769–1839) systematized geological mapping to create a four-dimensional view of the Earth's crust (Winchester 2001). By the early nineteenth century, the growing resource needs of the industrial revolution, coupled with this new appreciation of the underlying science, resulted in an enormous upsurge of interest in geology, the foundation of the Geological Society of London in 1815, and, in 1835, the establishment of the Geological Survey (now the British Geological Survey, BGS) under its first director, Sir Henry de la Beche, possibly the world's first example of a public science organization (and one that I was honoured to serve as its seventeenth director from 1990 to 1998). The BGS is a fine example of how a public scientific organization can respond to the changing needs of society while still maintaining its core values, skills, and data sets.

Another early example of 'public science' that has also stood the test of time, the British Meteorological Office, was first established in the mid-1800s under its director Robert Fitz Roy (Gribbin and Gribbin 2003) in response to the needs of the Royal Navy and the requirements of merchants for accurate weather forecasts to ensure that their exports would safely reach their destination.

Both the Geological Survey and the Meteorological Office were mirrored by the establishment of similar organizations in most countries, leading to a high level of international collaboration, whether in the naming of the geological timescale, the collection of weather observations, or the maintenance of seismic recording stations. These national scientific organizations also led to the development of international public science organizations, as exemplified by the World Meteorological Organization (WMO).

Therefore, the earliest 'public science organizations' arose not from curiosity regarding the natural environment but rather from a public need to more effectively exploit natural resources or minimize the financial impact of adverse weather. Public organizations such as the Geological Survey or the Meteorological Office were and are characterized by the strategic nature of their mandate, their obligation to collect data on a systematic national or international basis, and the broadly based public good nature of their work. An initial characteristic of such organizations was that they were fully funded by government. In more recent times, however, many have had to rely in part on other sources of funding, and one of the issues touched on in this essay is whether or not this damages their public role.

Public Science and Basic Science

To the extent that almost all basic science is supported by government, it could be regarded as 'public science.' Similarly, most universities get much of their general funding and most of their research funding from government. At the same time, most university researchers would shrink from the idea that they are involved with 'public science.' A distinguishing feature of basic science is perhaps the extent to which it is curiosity-driven and structured in many cases around one talented individual or a small group of talented individuals. In contrast, 'public science' very often has a strong strategic component that is frequently driven by a very clear long-term mandate for an institution such as to 'geologically map the country' or 'forecast the weather.' Not uncommonly, an institution will undertake some basic research in order to pursue its more strategic mandate, but seldom is basic research a core mandate. In other words, public science is seen as strategic and situated at the tactical or applied end of the science spectrum. That is the basis of the discussion in this essay. Specifically, this chapter will not deal with basic science. At the same time, it acknowledges that good strategic science (public science) is very dependent on a vigorous and well-funded basic science sector.

Public Science and Government

The early 'public good' model of public science outlined above is still of crucial importance to many national and international scientific endeavours, but the concept and the conduct of public science have undoubtedly become more diverse in recent decades. The Second World War and the subsequent Cold War produced an enormous increase in the level of government support for public science. Some countries established huge national laboratories which focused on defence-related science. With the end of the Cold War many of these laboratories moved into areas of broad national concern such as energy security, nuclear power, and greenhouse gas issues. Many of them established collaborative arrangements with particular industry sectors in order to strengthen national competitiveness and became progressively more reliant on industry funding.

Some countries maintained a centralized approach to their 'public science,' including France (through the Centre National de la Recherche Scientifique [CNRS]), Australia (through the Commonwealth Scientific and Research Organisation [CSIRO]), and the Netherlands (through the Toegepast Natuurwetenschappelijk Onderzoek [TNO]). For countries with a limited science base, this is a way to establish and maintain a critical mass in science. But such a high degree of centralization in public science does not always sit easily with the diversity of scientific ideas that are now needed to tackle some of the more intractable problems facing government today. As a result, in recent years we have seen a more broadly based and inclusive approach to public science in most democracies in an effort to ensure that scientific ideas are less constrained by institutional or government–private sector boundaries.

This public–private engagement has not always worked in the public interest. The tobacco and asbestos industries are two examples that spring to mind. The experience of BSE (mad cow disease) in Britain starkly illustrates the problems that can arise for scientists in a government department charged with ensuring the economic viability of an industry (the beef industry) while also being responsible for the public health aspects of industry. It was unclear whether the UK Department of Agriculture was the poacher or the gamekeeper. The problem in the case of BSE was further compounded, it would seem, by the reluctance of the public scientific community to look objectively at their results and in particular to be prepared to take into account divergent views. Instead, their response appears to have been to close ranks and to silence external mavericks, not necessarily by the strength and objectivity of the scientific argument but in some cases by seeking to limit research funding to the dissenters. The result was a flawed BSE strategy, a compromised government, a loss of scientific credibility, a communication and media disaster, massive economic costs, and perhaps some avoidable deaths. Openness, transparency, peer review (especially international peer review) of research and data, consultation with a wide range of stakeholders, including those with divergent views, and a clear communication strategy might have prevented or at least mitigated some of the problems posed by BSE and saved money and lives in the process.

As the past head of a public science organization with a major role in nuclear research, I have on occasion found it difficult to work with the nuclear power industry in a manner that serves the needs of industry, government, and the public. The problem arises in part from the tradition of secrecy that pervades the defence and nuclear weapons industries from which the nuclear power industry grew. A great deal of public science has been undertaken for nuclear waste programs. I know from my own experience that most of that science has been excellent, some of it absolutely outstanding and reflecting positively on the scientists and the industry that supported the research. Despite this, the industry has often been reluctant to expose that excellent science to public view, resulting in frustration for the scientist and his or her organization, a suspicious public, and an industry struggling to develop a nuclear waste strategy that is acceptable to the community. Transparency would help not only in the acceptance of scientific results and an improved relationship between public science and the nuclear industry, but also in providing the opportunity to develop a more considered strategic decision-making process that places greater weight on scientific evidence and meets long-term community needs rather than the short-term needs of politics, which all too often in the case of nuclear issues has meant putting off decisions until after the next election.

How then can public science be structured so that it not only meets the needs of government and the community but also recognizes that industry is a key stakeholder? In Australia, the federal government's development of the Cooperative Research Centre (CRC) scheme in 1991 has been one of the most innovative and successful approaches to public science in the past twenty years. The purpose of the CRC program is to develop close and effective strategic cooperation between government scientific institutions, the universities, and industry (CRCA 2004). The government provides some direct funding to CRCs. In the case of my organization, the CRC for Greenhouse Gas Technologies (CO2CRC), this amounts to around 20 per cent of the total funding. This funding provides the glue that holds the centre together and allows it to raise the rest of the cash and in-kind contributions needed to fund the research. In the case of CO2CRC, A$21 million of government funding over seven years provided the seed corn that enabled us to raise an additional $100 million (and growing). The other key component of the CRC model for public science is that it is funded for seven years, thereby enabling a centre to tackle major strategic issues. But a centre does not have a guaranteed existence beyond seven years, thereby ensuring that it does not spend its money on bricks and mortar or on a massive organizational infrastructure.

CRCs are now seen in Australia as vital components of public science, involving universities and more conventional public sector organizations yet remaining somewhat outside the normal public sector system. Not all of the long-established government institutions or universities that act as the research providers to the CRCs are necessarily always comfortable with the CRC concept. Some see them as drawing away funds that might otherwise have been 'theirs,' although in fact the CRCs are able to draw on funding sources that are not available to conventional public sector organizations. Others see them as unduly complex, which at times they are. But they have nonetheless provided Australia with an effective way of drawing together researchers and organizations that have not necessarily worked together in the past.

CRCs in many, though not all, cases also provide a more commercial focus and commonly have much closer working relationships with industry than traditional public scientific organizations. Some might see this as running counter to the needs of the public and public science, but as pointed out earlier, public science has since its earliest days been expected to confer economic benefit. I deal with this particular feature of public science in the next section.

Public Science and Industry

How should public science (including CRCs) work in collaboration with industry? Will the acceptance of industry funding by a public science organization serve to taint the nature of scientific advice or result in a prioritization of science that is more oriented towards conferring private benefit than public good? Will public science be able to maintain 'impartiality' if it works with or is funded by the user of its research, particularly if that user is a for-profit enterprise? Obviously there are potential dangers, and it is necessary to put safeguards in place. Some of the safeguards may relate to the way in which the program or the organization is structured. In particular, there must be clear identification of what is publicly funded, what is privately funded, and what is jointly funded. There must also be good corporate governance – an appropriately structured board, including an independent chair. External peer review and audit and an open and transparent approach to prioritization, decision making, and scientific discussion are also key to appropriate engagement between industry and public science. In my experience, the diversity of a board (including perhaps companies that normally compete) is also an effective way of ensuring impartiality.

What about commercial arrangements involving a public sector research organization providing consultancy services to a company on a fee-for-service basis? This is a common situation these days and might involve providing analyses, national and international development of economic or other models, strategic advice, or any number of other activities. Perhaps the public body has unique expertise or the only analytical equipment capable of producing the required information. In such circumstances, surely it would be foolish to deny access to the company, which after all has paid its taxes. Provided the service is fully paid for and the commercial arrangement is transparent, then it could be acceptable for a public science organization to provide this service to industry. However, it is necessary to add some important qualifiers to that acceptance. First, there should be no public subsidy of the work; second, there should be no priority given to that work ahead of other agreed priorities; third, there has to be a clear organizational structure and an underpinning strategy that enables these safeguards to be in place. Further, there must be a transparent system that ensures that unfavourable as well as favourable results will be released by the commercial organization. In other words, if a commercial organization wishes to use the standing of public science to enhance the credibility of positive results, it must be willing to accept the consequences of negative results derived through public science.

Finally, there must be a system that clearly and unequivocally defines the following:

1. public-good core science funded by the taxpayer;
2. collaborative science funded by a consortium of organizations and the public in approximate proportion to the likely benefit gained; and
3. contract science fully funded by the user (whether industry or some other organization) that directly gains the benefit arising from the science.

Defined in this way, together with the other safeguards outlined above, it is possible for public science to work with and on occasion for industry. Indeed, it would be foolish to imagine that public science can or should be pursued in a policy or industry vacuum. Also, given the massive transfer of assets and funding from the public sector to the private sector (water, electricity, health, security, etc.) in recent years, would it be sensible to pursue science without also considering the needs of the private sector? Indeed, given the massive amount of scientific work undertaken by industry itself, in many cases far in excess of the amount undertaken by the public sector, there is obviously much to gain from public sector–private sector partnerships in science.

Communicating Public Science to the Public

By its very nature, public science obviously pertains to the community or the public as a whole. Its interface with the public is commonly handled through 'government' whether through politicians, public servants, or a 'government spokesperson.' Often there is some inhibition on the scientist or even the scientific organization from talking directly to the community. Some scientific organizations have clear rules about not publicly discussing anything other than science. Perhaps this derives from the principle espoused by Winston Churchill that scientists should be on tap not on top! More likely it arises from the reasonable principle that scientists should not comment on areas beyond their area of knowledge. As a scientist, I find it frustrating to see undue weight placed on the utterances of scientists who comment on matters far beyond their area of expertise. At the same time, the community often does not just want the facts delivered in an impartial and totally objective manner (though they do want this); they also want to hear a scientific opinion based on those facts. But the public may find it impossible to judge whether an opinion is expressed as a 'spokesperson' of the organization (or government), as an 'informed scientist,' or as an 'individual' whose opinion should carry no more or less weight than other individuals.

If the scientist clearly identifies with an organization then the public will reasonably assume that he or she speaks with the authority and impartiality of the organization.

It is difficult to satisfactorily resolve the dilemma of ensuring that a scientist talking to the media does not assume, or is not perceived to assume, the role of a spokesperson for the organization when the view he or she expresses may be at variance with the considered view of the organization as a whole. One way of ensuring this is to allow only the 'official' spokesperson to speak, but this can adversely affect the credibility of the organization and give the impression of an inhibited, monolithic organization that does not allow scientific dissent. On the other hand, the employing organization can expect a degree of loyalty from an employee, with a dissenting view being offered in private to fellow scientists but not necessarily to the public. But could such a culture place public science (and the scientist) in the impossible position of being expected to have a greater degree of loyalty to an organization than to the public? In the example of BSE, few would argue with the view that public science has a greater obligation to public good than to the good of an organization. Indeed, a number of countries now have 'whistleblower' legislation to ensure that public good is recognized.

But other than in cases of falsifying data, or ignoring contrary data, many scientific issues, particularly in broad areas such as the environment, have no clear answers and in many cases no single option for resolution. In such circumstances, so-called 'consensus conferences,' 'town meetings,' or other forms of public debate that bring together the scientists and the community can be useful. Such meetings usually have the express purpose of ensuring that the community understands the limits to scientific certainty and that decisions must sometimes be made when the evidence is equivocal. It also provides the opportunity for the community to understand not only the cost of taking action but also the cost of not taking action. It also offers a vehicle for discussing that while an issue may be global, the actions and the impacts may be local. This in turn can help to minimize the prospect of a community taking a NIMBY (Not in My Back Yard) or BANANA (Build Absolutely Nothing Anywhere Near Anything) point of view.

Public science has a clear role to present the known facts concerning the impact of a particular action on an individual or a community and the likely environmental benefits of that action on the broader community. This of course inevitably casts public science in a sensitive and, on occasion, confrontational role, but nonetheless, it is a role that has to be accepted. What also has to be accepted is that on such occasions, the final decision on where the balance lies and what action will be taken rests not on science but on commercial issues, property values, votes, jobs, or politics, with these issues playing out at the local, regional, national, and international levels.

The greenhouse gas debate is a prime example of a complex issue that involves public science and the community in a multiplicity of interactions. For that reason, it is considered in some detail in the next section. I use my direct experience of one option for mitigation – geosequestration – as an example of the complexity of the debate.

Public Science and the Greenhouse Gas Debate

Evidence of past climate change was first apparent to geologists in the early nineteenth century, when the first geological surveys were being established. The prospect of global warming resulting from rising concentrations of CO_2 was first suggested by Arrhenius (1896). By the 1960s there was clear evidence of changes in atmospheric CO_2 concentration, notably from the spectacular atmospheric chemistry record obtained by the Mauna Loa monitoring station on Hawaii. This in itself is one of the classic examples of the value of long-term public funding of scientific monitoring: when monitoring commenced in the 1950s, there was no real concern regarding the 'greenhouse effect' and no obvious scientific attraction in monitoring the concentration of such a mundane molecule as CO_2. During the 1970s, the possibility of a 'nuclear winter' was regarded as the major climate concern. However, by the early 1980s, scientists and governments started to take note of the real prospect of a greenhouse effect resulting in global warming and climate change.

In 1993, 166 nations signed up to the United Nations Framework Convention on Climate Change (UNFCCC), which came into force in 1994. The Convention's objective was to achieve 'stabilization of greenhouse gas concentrations in the atmosphere below a level that would present dangerous anthropogenic interference with climate system' (UN 1992). In fact, there was, and still is, no agreement on what that 'dangerous concentration' might be. Nonetheless, most (though not all) scientists and the Intergovernmental Panel for Climate Change (IPCC) consider that the weight of evidence suggests that global warming is underway, and few would argue with the aim of decreasing global emissions of greenhouse gases, particularly CO_2, in the atmosphere.

The UNFCCC was followed in 1997 by the more contentious Kyoto Protocol (UNFCCC 1997). The strength of the protocol lies in the symbolism of a number of countries agreeing to take collective action to limit the increase in greenhouse gas emissions, especially CO_2. The weakness of the protocol is that it shows every indication that it will be ineffectual in bringing about a real decrease in CO_2 emissions by 2008–12. It also focuses essentially on one 'technical' response, the use of so-called 'Kyoto forests' (forests planted after 1990) as an enhanced CO_2 sink. Encouraging the planting of more trees (coupled with improved land use practices) is a commendable objective, as is setting targets for reducing emissions. But the protocol places no constraints on developing economies, which will soon provide a major proportion of total emissions. In addition, a number of developed countries, notably the United States and Australia, have chosen not to ratify it. Therefore, while the protocol has great symbolic significance to many people, there is also the danger that some people will believe that simply signing the protocol will ensure CO_2 emissions will decrease, when that is demonstrably not so.

Since the start of the industrial revolution, the atmospheric concentration of CO_2 has risen from about 270 ppm in 1750 to about 380 today (IPCC 2001), with the possibility that it could reach over 700 ppm by 2100, largely as a result of the increased use of fossil fuels. Steps such as greater energy efficiency are very important, and the long-term solution must be to move to an economy based on renewable energy rather than fossil fuels. There are those who would argue that we must do this now, without any delay. But for the moment it is not possible to meet all our energy needs from renewables, which still have to meet the challenge of being able to continuously provide large amounts of low-cost base load power, or to cheaply store large quantities of energy.

In the shorter term, moving to less carbon-intensive fuels will help, as will improved energy efficiency. The IPCC's climate change Model IS92a (IPCC 1992) factors an ongoing energy efficiency dividend of 1 per cent per annum for the remainder of this century. But the global increase in energy usage is projected to be approximately 2.5 per cent per annum, so energy efficiency alone will not keep pace with rising energy demand.

For some people the obvious step is to move increasingly to nuclear power, including building new nuclear power stations. But no country (with the possible exception of Finland) is currently willing to embrace this choice. The nuclear accidents at Three Mile Island and Chernobyl have profoundly and adversely affected community perceptions of the risks associated with nuclear power, whatever its benefits might be. Increased use of large-scale hydroelectric power is also off the agenda of almost every country because of community concerns regarding the environmental impact of large dams.

Large-scale moves to biomass for electricity generation have also been advocated. Although biomass does have a role to play, in many countries this would be at the expense of producing food; in other countries, more intensive agriculture coupled with major land use changes would not be acceptable, and in some countries with fragile soils, such as Australia, it is questionable that long-term, large-scale production of biomass is sustainable. Conversely, some of these countries have sufficient reserves of fossil fuels (particularly coal) to last a century or more – longer than some of the soils will last. Wind power is opposed by communities on the basis of its visual impact, coupled in some cases by concerns regarding bird kills or noise.

Long-term sequestration (storage) of CO_2 offers an alternative approach including ocean storage of CO_2 or long-term geological storage of carbon dioxide (geosequestration). Both technologies involve an increasingly vigorous public debate on their merits.

The oceans are the ultimate sink of most CO_2 in the atmosphere, but the atmosphere–ocean interaction is slow, and it will take many hundreds of years for the excess CO_2 in the atmosphere to re-equilibrate with the ocean. The process can potentially be speeded up. One option, tried on a small scale, involves artificial fertilization of the ocean by the introduction of finely divided iron in iron-depleted parts of the ocean to produce increased algal growth (Martin et al. 1994). In fact there appears to be little if any long-term enhancement of carbon storage, and this, coupled with a concern about the unforeseen effects of 'tinkering with nature' on the scale needed to achieve significant decreases in atmospheric concentration of CO_2, probably makes ocean fertilization option a non-starter.

The alternative approach to ocean storage is to directly inject the CO_2 into the water column at various water depths (Herzog, Caldeira, and Adams 2001). A key problem here is the significant impact on the marine fauna in the vicinity of the injection site. Also, implementing ocean storage on a large scale will change the pH of the ocean. But taking no action to curb CO_2 emissions will also result in a decrease in the pH of the ocean. Indeed, there is already evidence that over the last two hundred years, the pH of the ocean has decreased by 0.1 pH units – a significant change indeed. To date there have been two scientifically well-considered proposals to undertake small-scale ocean storage experiments: one off Hawaii, the other off Norway. Both were abandoned following community pressure and ultimately government opposition. Therefore, while there is real community concern regarding climate change, this does not translate into a willingness to accept any mitigation option whatsoever, particularly one that is perceived as impacting on 'public goods' such as the ocean. It is unlikely that ocean storage will be accepted by the community.

How can geosequestration be made more acceptable? Naturally

occurring CO_2 is commonly found trapped in the deep subsurface in many parts of the world. Geosequestration involves capturing CO_2 from major stationary sources of CO_2, particularly fossil fuel–fired power stations, compressing CO_2 to a dense supercritical state and then pumping it down into suitable geological formations at a depth of around 800 metres or greater, where the CO_2 will remain stored indefinitely. The technique has already been used at an industrial scale in a few places in North America (Moberg, Stewart, and Stachniak 2003) and Europe (Torp and Gale 2003) for a number of years. So is this the silver bullet we need so that we can continue to enjoy the benefits of low-cost fossil fuel–based energy without the looming problem of climate change? Can we have our cake and eat it? Well, no! However, geosequestration is an important component of the portfolio of responses that will be required if we are to make deep cuts to CO_2 emissions to the atmosphere, which is why an increasing research and demonstration effort is being directed at geosequestration by a number of publicly funded research organizations, including my own.

Based on preliminary investigations, it does appear that the world's geosequestration capacity is large and that it has the potential to allow us to make deep cuts in our CO_2 emissions (Bradshaw and Dance 2004). So why are we not doing it on a large scale now? There are two reasons. First the cost of geosequestration needs to be brought down; that is a technical issue that is outside the scope of this paper. Second, for the most part, there is no legal or regulatory framework to cover geosequestration; the community knows little about it and is not necessarily convinced that it is safe or sustainable. 'Won't it all leak back to the surface?' is a frequently asked question. Based on strong scientific evidence, the answer is no. But to counter this, opponents will cite the example of Lake Nyos in the Cameroons (Africa), where volcanic CO_2 was released catastrophically through a natural event from a crater lake, killing 1700 people (Stager 1987).

So whom should the public believe, how can they form a considered opinion on geosequestration, and what is the role of the scientist in the debate?

Given the evidence that geosequestration has a potentially important role to play in decreasing greenhouse gas emissions, how does the scientist go about persuading stakeholders (government, industry, community) that this technology should be part of the response to greenhouse gas concerns? Or is such advocacy inappropriate for the scientist? Should the scientist merely present the facts and then let the community make up its collective mind? This might be an appropriate course of action in some circumstances, but unfortunately the reality is that the scientists working on renewable energy feel threatened by geosequestration and on occasions feel bound to advocate that their technology is better. Such polarization of the scientific debate is unhelpful in terms of working with the community and politicians to develop a logical and considered approach to the problem of decreasing emissions of CO_2 to the atmosphere while maintaining all the good features of our liberal democratic system.

It is instructive to look at the manner in which ideas on geosequestration have evolved and how public science has responded nationally and internationally. My interest in geosequestration was first sparked in 1991, when a handful of researchers in the UK, Canada, the United States, Japan, the Netherlands, and Norway started to develop some of the early geosequestration ideas. But for much of the 1990s, funding was modest, and I well remember the frustration of trying to convince UK government departments and major electricity companies to consider geosequestration as a future mitigation option.

By the late 1990s the interest in geosequestration started to increase, as the petroleum companies and then the coal companies saw it as a potential way of continuing to use fossil fuels in a carbon-constrained world. The power generation companies were reluctant to consider geosequestration, but even they are starting to change. The implementation of CO_2 storage at the Sleipner gas project in the North Sea and the Weyburn enhanced oil recovery project in Canada also demonstrated technical feasibility on a fairly large scale.

Governments started to take an interest in geosequestration around 2000, as it became evident that renewable energy and improved energy efficiency were unlikely to deliver the necessary deep cuts in CO_2 emissions. Countries such as Australia and the United States, which are very dependent on fossil fuels for electricity generation, began to consider geosequestration more seriously. However, the fact that there was no international agreement relating to geosequestration was an inhibition; geosequestration is not a mitigation response recognized by Kyoto, due to the fact that the science has moved more rapidly then the politics.

This situation started to change with two important international initiatives. In late 2002, in Regina, Saskatchewan, the first meeting of the IPCC was held to consider CO_2 capture and storage (CCS). The second initiative took place in June 2003, when the United States convened the first meeting of the Carbon Sequestration Leadership Forum (CSLF) in Washington, DC. Let me briefly examine these two international initiatives, describe how they are evolving, and outline the role of the scientist, to show how public science can operate in an international setting.

The IPCC enjoys extraordinarily high stature as the authoritative body on climate change issues not only in the scientific community but in the community at large. It has a process that is extremely rigorous and bureaucratic, but also inclusive and transparent. For example, the Carbon Dioxide Capture and Storage (CCS) special volume involves one hundred lead authors from thirty countries and over one hundred contributing authors. The first draft of the CCS Special Report was sent to several hundred peer reviewers (including self-nominated reviewers), and every reviewer comment received is addressed. In the case of the chapter on geological storage, of which I am the coordinating lead author (with Sally Benson), we received over one thousand comments, each of which was addressed. The second-order draft received even more comments. The outcome of this painful process was the production of the IPCC's definitive CCS Special Volume in late 2005. The volume summarized our state of knowledge of CCS to early 2005 and assessed the technology in line the the IPCC mantra of being policy-relevant but not policy-prescriptive. Unlike other sections of the report which solely reflect the consensus of the authors, in conformity with the normal practice of the IPCC, the development of the 'Summary for Policy Makers' was approved line by line by all the participating countries before its acceptance by the UNFCCC. The *Special Report* has proved to be influential to other deliberations of the IPCC as well as to documents such as the *Stern Report*, and it is likely to be very influential in the discussions and negotiations that will take place regarding the post-Kyoto system beyond 2012. Added to this formality is the necessity to deal with scientific colleagues (and, on occasions, opponents) with a diversity of views.

In many ways the IPCC is a textbook example of how to develop scientific consensus at a global level. For those of us intimately involved, it is a tortuous, time-consuming process that most would only wish to do once in a lifetime. Also, it is not a mechanism for developing new knowledge; indeed, one of the underlying principles is that it is only concerned with the assessment of existing knowledge, and preferably knowledge cited in peer-reviewed journals.

The IPCC reports have their detractors, who on occasion question the conclusions, the models used, or the credibility of authors. Certainly the IPCC process is not the only way to take international science forward, but there is no question that its transparency, inclusiveness, rigor, and truly international approach give it exceptional credibility and, as a consequence, enormous influence at the international and national levels and throughout the community.

The CSLF is a different sort of vehicle for considering CCS technologies, particularly geosequestration as a mitigation option. Unlike IPCC, which studiously avoids developing policy responses, the CSLF has at its core the objective of ensuring that technology plays an appropriate role in addressing climate change concerns. It also seeks to ensure the right policy settings to take the technologies forward. Unlike the IPCC process, which involves the bringing together of individual leading scientists from around the world under the UN/IPCC umbrella, the CSLF is a ministerial-level body involving sixteen countries plus the European Commission with scientists nominated to CSLF by their countries, although in fact many scientists are also involved in the IPCC. The CSLF has its detractors: some see it as a ploy to counter Kyoto; others see it as an exclusive club dominated by a fossil fuels agenda. Some stakeholder groups, notably NGOs, have indicated unhappiness at what they perceive as their lack of engagement or recognition in the CSLF process, contrasting this with the fact that there are formally established technical and policy groups for CSLF but no stakeholder groups. Does this then cast a shadow on the involvement of scientists in the CSLF compared with their involvement in IPCC? In my view the answer is no. CSLF offers the opportunity not offered by IPCC for close dialogue between scientists and policymakers. This dialogue is essential if sensible policies are to be developed for an area of rapidly evolving technology that could be of profound community importance (and concern) in the future. Because it is so new, geosequestration lacks a clear regulatory and legal framework. Left to their own devices, governments and over-zealous regulators could put quite inappropriate and impractical regulation in place that would ensure that geosequestration was never ever able to play a role in greenhouse gas mitigation. There cannot be a laissez-faire attitude to a new technology such as geosequestration. But if climate change has disastrous consequences for the human race, as predicted by some, it would be foolish to preclude the use of any mitigation option – including geosequestration – because of inappropriate early regulation at the national or international levels.

Therefore, the CSLF is an important body in that it provides the science–policy interface that is so often lacking. It is for this reason that I and other scientists are prepared to contribute to CSLF, not because we believe it will take the basic science forward but because we believe it will take the application of the science forward and ensure that geosequestration policy and regulation are not developed in a technical vacuum.

Is there a downside to policy and science getting so close? Can policy exert undue influence on the direction and objectivity of the science? Maybe, but at this stage it is not evident. In some ways, CSLF and IPCC are at opposite ends of the scientific spectrum in terms of modus operandi and outputs. In reality they are complementary in that one is focused on science and the other on policy, but both are concerned with determining the extent to which geosequestration might be deployed as a CO_2 mitigation option – not as a single response but as part of a portfolio of responses.

This all seems all very sensible: The scientists are getting along well at the international level in IPCC, and the technical policy people similarly are comfortable with progress to date in CSLF. But what about the public? What about the media and the NGOs? What do they think?

The media have difficulties in dealing with a complex issue like geosequestration; even use of the name presents challenges. Conversely, geosequestration scientists have problems in dealing with the media. The nature of the media in a liberal democracy is to ensure the greatest number of readers or viewers. There are notable exceptions, I am pleased to say, who seek to ensure that their articles are informative, accurate, and non-sensational. But for some, the thirst for 'readership' is seen as requiring articles that talk not about 'geosequestration' or 'storage of CO_2' but 'CO_2 dumps' or, even more emotively, 'toxic dumps.' Alternatively, they depict the issues in terms of confrontation between competing technologies or, even better, between competing scientists.

In my many dealings with the press, I always emphasize that geosequestration should be seen as part of a variety of responses and technologies that will be needed to address greenhouse gas concerns. The only issue is the future mix of technologies that will be required, not which technology is required. But seldom is this reported, perhaps because it all sounds too reasonable and non-confrontational. However, it also has to be acknowledged, at least in Australia, and probably elsewhere too, that there are scientists who fervently believe that the only solution to greenhouse gases lies in renewables. Countries such as Denmark are cited for their commendable policy of encouraging wind power. What is not usually mentioned is that when the wind does not blow, Denmark uses its own conventional fossil fuel–based power stations or obtains electricity from the European grid, including electricity generated from German or Polish coals or French nuclear power. I observe this not in any way to denigrate wind power or any other form of renewable energy, for I hope that eventually it will be able to meet all of our electricity and transport needs. Rather, I wish to illustrate the need for a range of responses and technologies. But contrast this position with a statement taken from a recent submission of the Environment Business Council of Australia (7 August 2004):

> Assuming that geosequestration will be proven – and be cost effective is a bit like going for a drive in a car with no brakes, and hoping that someone will invest in brakes before you get to a corner. It's rather ironic that people who think that we can't assume that renewable technology will evolve and be more cost effective are quite happy to make a heroic leap of faith that geosequestration will be proven in the next five minutes. (Lowe 2004: 17)

There is no mention here of the fact that geosequestration has now been underway (at Sleipner) for seven years. And who are the people who think that geosequestration 'will be proven in the next five minutes?' Nobody to my knowledge. Years of research lie ahead.

What about the 'scaremonger' approach? Consider this recent statement from the organization Futureenergy (Whitehead 2004): 'With huge volumes of CO_2 stored underground, leakage of a significant volume of CO_2 through accident, mismanagement, geological instability, or terrorism could result in mass asphyxiation' (46). There is no scientific basis whatsoever for this doomsday scenario painted for geosequestration, and the concept of using it as a terrorist weapon is extraordinary! The statement goes on: 'To combat global warming and climate change we need to eliminate the production of greenhouse gases. Applied to fossil fuels, geosequestration will at best reduce CO_2 emissions by 80%, but energy produced using fossil fuels and geosequestration will still be adding CO_2 to the atmosphere. It's simpler to stop using fossil fuels altogether' (46).

To some people this is compelling stuff. It is also very misleading. If asked, most members of the public would express strong support for the idea of replacing fossil fuels with renewable energy right away. But put the question another way – would you be willing to stop using fossil fuels now if you had to give up air conditioning in the summer, only use electricity for a limited number of hours a day, do without light metals such as aluminum, have a less reliable water supply because the electric pumps could not be guaranteed to run all the time? I suspect few people would now say yes to stopping all use of fossil fuels. Yet that is the choice people would be facing if we moved too quickly to give up fossil fuels. In other words it is as easy (and as unproductive) to create doomsday scenarios for a 'non–fossil fuels world' as it is for a fossil fuel 'geosequestration world.' All this leaves the community and governments unwilling to take action on greenhouse gases because they hear no clear messages coming from the scientific community. The net result is that CO_2 emissions to the atmosphere continue to increase and the rate of global climate change accelerates – hardly a satisfactory outcome.

So what lessons are to be learned? First, the media have a key role to play in conveying scientific issues such as greenhouse gases to the public. But all too often the message is lost because of the need to represent those issues in terms of scientific conflicts or horror scenarios. Nor are scientists necessarily immune from making sensationalist claims and unsubstantiated statements when they perceive their area of science to be under threat. Such conflicts are wonderful for the media but can border on scientific irresponsibility or constitute a misuse of the trust that the public places in scientists and their pronouncements. My plea then is that scientists jealously guard the credibility and standing that they have with the public. They must ensure that they cannot be accused of spreading sensational tales of climate change in order to ensure attention for their area of science.

What are renewables or geosequestration capable of doing today? The scientists may well have put qualifiers into their original pronouncements (for example, about the 1 per cent probability or the ninety-ninth percentile or the risk profile), but such qualifiers are too complex to be picked up by the media and too difficult (or subtle) to be appreciated by the public.

There are certainly hazards for scientists in dealing with the media, and there is a tendency to sensationalize the issues whether through films such as *The Day after Tomorrow* or headlines such as 'Future Megastorms to Kill Millions.' At the same time the scientist depends on the media for getting messages across and, let us admit it, for boosting the prospects that a topic (or a research organization) will receive public funding. Scientists must improve the manner in which they deal with the media, and they must accept the disappointment of being misquoted or even misrepresented on occasions. But they also have an obligation to explain greenhouse gases and other complex issues to the public and to be honest in outlining why there is no one simple solution to complex environmental problems.

Conclusion

Having wandered far and wide over the issue of public science in a liberal democracy, I would like to conclude by addressing the three questions posed by this volume:

1. *Can science retain independence and objectivity in the face of demands to meet commercial and public policy objectives?* On this discourse, I have outlined how independence and objectivity can be retained despite commercial and policy demands, but it is not always easy. I have described how the Cooperative Research Centre system attempts to address this issue, but the methods used are not necessarily universally applicable. Nonetheless some general safeguards can be widely employed, such as an external board with real powers of review and decisions made up of a diversity of members/directors. The chair should be independent. External expert reviewers and transparency of decision making are key features. There must be clear identification of program objectives, structure, and funding. A tripartite system is proposed:
   (a) a core program that is publicly funded and focused on the delivery of public goods;
   (b) a collaborative program jointly funded by the public and the user community and delivering benefits to the public and private (or user) sectors; and
   (c) a contract program where there is no broadly based public funding and where the user fully meets the cost of the work.

   Structured in this way, public and private benefits can be clearly defined, and funded, without jeopardizing the impartiality and standing of an organization. There is no magic formula for the relative proportions of these three programs, but of course it has to be recognized that any public science organization that receives most of its funding from (c) and little from (a) is unlikely to be perceived as independent, dispassionate, or even 'public.' Therefore, while tripartite structure, an independent board, openness, and above all excellent science are extremely important, the bottom line is that perceptions must also be carefully addressed.

2. *In what ways is scientific discourse privileged in the formation of public policy?* Is science privileged or is it just doing its job? There is no question that public science can and should have a major influence on public policy in many areas. The public and government expect science to influence policy, although it has to be recognized that the final decision on a complex problem such as greenhouse gas mitigation will be based not just on science but on social, economic, and political issues. But the public science must be good science, and the mad cow disease episodes in the UK demonstrates how questionable science and bad policy can feed on each other, to the detriment of the public and an industry.

   Scientists do enjoy privileged access to politicians and senior government officials and can have a reasonable expectation that they will receive a hearing. It is important that scientists not misuse this position by claiming an authority beyond their area of expertise. At the same time, many issues relating to public health or the environment are complex and should transcend conventional scientific and institutional boundaries. Our current structures often do not handle this well.

3. *How can scientific knowledge and methodology be made compatible with the interdisciplinarity and integration required of public policy discourse and formation?* As previously pointed out, our current scientific structures do pose problems for interdisciplinary issues and do not always handle integration well. The CRC model mentioned earlier can help but can never be the entire answer. The complex topic of greenhouse gas mitigation and climate change is a good example of the issues that can arise. These can include tensions between different areas of science, the proponents of different solutions, and the tendency of some to depict the issues in black and white when the reality is that there are many shades of grey. The media can be a very important component of any decision by government or the community. Consequently, a well-defined communication strategy must be part of a resolution of issues underpinned by science. It is also necessary to recognize that for some problems there is no obvious solution, and the scientist must on occasion be willing to state this, even when it is unpalatable.

   The global scope of some issues such as greenhouse gases adds an additional dimension to the problem of interdisciplinarity and public discourse. The IPCC in some ways offers a remarkably effective example of how to draw a huge community of scientists into a scientific assessment process and in so doing confer a remarkably high standing to their pronouncements. Some would question the basis of that standing, but having been deeply enmeshed in the process, I must say that despite its imperfections and bureaucracy, there is no obviously better system. Nonetheless it does not offer the entire answer. A strong interface between science and policy, such as that offered by CSLF is also very important and is complementary to a rigorous scientifically driven process that underpins IPCC.

REFERENCES

Arrhenius, S. 1896: On the Influence of Carbonic Acid in the Air upon the Temperature of the Ground. *Philosophical Magazine* 41: 237–76.

Bradshaw, J., and T. Dance. 2004. *Mapping Geological Storage Prospectivity of CO_2 for the World's Sedimentary Basins and Regional Source to Sink Matching*. Presented at the Seventh Greenhouse Gas Control Technologies Conference, 5–9 September 2004, Vancouver, Canada. http://uregina.ca/ghgt7/pdf/papers/peer/426.pdf.

Cooperative Research Centre Association (CRCA). 2004: About CRCs – Role of the CRC Program. http://www.crca.asn.au.

Gribbin, J., and M. Gribbin. 2003. *Fitzroy – The Remarkable Story of Darwin's Captain and the Invention of the Weather Forecast*. New Haven, CT: Yale University Press.

Herzog, H., K. Caldeira, and E. Adams. 2001. Carbon Sequestration via Direct Injection. In *Encyclopaedia of Ocean Sciences*, edited by J.H. Steele, S.A. Thorpe, and K.K. Turekian, 1: 408–14. London: Academic Press.

Intergovernmental Panel on Climate Change (IPCC). 1992. *Climate Change 1992 – The Supplementary Report to The IPCC Scientific Assessment*. Prepared for IPCC by Working Group I: J.T. Houghton, B.A. Callander, and S.K. Varney. Cambridge: Cambridge University Press.

– 2001. *Climate Change 2001 – The Scientific Basis*. Contribution of Working Group I to the Third Assessment Report of the Intergovernmental Panel on Climate Change, edited by J.T. Houghton, Y. Ding, D.J. Griggs, M. Noguer, P.J. van der Linden, X. Dai, K. Maskell, and C.A. Johnson. Cambridge and New York: Cambridge University Press.

Lowe, I. 2004. Interim Submission on the Energy White Paper to Prime Minister and Members of Cabinet. Environment Business Australia (EBA). 7 August.

Martin, J.H., K.H. Coale, K.S. Johnson, S.E. Fitzwater, R.M. Gordon, S.J. Tanner, C.N. Hunter, V.A. Elrod, J.L. Nowicki, T.L. Coley, R.T. Barber, S. Lindley, A.J. Watson, K. Vanscoy, C.S. Law, M.I. Liddicoat, R. Ling, T. Stanton, J. Stockel, C. Collins, A. Anderson, R. Bidigare, M. Ondrusek, M. Latasa, F.J. Millero, K. Lee, W. Yao, J.Z. Zhang, G. Friederich, C. Sakamoto, F. Chavez, K. Buck, Z. Kolber, R. Greene, P. Falkowski, S.W. Chisholm, F. Hoge, R. Swift, J. Yungel, S. Turner, P. Nightingale, A. Hatton, P. Liss, and N.W. Tindale. 1994. Testing the Iron Hypothesis in Ecosystems of the Equatorial Pacific-Ocean. *Nature*, 8 September, 123–9.

Moberg, R., D.B. Stewart, and D. Stachniak. 2003. *The IEA Weyburn CO_2 Monitoring Project*. Proceedings of the Sixth International Conference on Greenhouse Gas Control Technologies, edited by J. Gale and Y. Kaya, 1–4 October 2002, Kyoto, Japan.

Playfair, J. 1802. *Illustrations of the Huttonian Theory of the Earth*. Printed for Cadell and Davies, London, and William Creech, Edinburgh.

Sobell, D. 1995. *Longitude – The True Story of a Lone Genius Who Solved the Greatest Scientific Problem of His Time*. Walker & Company.

Stager, J.C. 1987. Silent Death from Cameroon's Killer Lake. *National Geographic*, September, 404–20.

Torp, T.A., and J. Gale. 2003. *Demonstrating Storage of CO_2 in Geological Reservoirs – The Sleipner and SACS Projects*. Proceedings of the Sixth International Conference on Greenhouse Gas Control Technologies, edited by J. Gale and Y. Kaya. 1–4 October 2002, Kyoto, Japan.

United Nations (UN). 1992. United Nations Framework Convention on Climate Change. http://unfccc.int/resource/docs/convkp/conveng.pdf.

United Nations Framework Convention on Climate Change (UNFCCC). 1997. Kyoto Protocol to the United Nations Framework Convention on Climate Change. http://unfccc.int/resource/docs/convkp/kpeng.pdf.

Westfall, R.S. 1983. *Never at Rest – A Biography of Isaac Newton*. Cambridge: Cambridge University Press.

Winchester, S. 2001. *The Map That Changed the World – William Smith and the Birth of Modern Geology*. New York: HarperCollins.

Whitehead, A. 2004. Keeping the Fossil Fuel Industry Alive. *Issues* 68: 46.

SECTION TWO

Solutions to the Problems: Philosophic

5 The Role of Humanities Policy in Public Science

ROBERT FRODEMAN

Plato's masterwork, *The Republic*, begins with a playful act of coercion, when Socrates' friends force him into a conversation on the nature of justice. At the end of Book 4, after finally offering his definition of justice, Socrates tries to conclude the conversation. He is forced to stay, however, and to further explain his thoughts concerning the nature of the ideal community.

This is the setting for Socrates' account of what he calls the three waves – a series of suggestions, each more odd than the last, that he hardly dares mention, out of fear that they will seem ridiculous to his companions. The first wave is concerned with the proposition of sexual equality. The second discusses the virtues of eugenics and the communal raising of children. But as radical as these first two suggestions are, Socrates shows his greatest trepidation about the reaction to his third wave – that a just society is only possible once philosophers become kings, and kings philosophers.

It is unclear why Plato considers this to be the greatest challenge, and the greatest absurdity, of them all. It may reflect his sense that communities can never accept the radical insights of the philosopher. Or Plato may feel that philosophy is inherently so distant from everyday concerns that governance by philosophers is a practical contradiction. We do know that Plato's own experiment in policy advising with the tyrant Dionysus went badly awry, on some accounts with Plato being sold into slavery, only to be rescued later by friends.

I bring up the example of Plato in order to emphasize the challenges and opportunities that lie before us. For the relationship between philosophy and the community has become relevant again. Strikingly, it has been the government itself, in the form of public science agencies, which has turned to philosophy and the humanities for help, rather than vice versa. Since 1990, U.S. federal science agencies – such as the National Institutes of Health and the National Science Foundation – have steadily increased their support of social science and humanities research. Indeed, as I can attest from my own experience as a consultant for the U.S. Geological Survey from 1993 to 2003, public science agencies are increasingly being driven, in some cases despite their own inclinations, to include concerns about ethics and values in their research.

This support is all the more striking in that it has happened at a time when federal support for *direct* humanities research, through the National Endowment for the Arts and the National Endowment for the Humanities, has declined. Indeed, in the mid-1990s these two agencies came close to being eliminated. At the same time, humanistic and social science research was growing within public science agencies. These social science and humanities research programs go by a variety of names – for instance, the Human Dimensions of Global Change Research within the Climate Change Science Program, or the Office of Environmental Justice within the U.S. Environmental Protection Agency – but the most common names for such programs are EVS (ethics and values studies) and ELSI (the ethical, legal, and societal implications of science and technology). Moreover, these trends are continuing: for instance, the new National Nanotechnology Initiative has set aside 3 per cent of its budget for ELSI-type research.

There are at least two reasons for the attention being given to ethics and values within scientific and technological research. First, it has become increasingly obvious that the positivist assumptions that have underlain science policy are inadequate. For quite a long time, and still in some circles, it has been assumed that science was by its very nature relevant to policy considerations. This belief, sometimes called the linear hypothesis, assumed that a straight line could be drawn from scientific information to policymaking. Contentious political debate would be informed and directed by the calm conclusions of the scientific process.

Increasingly, however, there is a sense within scientific and policy circles that this assumption is inadequate. So, for instance, when the United States created the Global Change Research Program in 1989 the program was justified in terms of the role that science can play in addressing the policy implications surrounding climate change. To quote:

> The US Global Change Research Program was conceived and developed to be policy-relevant and, hence, to support the needs of the United States

> and other nations by addressing significant uncertainties in knowledge concerning natural and human-induced changes in the Earth's environment ... The USGCRP is designed to produce a predictive understanding of the Earth system to support national and international policymaking activities across a broad spectrum of environmental issues. (Committee on Earth and Environmental Sciences 1993: 5)

Since 1990 the U.S. government has spent more than $20 billion on climate science. Across this same span, however, the range of predicted globally averaged temperatures for the year 2100 has *increased* – from 1.4–5.4 degrees Celsius to 1.4–5.8 degrees Celsius. Simply put, science is as likely to discover hidden complexities and uncertainties as it is to identify a clear and straightforward answer to a problem.

A second issue driving the integration of the humanities with scientific research is the fact that science and technology today have become cultural and philosophical events. Science and technology have been so successful in uncovering the nature of and manipulating the world that they have transcended the boundaries of their disciplines. Originally posited as the opposite of and the repudiation of metaphysics, science and technology today have become metaphysical events in their own right, raising fundamental questions about what it means to be human and about the nature of the good life.

It seems clear that the times demand a corollary to the field of science policy. Just as science policy seeks to offer a systematic evaluation of how science contributes to decision making, humanities policy can methodically investigate how the humanities can better contribute to policymaking and how it can help science and technology take better account of societal values. It is worth emphasizing that this represents a signal opportunity for the humanities to contribute to society while at the same time improving their own lot in terms of stature and funding. But I will argue, only partially in jest, that this opportunity can only be seized if we abandon our old dreams of becoming philosopher-kings, or even advisers to presidents, and instead seek to become philosopher-bureaucrats.

Re-envisioning Values in Policy Debates

What is at stake here, of course, is something more than just the future of a given policy school (scientific or otherwise) or scientific research program. The real problem is our society's over-reliance upon technical solutions to our problems – fixes involving a new tax policy, economic mechanism, or scientific or technological breakthrough that allow us to overcome a policy impasse without making a change in ourselves. Of course, some problems are amenable to technical solutions, but most live in a grey area that requires a mix of science, technology, and values analysis. For example, wildfire policy has scientific (e.g., fire ecology), technical (e.g., fire retardant), and axiological (e.g., the meaning of a healthy forest) aspects. Stephen Pyne (1999) remarks that in wildfire management, 'humanistic scholarship' is necessary, because the 'technology could enable but not advise, [the] science could advise but not choose,' and that ultimately the world of political economy needs 'the vitality and rigor of philosophy, literature, and history if it [is] to choose wisely' (author's note: no pagination). In the end, complex problems like those presented in wildfire management present a bewildering mix of facts and values, and we are forced to examine, and perhaps alter, our beliefs about the right course of action. This requires public forums capable of fostering greater openness to self-improvement, better-tempered conversation, and deeper reflection upon the meaning of the good life within a technoscientific world. These goals have traditionally belonged to the humanities; progress in our public debates requires that they become part of our policy processes.

Informed by the concepts, tools, and methodologies found in the humanities – for example, the wider perspective offered by history, the empathic understanding generated by literature, poetry, and art, and the logical clarity offered by philosophy – the humanities aid policy context analysis and enhance reflective dialogue among stakeholders in the policy process. They supplement the values-mapping efforts of the social sciences by providing new categories of description and alternative methods of evaluating policymaking. As a means of policy resolution, humanities policy generates opportunities for values education, clarification, enhancement, and transformation.

Granted, the humanities are not widely celebrated for their practical utility. For over a century now they have been justified largely on romanticist grounds, their worth a matter beyond basic necessities, consisting in the distinctive pleasures of the life of the mind. This worthy point should not blind us to the fact that since ancient times what we today call humanistic reflection was considered essential to a good life.

Narrow and Wide Humanities Policy

Critiques of current science policy – or more simply, the dawning recognition that the simple application of the natural sciences alone is unlikely to solve the societal problems in whose name they are justified – have led public funding agencies to make modest investments in social science. For example, research into the social and political aspects of climate change – known as 'human contributions and responses to global change' – receives around 2 per cent of the U.S. Global Climate Change Research budget, totalling $50 million. Even here, however, the overwhelming majority of this investment goes toward quantitative (often economic) research. The investment in the humanistic aspects of issues such as climate change has remained quite small. The human genome project co-sponsors – the National Institutes of Health (NIH) and the Department of Energy (DOE) – have devoted 5 and 3 per cent of their respective budgets to societal impacts research.

There is of course some overlap between the fields, but to draw out the differences: the social sciences describe values while the humanities seek to improve them. Drawing from fields such as philosophy, literature, art, history, and religion, humanities policy applies humanistic knowledge and perspectives to problems in order to clarify, explore, challenge, and redefine patterns of thought among stakeholders in the policy process. This integration of the humanities into policy deliberations can take different (and complementary) paths, which may be summarized in terms of narrow and wide humanities policy.

The narrow approach to humanities policy is already present today in a variety of federal contexts, such as ELSI programs within the human genome project, the National Nanotechnology Initiative, and EVS programs within the National Science Foundation's Social Science Directorate. This approach is characterized by a predominant emphasis on questions of ethics and epistemology. Bracketing areas of philosophic concern such as metaphysics and aesthetics, this approach focuses on questions of logic and knowledge within issues such as the reliability of genetic testing for susceptibilities to various medical conditions, and issues such as privacy, autonomy, and prior and informed consent. Similarly, issues such as patient and research volunteer safety and fairness in the use of genetic information by insurers, employers, and the courts have loomed large.

Narrow humanities policy can also be defined in terms of its focus upon process rather than product. It takes a proceduralist approach to questions of values, emphasizing that the right result is the one that comes from following the proper procedures: open deliberation, prior and informed consent, and opportunity for dialogue. This perspective urges decision makers and participants to overtly pronounce and defend their value interests rather than treat them as personal preferences or as purely given. Practitioners are urged to be open and honest about their value commitments and make values an explicit part of their rationale for decision making, just as scientific facts are.

In seeking to uncover and clarify motivations, humanities policy can proceed by means of analysis or by shared dialogue. In the former case, humanities policy compares the stated ('formal') goals of an agency with its actual ('effective') goals, and incorporates some tenets of the policy sciences. Among the science policy research community, the policy sciences may come closest to the value-critical analysis promoted by humanities policy. By drawing out logical implications and, in some cases, contradictions, we can uncover philosophic values and assumptions that underlie more visible actions and decisions. In this respect, humanities policy reveals the existing, if otherwise invisible, motivating values within an agency or science policy. While such values, once revealed, may then become open to public or private critique, the specific context will determine whether they are then submitted to evaluation and possible refinement, or whether the analysis will simply be meant to lead to greater transparency and more efficiently focused energies.

In contrast, wide humanities policy highlights two additional factors to those covered by narrow humanities policy: drawing upon a wider set of humanities perspectives and emphasizing values education, evaluation, and modification. Humanities policy not only should be concerned with seeing that actions are consistent with values; it should also determine, as far as possible, which values are the best ones. Humanities policy in this stronger form seeks not just an accounting of values but an active role in shaping this landscape.

Wide humanities policy attempts to reshape the fundamental landscape of policy discussions: it is an attempt at world-making, not just map-making. Of course, the new landscape envisioned by wide humanities policy is not pre-formed; its shape and nuance will result from active dialogue on the values and goals of participants and decision makers. Humanities policy rejoins the battle to identify and promulgate values that improve society and create good policy. It is a rejection of the view that sees values as inevitably subjective.

Moreover, wide humanities policy takes up traditional areas of philosophical reflection that have fallen into disfavour, investigating questions such as what it means to be human. It believes that many of the issues being brought up by science and technology today return us to traditional aesthetic, metaphysical, and theological questions. For instance, possible future advances in biotechnology do more than simply raise issues of safety and prior consent; they also go to the heart of what it means to be human. What would be the consequences for our sense of ourselves if we could consciously design children? How would our sense of accomplishment be affected if our skills and achievements were picked by someone else? (see McKibben 2003 and Sandel 2004).

Aesthetics provides a prominent example of the possible contribution that the humanities can make to policymaking. While the analysis of beauty has long been ruled by romanticist assumptions that see art as predominantly a means of self-expression, aesthetics has also been understood as tasked with forcefully bringing the reality of a situation home to people (Heidegger 1971). On this view, aesthetics consists in *realization*, making something real and relevant to people's lives, whether it is a scientific fact or a perplexity that a community finds itself in. Pictures, paintings, and fictional narrative become bridges between bare fact and poignant meaning, places where people 'get it,' fully grasping the importance of, say, scientific insights to their daily lives.

Aesthetics already plays a constitutive (if usually unacknowledged) role in the framing of public policy. Take the example of acid mine drainage. Acid mine drainage (AMD) is a water quality problem common to rivers and lakes affected by water draining from mine sites. It is a critical water quality issue around the world, affecting nations from the Far East to Europe and the Americas. Estimates of the costs of cleanup within the United States alone are in the tens of billion of dollars. As a matter of ongoing public policy, the beauty and popularity of a damaged area is factored into the decision process (along with other criteria such as cost, proximity to population areas, and degree of damage). Humanities policy can help improve such deliberations by making them more honest, systematic, and self-aware, as well as help them appreciate the ways in which aesthetic judgment are susceptible to reasonable discussion (Frodeman 2003).

Religious thought provides another notable example of the possible contribution of humanities policy. Part of the reason why values education has been passed over within the policy movement lies in our lack of appreciation of the spiritual dimension of scientific practice, whether it be natural, social, or policy science. The point here has nothing to do with sectarian religion. Becoming a scientist requires much more than technical skill at memorizing congeries of facts or manipulating formulas, equipment, or methodology. It also requires more than the mysterious spark of creativity that seizes upon a problem in an original way. Becoming a scientist requires disciplining the soul as well as the intellect. The patient sifting of facts, the willingness to set aside personal desires to follow evidence wherever it leads, the fair-mindedness that helps an opponent improve his or her own argument to the detriment of one's own, the ability to live with uncertainty as a permanent fact of life: these qualities constitute what can be identified as the spiritual element lying at the heart of science.

This point has real implications for humanities policy. Consider, for instance, a Buddhist perspective on policy. At its root Buddhism is concerned with the management of desire, offering a psychological and philosophical reading of our troubles as being less based in the lack of possessions and more rooted in our unwillingness to place limits on our wants. Buddhist practice – for Buddhism is primarily a set of practices rather than a system of beliefs – focuses on loosening our attachment to our own wants. Suffering results from the attachment to what we want; lessen this, and we lessen our heartache.

Such points have generally been taken as a matter of personal philosophy. But as an example of a humanities policy, a Buddhist-influenced science policy could complement our predominantly scientistic approach to problems by recognizing the folly of dogmatic devotion to technological fixes (see Sivaraksa 1992). This approach toward humanities policy could thus help educate us to be more judicious in the pursuit of our own desires within policy debates.

As suggested earlier – and not without a touch of irony – the most effective way to promote such practices may be to extract and generalize the set of skills found within scientific practice, adapting them for the world of policymaking and political debate. But if an education in personal values is possible within scientific practice, why not within the practice of policymaking and political debate? This would not, of course, mean an education in what is the 'right' opinion about, for example, welfare payments or the size of government, but rather an increased attention to improving the process and demeanour of political debate through personal transformation. This transformation also hearkens back to the idea of *Bildung*, a German term that defines education as largely consisting in the development of a self that is more aware, empathetic, and controlled.

The outstanding current example of a wide approach to humanities policy is the President's Council on Bioethics, which uses a wide range of humanities materials (philosophy, literature, religion, etc.) to inform its deliberations on issues such as stem cell research, cloning, genetic enhancement, and aging. The field of bioethics, with its origins in the 1960s, is an exemplary case of narrow humanities policy, focusing on various questions of ethics and epistemology such as the autonomy and rights of patients, and devising more nuanced definitions such as that of brain death. In contrast, the President's Council has been distinctive in expanding the range of topics to include the full range of the humanities. Its recent compilation of readings, *Being Human*, draws from a wide variety of poetry, sacred books, history, philosophy, science, and personal essays (President's Council on Bioethics 2003).

The reactions that the Council's deliberations have elicited have been telling. On the one hand, the Council's attempt to bring an expanded sense of the humanities to bear in policy formulation has been criticized for its technological pessimism and perceived politically conservative agenda and for its lack of explicit policy recommendations: 'There are times for getting to the damn point' (Brainard 2004: A22). But on the other, *Being Human* has sold out its initial printing of 5000 copies, and its work has been praised in a number of publications as a groundbreaking effort in alerting the public to the opportunities and dangers of biotechnology (e.g., Schaub 2004). The Council's goal of informing rather than directing public conversation exemplifies the possible contributions of humanities policy.

Conclusion

This essay constitutes only a prolegomenon for a future humanities policy. The only real way to tell whether the claims made here are cogent will be to test them through a series of case studies (e.g., Frodeman 2003). Only through a sustained exploration of issues such as climate change, biotechnology, and nanotechnology will we be able to identify the consequences of a serious commitment to humanities policy.

Nonetheless, this essay does serve a modest purpose. For even the bare introduction of the idea that the humanities have significant contributions to make to policy debates serves as an invitation to keep an eye out for neglected dimensions of societal issues. The development of a more humanistic approach to science policy research will best occur through a thousand inquiring thoughts and incremental actions in as many situations.

REFERENCES

Brainard, Jeffrey. 2004. 'A New Kind of Bioethics,' *Chronicle of Higher Education* (21 May): A22.
Committee on Earth and Environmental Sciences. 1993. *Our Changing Planet: The FY 1994 U.S. Global Change Research Program*. Washington, DC: National Science Foundation.
Frodeman, Robert. 2003. *Geo-Logic: Breaking Ground between Philosophy and the Earth Sciences*. Albany, NY: SUNY University Press.
Frodeman, Robert, and Carl Mitcham. 2004. Philosophy of Science Policy. *Philosophy Today.* Special supplement. 3–14.
Heidegger, Martin. 1971. The Origin of the Work of Art. In *Poetry, Language, Thought*, translated by Albert Hofstadter, 17–87. New York: Harper and Row.
Kass, Leon. 1997. Testimony before the American Bioethics Advisory Committee, 14 March. http://www.all.org/abac/clon-sec.htm.
McKibben, Bill. 2003. *Enough: Staying Human in an Engineered Age*. New York: Henry Holt.
President's Council on Bioethics. 2003. *Being Human: Readings from the President's Council on Bioethics.* New York: Regan Books. http://www.bioethics.gov/bookshelf/.
Pyne, Stephen J. 1999. Smokechasing? *NHC Home Publications Ideas* 6, no. 2. http://www.nhc.rtp.nc.us/ideasv62/pyne.htm.
Sandel, Michael. 2004. The Case against Perfection. *Atlantic Monthly*, April.
Schaub, Diana. 2004. Methuselah and Us. *The New Atlantis* 2, no. 4. http://www.thenewatlantis.com/archive/4/schaub.htm.
Sivaraksa, Sulak. 1992. *Seeds of Peace: A Buddhist Vision for Renewing Society.* Berkeley: Parallax.
Stehr, Nico, 2005. *Knowledge Politics: Governing the Consequences of Science and Technology.* Boulder, CO: Paradigm.

6 Science Studies Encounter Public Science: Mertonian Norms, the Local Life of Science, and the Long Duré

GORDON McOUAT

Public Science and the Liberal State at the Beginning of the Last Century

When we talk about 'public science,' we must not confuse it with the related notion of the 'public understanding of science' (or 'PUS,' that ugly term used by British administrators and those earnest popularizers of received images of science).[1] PUS itself may have had its origin in the eighteenth century – with notions of the spread of 'Enlightenment' – right at the very time of the rise of 'public science.' The two were closely related, but in unexpected ways. Science, as we have seen in the works of Larry Stewart and Margaret Jacob, was public from the very start (Stewart 1992).

One way of founding the relationship between the political, the social, and the public was established at the beginning of the last century by the great positivist, eugenist, and founder of modern statistics, Karl Pearson (1857–1936), in his monumental *Grammar of Science* of 1892 (Pearson 1937) and especially in his programmatical 'The Function of Science in the Modern State,' published as the prefatory essay of the supplemental eight volumes of the tenth edition of the *Encyclopaedia Britannica* (1902–3) (the one just before the much-admired eleventh edition). Pearson set out one set of parameters for *that* long century's public place of science and the related role of science in the liberal state.

Simply – perhaps too simply – for Pearson science provided us with the intersubjective means to mediate and militate against prejudice, opinion, and the individual persepective. For Pearson, the scientific 'frame of mind' was an 'essential of good citizenship.' These were all matters of state. The teaching of any particular branch of science (it did not matter which) would instil into any citizens the methodology of transcending their own particular subjective view. This was the very idea of a *liberal* education.

Pearson's view was not the only option available for discussing the relationship between science and politics. The nineteenth century had given us the 'value-freedom' controversy (Proctor 1991), and the relationship between the structure of nature and the structure of the state in the Ernst Haeckel/Rudolph Virchow dispute over the 'freedom of science' in the late nineteenth century.[2] But Pearson spoke to the next century. The twentieth century is dotted with various entanglements between science, the public, and the state – with Whiggism, against totalitarianism (Arendt), for socialism (Bernal), for liberal democracy, against 'enframing' (Heidegger), for enframing. In Pearson's view – where perhaps we might most expect it – we find very little of the kind of 'instrumentalism' that would mark later technocratic debates over the meaning of scientific education and its benefit to the material well-being of citizenry – debates wracking the second half of the twentieth century. Rather, for Pearson, science provides *a frame of mind* – a true ground of the *possibility* of being a modern citizen in a liberal democracy.[3]

'I assert,' wrote Pearson, 'that the encouragement of scientific investigation and the spread of scientific knowledge by largely inculcating scientific habits of mind will lead to more efficient citizenship and so to increased social stability.' 'Minds trained to scientific methods are less likely to be led by mere appeal to the passions or by blind emotional excitement to sanction acts which in the end may lead to social disaster' (1937: 13). For Pearson, we do not turn to science for any specific instrumental ends. Rather, science is a kind of control on the passions and divisive interests.[4] And, moreover, science is *an aid to stabilization of the state*. (No wonder why Karl Pearson was called a 'social imperialist' [Semmel 1960].)

Pearson's very own mathematical science – statistics – was 'public' from the very beginning; its very objects – population censuses, surveys, laws of large numbers – were built in concert with the rise of the modern state, the modern population, and the modern citizen (Hacking 1990; Porter 1986). For Pearson, science was public not only because its object was public, but also because it presented the very possibility of *being* a public.

As such, Pearson was not so concerned about the *instrumental* effects of science, technological or factual – it was *methodology*, the holy grail of a modern philosophy of science: 'the value of science for practical life turns upon the efficient training it provides in *method*' (Pearson 1991: 13). In Pearson (1991) he follows with what he considers the 'essentials of good science.' (Be cautioned, they do *not* exclude a vivid imagination or a deep passion and aesthetic sense.) Pearson's essentials of good science contain all the usual suspects in our modern understanding of the so-called 'scientific method' – intersubjectivity, testability, critical reason, scepticism, limits upon bias. He meant this route to be a one-way street: a good citizen learns the right methodology *from* science.

Pearson's notion of the public good of science at the very end of the nineteenth century set much of the form for the study of the public function of science during the first half of the twentieth century. The sociology of that period concerned itself with the grounds of 'good science' – under what conditions could it prosper? (Ben-David 1971; 1991). Its main champion was the American sociologist of science Robert K. Merton (1910–2003). He was mainly concerned with norms and values.

End of Mertonian Values?

In the 1980s and 1990s, radically sceptical critics of the received notion of science began to appear (mainly out of Edinburgh, and Bath in England). The modern field of 'social studies of knowledge' (SSK, or, perhaps, science and technology studies – STS) cut its teeth on deconstructing and laying waste to those 'essentials of good science' so well reified and put to similar use in the mid-century in Robert Merton. His famous 'four cultural values' (both descriptive and normative) of the scientist were grounded on Pearsonian norms: universalism, communism, disinterestedness, and scepticism (Merton 1942). Here is how one latter-day commentator describes the Mertonian values:

> Universalism requires that science be independent of race, color, or creed and that it should be essentially international. Communalism requires that scientific knowledge should be public knowledge; that the results of research should be published; that there should be freedom of exchange of scientific information between scientists everywhere, and that scientists should be responsible to the scientific community for the trustworthiness of their published work. Disinterestedness requires that the results of bona fide scientific research should not be manipulated to serve considerations such as personal profit, ideology, or expediency, in other words they should be honest and objective; it does not mean that research should not

> be competitive. Organized skepticism requires that statements should not be accepted on the word of authority, but that scientists should be free to question them and that the truth of any statement should finally rest on a comparison with observed fact (Brown 1986).

On each 'value' our Edinburgh (and Bathist) friends in the sociology of knowledge have gnawed profusely (Bloor 1991). Knowledge, we have learned from the sociological turn, is *always* local, fractured, interested, and biased.

We need not rehearse the phenomenal rise and petering of the brief 'science wars' of the 1990s – a dispute arising out of the quick ascendancy of radical sociological schools and the opposition they generated among some practising natural scientists and their philosophical supporters (Parsons 2003). That dispute revolved almost exclusively around the problems of Mertonian/Pearsonian norms. In the 'science wars,' the main defence of the *received* ethos of science seemed to come from scientists (and perhaps some analytic philosophers and a good many political scientists to boot) who had so much invested in telling that internal story about their particular sciences (Boghossian 1996, Gross and Levitt 1994). From that debate we have learned a lot about embodied objectivity, partial perspectives, the disunity of science, and the networks of negotiations (now between people and things) that make up a scientific fact (Golinski 2005). But very little was said about the other side of the equation: the political.

This concentration on the one side of the Pearson/Mertonian legacy may have blinded us to other options available in the debates over science and politics. We have been warned about this before: that we may indeed have missed something about the very thrust of Pearson's argument and its Mertonian version. We *may* have missed something about the importance of those regulating sets of norms for the very self-construction of scientific possibility – about the relationship between science and a liberal democracy, about the ecology of scientific space. This chapter is about that missing place. It brushes lightly over many of the issues that could be raised. Instead, it examines the black-boxed side of the equation: the internal norms of scientists and how those might be themselves changing. Near the end I will introduce a research project – still in the works – to examine a changing ecology of science, using many of the tools of the new social turn without, however, losing sight of a bigger picture. For there is a bigger picture: namely, the one addressed by the essays in this book.

Kuhnification?

Fringe sociologists of knowledge, such as the indefatigable Steve Fuller, have lamented something they have called the 'Kuhnification' of 'science studies' (2000). For them, while the new sceptical sociology of knowledge was self-congratulatorily 'revolutionary,' it masked a stodgily reactionary core – a core that pointed in the opposite direction to the deep social interests motivating those first waves of the public debate about science in the nineteenth and early twentieth centuries. The pedigree did seem clear for those who care to look (although the new professionalized SSK, as any good revolutionary does, was clever at concealing family roots). Provocatively, Fuller claims that Kuhn and his sceptical followers are really only the latest in a long line of conservative theorists who emphasized the 'tacit,' the 'traditional,' the 'conventional,' and the 'irrational' over the rational and the critical.[5] The genealogy is a great deal more complicated than Fuller lets on, although there is indeed a noticeable lineage. Edmund Burke's great eighteenth-century poke at Paine and the French Revolution – *Reflections on the Revolution in France* (1790) – was a caution against the Enlightenment expansion of scientific critical reason into *social-cultural* and constitutional matters. For Burke, in the social-cultural domain, meaning and grounds are not established by perpetual scientific critique and 'objective' analysis but by 'tradition' and by 'skill' (in this case the aristocratic skill of governance), by local knowledge, by interest and 'bias' (Burke happily used the word 'prejudice'). Applying the critical reason of science in social-cultural matters would at best stultify culture and at worst lead to political tyranny, Burke warned. That is why the British constitution was so exemplary (!). It was tacit, and aristocratic.

In the mid-twentieth century, British chemist and philosopher Michael Polanyi took up Burke but gave him an ironic twist by expanding the traditional critique of reason into the *very core* of science itself (where Burke always feared to tread). For Polanyi, science itself was constituted by just such a Burkean organic unity of tacit, traditional, conventional skills, incapable of explication and examination by the strictures of critical judgment.[6] Remarkably, Polanyi intended his turn to 'tacit' and 'traditional' knowledge as a way of *blocking* social and political intrusion *into* science – and rescuing pure science from the imperialism of utilitarianism and instrumental knowledge. If science was a 'culture' unto its own, with its own unwritten (unwritable!) rules, traditions, skills, romantic intuitions, and Truth a kind of transcendental attractor point of a skilled set of practitioners, then any intrusion of the 'social' (political, instrumental) into science would diverge it from its noble and proper goal. Such an intrusion would lead to a kind of Burkean described tyranny within the very core of science. Moreover, it would lead to stagnation within science. 'Any authority which would undertake to direct the work of the scientist centrally would bring the progress of science virtually to a standstill' (Polanyi 1969: 51). There was nothing to be communicated across the border from either side. Keeping them (i.e., science and the state) apart was the very ground of the liberal state.

Polanyi's new gestalt notion of science was directed against a particularly Baconian/Marxist engagement with the social place of science – mainly in the work of the crystallographer-sage J.D. Bernal (1939; Werskey 1978). Bernal thought that the social ills associated with recent science came not from its instrumentalism and its political involvement (for Bernal, science was 'instrumental' and political all the way down), but rather from its deflection into *certain* personal and destructive directions by the interests of capital. Capital, not instrumental knowledge (nor 'Gestell,' nor the 'masses), deflected science, said Bernal.[7] Polanyi had very little to say about capital. On that score he was indeed a classic liberal.[8]

Fuller claims that the origin and élan of recent post-Kuhnian 'science studies' is to be found not in a wholly politically engaged study of, say, science in a liberal democracy – such as that offered by Bernal – but in the isolation of science as a separate elite culture, held together as a way of life. Ironically, while Bernal may have won on the funding side of things (that is, we have become, in a sense, entirely Bernalized), Polanyi may have won on the intellectual side of things (at least with those engaged in the social examination of science) (Young 1980). And so the social study of science has become largely ineffectual on the political scene.

Here Fuller tried (rather unsuccessfully, one might say) to do for Kuhn what Kuhn purportedly did for the 'received view' of science – he contextualizes him as part of the Cold War attempt to have one's scientific cake (of the instrumentalism of weapons research and the enormous growth of science in the postwar years) and eat it too (that is, claim that it was entirely *above* social issues, instrumentalism, engagement). For Fuller, Kuhn's theory of scientific change tended to insulate the development of scientific ideas from their wider context. A Kuhnian 'paradigm' is a paradigm because it is wholly self-enclosed, incapable of *scrutiny from the outside*. Moreover, and notoriously, the post-Kuhnians refused to take a stand on the value of their object of study. They held it with great pride that they were 'value free' (of course, missing the irony). As such, we find a diffusion of responsibility – the cleverest trick of modernity, if you like (Latour 2004). Here is where the trouble begins.

I hold no brief for Fuller's rather peculiar and procrustean reading of Kuhn and post-Kuhnian social study of science. I cannot quite understand why he wants to lay the entire recent social turn at the door of Thomas Kuhn (after all, the social study of science is a very diverse, dare I say 'disunited,' movement, representing a spectrum of engagement, most of which is highly critical of Kuhn) (McOuat 2001). But in the remainder of this paper I want to examine something that may have indeed been missing in the recent constructivist turn – a turn that, while I wholly endorse it, I worry avoids some of the questions raised by earlier understandings of science and its place as discussed by Pearson. Perhaps we can use certain skills and tools acquired by those in SSK to revisit those issues raised by Pearson, by Polanyi, by Bernal – by the older views of science, its public, and its norms. Moreover, we may be able to use the recent encounter with social studies of science to reintroduce discussions about norms and methods so very absent from that turn. I want to talk about something that has been introduced by recent (well, perhaps not so recent) science studies – the study of the very local, the local that seemed to arise from studies of laboratory practices.

The End of Laboratory Life as We Know It

One of the great canonical texts of the social- (later just plain) constructivist turn in the social studies of science is Bruno Latour and Steve Woolgar's *Laboratory Life* (1989). First published in 1979, *Laboratory Life* was the pioneering exploration of the *interior* of that great authoritative site of modernity, the large-scale research laboratory. For earlier studies of science, including the first wave of those social turners, the laboratory remained off-limits – a black box where truly miraculous things get done. We all knew what went on in *there*: the discovery of facts and the testing of theories. The real task, for earlier social studies of science, was to expose the social interests and skill going *into* the very possibility of a laboratory. In good phenomenological manner, Latour and his followers bracketed all that. They were interested in the construction of a scientific fact, on the factory floor if you will. They discovered that it was much more complicated than the usual stories of 'scientific method' and norms.

Almost immediately, post-*Laboratory Life*, laboratory studies became all the rage (albeit with rather diminishing returns). Science studies turned to laboratories and what were called ethnomethodologies of scientific practice. German social theorist Karin Knorr-Cetina (1981) explored knowledge as it was 'manufactured' on the floor; Sharon Traweek (1988) spent time in a large-scale particle accelerator in Japan and there found much social construction, and even more masculinity. Others followed suit. The philosophy of science even turned to the laboratory (Hacking 1983).

What did they find there? Our new intrepid anthropologists of science discovered that there was no real 'inside' and 'outside' of the laboratory at all – just long networks of negotiation, exchange, and translation, involving social actors, political actors, aesthetic actors, rhetorical actors, technological actors, and so on. What they did *not* find is a simple collection of Mertonian norms exemplarily guiding the hand of the laboratory actors. (It should be noted that they went in there not believing it in the first place, so it is not surprising.) Old questions that motivated Pearson, Polanyi, Bernal, and Merton seemed no longer applicable, or just to melt away. Science was not 'exemplary' in any sense. It was not separate and pure. It was not any of those old things.

Now, with such insights, the new social studies of science *could* have and indeed *did* address some of these older questions – for example, the great problem of the 'expert' in a modern democratic state could now be reformulated. Expertise was now distributed – opening up, perhaps, a new conception of responsibility and public participation in science. The jury is still out on that one, though – and few have taken it up (Jasanoff 2005). But the study of the life of the laboratories and the local was always extremely short-term and rather internalized to the life of the lab and its network – the typical in-out of the early anthropological ventures of, say, the mid-nineteenth-century cultural anthropologist.[9] With that sort of encounter, any long-term 'ecology' of the laboratory and its place was out of the question.[10] Science studies have been (and are) notoriously uninterested in the long duré. They were concerned with the stabilization of a scientific fact. But perhaps it is in the long duré that different questions can indeed be asked – questions ignored in our local turn. And it is just that sort of ecology that is demanded of the new situation in science, where we *do* face radical changes in funding structure, in the place of science, where the liberal state *is* re-evaluating its relationship with science (in part as a result of the success of science studies). Let us pause for a moment to look at that new situation. These thoughts are entirely preliminary – even impressionistic. But they do impinge upon an environment within which science gets done.

Recent studies in the structural funding of science have noted a (for some, disturbing) trend. In Canada we can point to the shift from 'centres of excellence' to 'centres of innovation.' In rhetoric echoing the 'decline of science' movement of nineteenth-century England (Babbage 1830), policy analysts of scientific research trends either praise or lament this particular shift in science funding.

I will prejudice the matter by quoting only one or two studies. (The amount of data on these developments is huge, and most of the work done on this has been the result of policy analysis and bibliometric studies. Much of it remains undigested.) In her address to the 26th Annual AAAS Colloquium on Science and Technology Policy, 2001, Diana Hicks, senior policy analyst at CHI Research, Inc., analysed the 'Changing Science and Technology Environment.' She identifies three areas of interest: the phenomenal growth in 'innovation' as a context of research, the change in the research enterprise itself at both national and international levels, and the close connection between innovation and research (focusing on university patenting and patent-to-paper citations) (Hicks 2004). It is as we would expect – we are witnessing a large-scale shift in funding, from local to global, from what Polanyi might recognize as 'pure' research to engaged research oriented towards a particular instrumental or social end, and a large growth in public/private partnership. (However, it should be noted that this notion of an entirely recent and exclusive engagement between science and the private sector has been challenged by some historians of science.)[11] In May 2002, the OECD conducted a workshop concerning 'Science Funding in Transition,' coming to many of the same conclusions as the Hicks report (OECD 2002; O'Malley, Doolittle, and McOuat forthcoming). The situation seems clear: we could do well to examine the shifts in funding provided by the construction and renovation of 'Genome Canada' – the major funding body and knowledge clearinghouse for genetic research in this country, sticking with a close examination of the *types* of competitions and the nature of awards.[12]

Policy analysis of science usually remains at this level. But I propose a different route. Over the last few years, students of mine have been tracking the *internal* development of laboratory life in a series of laboratories devoted to the study of genetics and genomics. Those laboratories were chosen especially for their proximity to the debate about the nature of pure versus applied genetics. (One of the key laboratories is that of the world-renowned microbiologist Ford Doolittle. Doolittle has been a continuous gadfly – albeit a well-connected gadfly – around the exposed parts of genetic science. Having developed many of the techniques for mitochondrial DNA phylogenetics, Doolittle likes to think of himself as a constant innovator, indeed a revolutionary. His laboratory is strikingly and decidedly non-applied. Phylogenetics may or may not have particular economic benefits. However, their projects concern issues such as the evolution of complexity and the prokaryote/eukaryote transition, lateral gene transfer, within-'species' genomic diversity, and environmental genomics.)

Our methodology combines classic ethnomethodology of participant/observer laboratory studies with a more standard bibliometric study of citations, publishing, and other sites of scientific legitimacy. We propose to target a spectrum of laboratories, some actively involved in Polanyi-esque 'pure science,' others very much centres of innovation and public/private integrations. We chose these laboratories in order to examine the way the changing ecology of science and its funding changes *the internal life* of laboratories (if, indeed, there is such a thing). One thing that we *cannot* rule out is the change in Mertonian/Pearsonian norms.[13] Practising scientists put much truck in the Mertonian norms. In a sense, these are their self-legitimization. But, as new forms of scientific funding and research press in upon these laboratories, we hope to examine the long-term nature of these norms, of the place of expertise and legitimacy (objectivity, the rest) in a changing environment. We will do so by closely following those changes, not in national policy, not on the distribution of ever-scarce funding, but on the very life of laboratories over the long term and the short term.

The Life of Science in a Changing Liberal Democracy

Perhaps before we ask ourselves about the nature of the scientific expert, the role of the scientist in public decisions, the very public of science, we have to ask ourselves this: in what way are the very forms of life of scientific practice being changed in late modernity? What does this evolution mean to science? Is this merely more of the same from the very beginning of modern science? These are perennial questions. We can bring the formidable tools of recent close examinations of science to explore the very ecology of that life. If indeed a new ecological landscape changes even the propagandistic/rhetorical nature of Mertonian/Pearsonian norms, we might be able to track it – on the ground. 'Will scientific knowledge itself become politicized?' It already has.

NOTES

1 In Britain they now have 'CO-PUS,' established by the Royal Society and the Office of Science and Technology (Miller 2001).

2 The leading German cellular pathologist and anthropologist Rudolph Virchow in his speech before the fiftieth conference of the German Association of Naturalists and Physicians, 'Die Freiheit der Wissenschaft im modernen Staate' (1877), famously spoke against the teaching of evolution in the German school system. Although a political liberal and philosophical naturalist, Virchow believed that Darwinism was at once non-scientific (an unproved hypothesis) *and* dangerously political (i.e., associated with the rising Social Democratic Party's view of historical development and politics). The great German phylogenist and Darwinist Ernst Haeckel responded immediately that evolution should be taught in the schools because it was indeed a science and, *as a science*, it was necessarily politically neutral (Haeckel 1879). In spite of their disagreement, Virchow and Haeckel agreed on virtually everything about the relationship between science, the public, and the state.

3 Now, it is not clear how democratic or liberal Pearson actually was. He speaks of an 'aristocracy of science' with almost Carlylean enthusiasm and imagines the construction of a scientific elite of experts to replace the old oligarchy. Yet, he was ambiguous: 'We must aristocratise government at the same time as we democratize it' (Pearson 1991: 306). Perhaps Theodore Porter (2002) overstates the position when he writes about 'Statistical Utopianism in an Age of Aristocratic Efficiency.'

4 Here Pearson presents a wonderful social-Darwinian take on a Cartesian theme. See Pearson (1902–3), where he begins his discussion of the 'function' with a nod to Darwin, and this astonishing claim: 'By observation on the death-rate of relatives – i.e., of groups of individuals possessing like characters or combinations of characters – it has been demonstrated that 50 to 80 per cent of the deaths among mankind are selective' (viii). On the passions and the virtues, see Albert Hirschmann (1977).

5 Habermas (1989) made the same move against certain French 'radicals.'

6 Perhaps the first to point out the importance and irony of Polanyi's take on Burke was the communitarian philosopher Alisdair MacIntyre (1977).

7 Here, for example, he would locate the evils of Monsanto in the structure of private production and research.
8 Of course I am vulgarizing here. Bernal's position on the social function of science is at once very nuanced and also quite naive.
9 Latour stayed for under two years, long by the standards of laboratory studies.
10 In a sense, the anthropologists were pre-Humboldtian.
11 For example, the papers presented at the Dibner Institute summer seminar, 'The Business of Life: Life Sciences and Industry in the 20th Century,' 15–22 May 2002.
12 See www.genomecanada.ca (accessed 1 October 2004). The Genome Canada website contains lots of PUS. Pessimists such as Fukuyama think that the shift in genomic research is a mark of our 'posthuman' future (2002).
13 The late John Ziman was working on just this issue when he died in January 2005. See his re-evaluation of the Mertonian norms under new 'post-academic' science (2000). Pessimistically, Ziman worried that the old norms were being replaced with the norms of industrial science: 'proprietary, local, authoritarian, commissioned, and expert' (81). It is still unclear whether scientists have explicitly adopted these as their operative values.

REFERENCES

Babbage, C. 1830. *Reflections on the Decline of Science in England.* London: B. Fellowes.
Ben-David, J. 1971. *The Scientist's Role in Society: A Comparative Study.* Englewood Cliffs, NJ: Prentice-Hall.
– 1991. *Scientific Growth: Essays on the Social Organisation and Ethos of Science.* Berkeley: University of California Press.
Bernal, J.D. 1939. *The Social Function of Science.* London: George Routledge and Sons.
Bloor, D. 1991. *Knowledge and Social Imagery.* 2nd ed. Chicago: University of Chicago Press.
Boghossian, P. 1996. What the Sokal Hoax Ought to Teach Us. *Times Literary Supplement*, 13 December, 14–15.
Brown, R.H. 1986. *The Wisdom of Science.* Cambridge: Cambridge University Press.
Fukuyama, F. 2002. *Our Posthuman Future.* New York: Picador.
Fuller, S. 2000. *Thomas Kuhn: A Philosophical History for Our Times.* Chicago: University of Chicago Press.

Golinski, J. 2005. *Making Natural Knowledge: Constructivism and the History of Science.* Chicago: University of Chicago Press.
Gross, P., and N. Levitt, eds. 1994. *Higher Superstition: The Academic Left and Its Quarrel with Science.* Baltimore: Johns Hopkins University Press.
Habermas, J. 1989. *The New Conservativism.* Cambridge, MA: MIT Press.
Hacking, I. 1983. *Representing and Intervening.* Cambridge: Cambridge University Press.
– 1990. *The Taming of Chance.* Cambridge: Cambridge University Press.
Haeckel, E. 1879. *Freedom in Science and Teaching.* Reprint edition. Honolulu, HI: University Press of the Pacific, 2004.
Hicks, D. 2004. The Changing Science and Technology Environment. *AAAS Yearbook 2002.* http://www.aaas.org/spp/yearbook/2002/ch33.pdf (accessed 1 October 2004).
Hirschmann, A. 1977. *The Passions and the Interests: Political Arguments for Capitalism before Its Triumph.* Princeton: Princeton University Press.
Jasanoff, S. 2005. *Designs on Nature: Science and Democracy in Europe and the United States.* Princeton: Princeton University Press.
Knorr-Cetina, K. 1981. *The Manufacture of Knowledge.* Oxford: Pergamon Press.
Latour, B. 2004. *The Politics of Nature.* Cambridge: Harvard University Press.
Latour, B., and S. Woolgar. 1989. *Laboratory Life.* 2nd ed. Princeton: Princeton University Press.
MacIntyre, A. 1977. Epistemological Crises, Dramatic Narrative and the Philosophy of Science. *The Monist* 6: 453–72.
McOuat, G. 2001. The Mistaken Gestalt of Science Studies: Steve Fuller Takes on Kuhn. *Canadian Journal of History* 36: 523–7.
Merton, R. 1942. Science and Technology in a Democratic Order. *Journal of Legal and Political Sociology* 1: 115–26.
Miller, S. 2001. Public Understanding of Science at the Crossroads. *Public Understanding of Science* 10: 115–20.
OECD. 2002. Summary Report on Workshop 'Science Funding in Transition – Changing Paradigms and First Experiences of Implementation.' http://www.oecd.org/dataoecd/44/43/2506954.pdf (accessed 1 October 2004).
O'Malley, M., F. Doolittle, and G. McOuat. Forthcoming. Evaluating the Triple Helix Account of Innovation.
Parsons, K., ed. 2003. *The Science Wars: Debating Scientific Knowledge and Technology.* Amhurst: Prometheus Books.
Pearson, K. 1937. *The Grammar of Science.* London: J.M. Dent and Sons.
Pearson, K. 1902–3. The Function of Science in the Modern State. *Encyclopaedia Britannica.* 10th ed. Vol. 8: vii–xxxvii. London: Adam and Charles Black.

– 1991. The Moral Basis of Socialism (1887). *The Ethic of Freethought*. London: Unwin.

Polanyi, M. 1969. The Republic of Science: Its Political and Economic Theory. In *Knowing and Being*, edited by Marjorie Grene, 49–72. Chicago: University of Chicago Press.

Porter, T. 1986. *The Rise of Statistical Thinking 1820–1900*. Princeton: Princeton University Press.

– 2002. Statistical Utopianism in an Age of Aristocratic Efficiency. In *Science and Civil Society*, edited by L. Nyhart and T. Broman, 210–27. Chicago: University of Chicago Press.

Proctor, R.N. 1991. *Value Free Science?* Cambridge: Harvard University Press.

Semmel, B. 1960. *Imperialism and Social Reform: English Social-Imperial Thought, 1895–1914*. Cambridge: Harvard University Press.

Stewart, L. 1992. *The Rise of Public Science: Rhetoric, Technology and Natural Philosophy in Newtonian Britain, 1660–1750*. Cambridge: Cambridge University Press.

Traweek, S. 1988. *Beamtimes and Lifetimes*. Cambridge, MA: Harvard University Press.

Werskey, G. 1978. *The Visible College*. London: Allen Lane.

Young, R.M. 1980. The Relevance of Bernal's Questions. *Radical Science Journal* 10: 85–94.

Ziman, J. 2000. *Real Science: What It Is, and What It Means*. Cambridge: Cambridge University Press.

7 The Democratic Deficit of Science and Its Possible Remedies

IAN JARVIE

There is an old adage that goes 'If it ain't broke, don't fix it.' Philip Kitcher, in *Science, Truth, and Democracy* (2001) brings forward as examples of brokenness the human genome project, the cancellation of the superconducting supercollider, and investigations of race and IQ. Martin and Richards (1995) bring forward the controversies over water fluoridation and over Linus Pauling's theories about vitamin C and cancer. At one level, all that these examples demonstrate is that the experts disagree and that funding sometimes goes to and sometimes is denied to one project or the other. They do not by themselves show what within science is broken.

Before I continue with a discussion of these cases, let me bring out some problems of the received sociology and philosophy of science as they manifest themselves in Kitcher's work. His intent is to find a *via media* between two 'images' of science: one positive and one negative. Using the language of images has the effect of trivializing the issues at hand by focusing on perception and not reality. It suggests as well that science and anti-science are on some kind of par and deserve equal time. It also suggests that scepticism about science comes from the anti-science side. All this received sociology and philosophy seem to me suspect. As long ago as the late 1930s the sociologist of science Robert K. Merton (1938) had identified institutionalized scepticism as part of the ethos of organized science. This scepticism is directed towards science, that is, towards scientific work – that of colleagues and of oneself. Thus a positive attitude towards science includes scepticism towards it and all of its claims. Such scepticism directed towards science and its claims is constructive; scepticism directed towards anti-science and its claims is annihilating. A sociology and philosophy of science that holds otherwise flirts with authoritarianism.

Kitcher's characterization of what he terms the image of the scientific faithful includes views that I endorse. However his choice of label, 'the scientific faithful,' strikes me as needlessly loaded. Kitcher's language insinuates that pro-science is a kind of faith, a substitute religion that parallels the way anti-science sometimes stems from religious sources.[1] But Kitcher names no names and lumps together what needs to be separated. Kitcher wants to abandon the Weberian philosophy of science as value-free, but he still seeks to legitimate a somewhat curbed authority of science rather than acknowledging that its scepticism is opposed to all authority, including its own.

This brings me back to the cases offered as evidence that there is something amiss in science. All the five cases at issue are of scientific matters that became matters of public interest and in some cases also public intervention – intervention that led to funding of the genome project, denial of funding to the superconducting supercollider, funding for and public debate about research on race and IQ. In the cases of Pauling's views on cancer and of water fluoridation we have another factor, namely scientific dispute and a kind of politics of dispute. Were Pauling's ideas treated sufficiently seriously? Was further research effectively discouraged? Does fluoridating water have severe side effects? None of these cases involves fraud; none involves science being trumped by superstition. Each involves policy, willingness to spend money or to undertake social engineering. At the interface of science with the wider society what else should we expect other than controversy, the deploying of power and influence – in short, politics?

How is it that the juxtaposition of politics with science is setting off alarm bells for the present universe of discourse? For its entire modern history science has proceeded from problem to problem by open dispute, the formation of schools of thought, parties of influence, centres of concentration. The issues have been thrashed out by several means including, of course, research. The outcome was not certain knowledge. Karl Popper, we may recall, characterizes science as conjectural knowledge, an edifice built by driving piles down into swampy ground. An alternative metaphor, of the house of science as a temporary structure, the best we can currently cobble together, was one used by J. Robert Oppenheimer (1954: ch. 6). These two giant figures of the mid-twentieth century never for a moment would have said that science was value-free, free of politics, somehow detached from the vagaries of the real world and politics in the wider sense. Each thought science should be respected but not romanticized as authoritative, unchanging, certain knowledge.

Neither thinker would have said that science needs to be insulated from the public, as Michael Polanyi (1969: ch. 4) argues. His case is compelling but flawed. The flaw is very simple and was pointed out by John Stuart Mill in the middle of the nineteenth century (Mill 1859: ch. 2). Arguing for freedom of thought and speech, Mill reasoned that if you silence anyone, you silence a possible source of truth. Exclusion by insulation is a version of silencing. The following, weakened, version of Mill's point should do for the present discussion: if you silence or exclude any voice you possibly silence a source of criticism. Polanyi tries to fend off the point by characterizing non-expert intervenors as cranks and dabblers whom science can ignore without loss. He wrote in the full knowledge that cranks and dabblers were sometimes the sources of good ideas and good criticisms. Only an ideal of efficiency could make sense of not listening to them: we haven't got time. But science is not the Manhattan Project. There is time, plenty of it. Debates should be open and not closed. (If that sounds too strong, then how about: scientific debates should be open, but not all debates, to anyone, all the time.) They need closure or the guillotine only when there is a compelling reason. We are, after all, hunting big game: how things are. Why not take time?

Part of the problem, as I indicated with my allusion to the Manhattan Project, stems from the uncritical mixing of science and technology. While it is true that they are sometimes indistinguishable, it remains useful to separate them in analysis. Time and hence urgency is much more characteristic of technology than of science. One characterization of technology would be innovation that we can clearly see will have practical effect on human life. This is vague, no doubt. Newton's theory of gravity and Darwin's theory of evolution eventually had tremendous practical effect on human life in and of themselves – but not immediately and not always clearly foreseen.

Technology is the Trojan horse in these discussions that allows Polanyi to smuggle in two features of science that he then accepts with alacrity: the community of science as a body of experts who apply the theories of science and thus wield legitimate authority. The authority of theory is legitimated by the expertise, and vice versa. Let us start with the expertise. First, a simple observation. When one listens to the experts one quickly learns that they do not agree; in the forefront of science consensus is rare. This is because expertise consists partly in awareness of what we do not know as well as awareness that there are others equally expert with whom one disagrees. From cosmology to medicine there are at best shifting coalitions, or summaries of the current consensus when there is one, and it is not difficult to find expert dissent and opposition. No one in science seems to think this is a bad thing. Indeed, putting forward new and better theories, bringing down current theories, pushing forward with the debate, are all intrinsic to science and its progress.

Experts, consultants, and the like seldom agree. They might consent to sign a compromise statement, but that is *pro tem*, always reserving the right to cancel past consent and with implicit or sometimes explicit reservations. In what, then, consists the authority of expertise? Is there such a thing? In technologies like airplane design, bridge building, and surgery, there are legal regimes that specify that authority and also its limits. These are matters of public law and regulation, not of internal decision by the practitioners. Planes and bridges sometimes come apart; operations are successful but the patient dies. Our usual response is to consider revising our public codes and standards. So to query expertise is not to say that any John or Jane can build an airplane or a bridge, or perform surgery. What it amounts to is a plea that we do not treat technological expertise as magic, that is, as perfect, as unchallengeable, as authoritative. Indeed, the category of technical expertise itself and the criteria enforced come from the wider society, from outside.

What of expertise in pure science as opposed to technology? Pure science involves the expenditure of money. Money goes to those who have gone through the educational system, which functions both to initiate and to exclude. It could hardly be otherwise. Does that legitimate excluding cranks (Linus Pauling on cancer) or dabblers (Erwin Schrödinger on biology)? I think not.[2] The problem of these cases is that they may show that scientific debate is not as open as it could be and that science contains vested interests and politics just as do all social organizations from the family to the legislature. Democrats must profess that: politics is not a bad thing. It is not a pollution of some pure state. Politics is the interaction of differing interests in search of cooperation, compromise, standoff, mutual understanding, and the like. It goes on all the time in science, though mostly out of the public view, and mostly not obedient to the political ideologies and coalitions of the wider society. Few scientific quarrels politicize along the lines of socialist and anti-socialist, liberal and conservative, Republican and Democrat. They politicize around scientific issues, and their partisans organize to gain money and power for the furtherance of what they see as the best science. What is the best science is contested in the best scientific circles.

Politics, then, is not a dirty word, and applying it to science does not demean that endeavour. Politics is the best system of coordination, negotiation, and compromise in a world where people differ in their goals, their values, and their assessment of means. I certainly agree that Max Weber's naive idea of science as value-free cannot be sustained (Mehta 1998). But I strongly dissent from drawing the conclusion that that means that science is on a par with other belief systems, a social construct like any other.[3] On the contrary. Most intellectual systems are closed. That is, they are dogmatic, intolerant, and, therefore, not intellectually serious. Science, by contrast, is an organized way to pool human resources in the quest to understand the world. This quest has, to be sure, an element of curiosity. But it also stems, in an evolutionary sense, from a drive to master our environment. The best way to survive in your niche surely has something to do with just how accurate is the information you develop about it. That the niche is the way it is and not another way is all that 'truth' means. The reason for stressing truth, then, is Pickwickian: in order that we may appreciate error when we make it. If food resources in the niche are limited, better that we know that than that we think they are unlimited.

Three Questions

To continue this discussion of what is allegedly broken in science I will address three explicit questions.

Question 1: Can science retain independence and objectivity?

The answer is yes, as long as officially we do not equate independence and objectivity with absence of public oversight, on the one hand, and with some sort of unbiased, value-freedom on the other.

The extent to which decisions affecting pure science and decisions within pure science lend themselves to democratization is problematic. The most obvious is public funding of research, especially for pure science. Governments usually hand over monies to boards of experts to distribute. Those boards can lobby governments, but the global sums involved are invariably a political decision. How could it be otherwise? In a democracy the expenditure of public money must be accounted for. But pure scientists working at their computers can hardly complain of being subject to excessive control, of having non-scientists make their scientific decisions. It is only when research requires substantial infusions of money that researchers may have to face the tribunal of others, peers and bureaucrats. That such systems of allocation could be improved goes without saying. The possibility of such improvement is not a sign of crisis, however; it is the outcome of having learned from experience: the conjecture that this is the best arrangement we can devise or implement has been falsified.

Objectivity in science is not about detachment or lack of bias. Commenting on an earlier version of social constructivism, Popper wrote:

> If scientific objectivity were founded, as the sociologistic theory of knowledge naively assumes, upon the individual scientist's impartiality or objectivity, then we should have to say goodbye to it. Indeed, we must be in a way more radically sceptical ... for there is no doubt that we are all suffering under our own system of prejudices ... that we all take many things as self-evident, that we accept them uncritically and even with the naive and cocksure belief that criticism is quite unnecessary; and scientists are no exception to this rule. (Popper 1945, 2: 23)

Objectivity is better construed as the friendly hostile cooperation of scientists, something that is induced by the artfulness of institutional design and proliferation – for example, the founding of new and dissenting journals; the creation of research centres financed by rival industries; the vigorousness of the learned associations in recognizing work that challenges the mainstream. Science has multiple ways to supervise and protect its objectivity and intellectual independence.

Question 2: In what ways is scientific discourse privileged in the formation of public policy?

The short answer is, in no way. In the formation of democratic public policy all inputs are equal, at least in a formal sense. Just as there are individuals and groups whose word, for various reasons, is given greater weight, so there can be the same effect with science. But this variable is social and individual. Remember that in the early years of the Second World War the nuclear physicists concerned with the possibility that Germany might do nuclear weapons research went to some trouble to get Einstein to sign their letter to President Roosevelt; his prestige was such that it ensured a hearing for the views expressed.

The way science is now organized should not be a fetish. It was given its present shape in the United States by a handful of senior scientist-bureaucrats, notably Vannevar Bush, James Bryant Conant, Ernest O. Lawrence, J. Robert Oppenheimer, and Glenn Seaborg.[4] They were driven by their experience with the Manhattan Project and then by Cold War considerations. What served then may not serve us now. Lesser countries were of course influenced by how the United States did things, although sometimes they could not and other times they would not do it the same way. I shall argue that diversity of science policy and institutions among states may be a saving grace.

The very fact that we speak of public policy means that it is a public matter and not one for the scientists to settle among themselves. Consider the case of stem cell research. A recent headline in the *International Herald Tribune* (*IHT*) read: 'Britain fills gap left by U.S. in stem cell research' (2004). The gist of the story was that because of tight restrictions on federal funding for stem cell research in the United States there had been a considerable slowdown in research there and the exit of at least one top researcher to the United Kingdom. In the United Kingdom a different regime had been worked out by negotiation between scientists and government that imposed central control but still facilitated rather than stifled research. By contrast, since the remit of the U.S. federal government does not extend to private companies, there was, in the United States, an unrestricted and possibly chaotic regime, tightly controlled on one hand, uncontrolled on the other. Thus, the *IHT* story goes on, there might be a shift of momentum in this area from the United States to the United Kingdom or elsewhere. Far from being a crisis of science, this may be a healthy development as it is a shift of the centre of gravity of some research.[5]

One lesson to be learned is that there are good and there are poor institutional solutions to issues of public concern. The *IHT* article intimates that the U.S. stem cell solution is poor because it drives research away from government supervision and into an unsupervised area. The British solution is good because it does not trammel research, yet it keeps it under close supervision. These claims are worth discussing, and their discussion must be much wider than the scientific community.

Question 3: How can opening closed scientific communities be done without scientific knowledge itself becoming politicized?

Polanyi, of course, paints scientific communities as closed to outsiders in order to exclude cranks and dabblers. The dangers posed by cranks and dabblers not being obvious, it is interesting that Polanyi does not specify them. Yet by Polanyi's own account, scientific communities are anything but closed to outsiders. Most are housed in universities, research institutions, or corporate laboratories, where they are seldom, shall we say, self-administered. At least, his view of them as self-administering will not hold up. Also, by their nature these communities are constantly recruiting and initiating newcomers, saying farewell to those that move on. Newcomers and those on the way out are liminal, seeking to become insiders or ceasing to be insiders. Yet dealings with them are unavoidable. Scientific communities or networks are also themselves constantly forming, reforming, and disbanding. It is not clear, during these processes, just who is insider and who is not. Commonsense sociology, then, shows us that Polanyi's picture is too simplified for the present day. If we turn to some collateral evidence, such as Watson and Crick's memoirs, we can see clearly that marginal scientists can get access to information from other groups and proceed to crack a problem without ever being acknowledged as insiders (Watson 1968; Crick 1988).[6] Meanwhile Polanyi's description of Britain's Royal Society looking into matters suggested to it by government offers a very clear case where the mandate and the reported outcome were called up by outsiders.

None of these sources reports anything like scientific knowledge becoming detrimentally politicized. What is a paradigm case of that? It is what Polanyi rightly combated, something like the Lysenko case in the USSR, when one party to a scientific dispute became aligned with the ruling party in the wider polity (Medvedev 1969). Thus their bad science was uncritically promoted and funded, and dissenting views discouraged, disallowed, or otherwise stifled. Charges that this sort of thing goes on in democracies are almost non-existent. Some have tried to argue that science was politicized in and by the Cold War, not just its militarizing, its acceptance of a regime of secrecy, but also in its acute dependence on government funding. The hangover of some of these arrangements well beyond the Cold War looks less and less like a blessing. Maybe a *prosperous, high-status*, and *authoritative* science is a science in tension with its own sceptical and conjectural constitution.

As with the previous questions, I do think impoverished sociology of science is partly the villain here. There is politics in science. Science in democracy is part of a society that governs itself by politics. Yet scientific institutions are structured so that the wider political commitments of scientists, the pressures from sources of finance, and the religious beliefs of scientists and the surrounding community are not shields that prevent scientific work from being exposed to the critical scrutiny of others. No doubt the system, like any institutional setup, is far from perfect, including far from being as readily self-correcting as we might wish it to be. Yet what else can we do except use our knowledge of such deformations as the Lysenko case to devise institutions that make such a development one that sets off alarm bells?

In a way, in the present social setup of science, scientists are their own watchdogs. Through the peer system they are their own specialist watchdogs. But through more representative scientific institutions, those who are not specialists, not insiders, yet still scientists, have a watching brief on the work of others. As Polanyi rightly indicates, the first audience beyond the specialists for scientific work is other scientists: beginning with those in close or cognate fields and gradually spreading outwards through such general journals as *Nature* and *Science* before reaching the general intellectual public through science journalists and such outlets as *Scientific American*, *Popular Science*, and science books for the general reader. Beginning with peer review, none of these audiences is subject to the control of the insiders. That is, the audiences' criticisms, if any, cannot be curbed. What we could use, it seems to me, is more of this diffusion, and better channels for feedback to the insiders.

By and large the relations between scientific peers are not governed democratically. The institutions of learned experts, the laboratories, government and private, the universities, are *all* organized hierarchically, not democratically. To some degree they are gerontocratic, loaded so that decisions are kept out of the hands of the newcomers, who often constitute the majority. You may say yes, power is in the hands of the knowledgeable. Would you have it otherwise? One answer is that this is the wrong question, since it is a version of 'Who should rule?' It ignores Popper's cogent argument that, regardless of who rules, the urgent question is how we control rulers and get rid of them when the need arises (Popper 1945).

Another answer is this: scientific institutions are institutions of certification as well as of research. Physicists, for example, are by and large people with advanced degrees in physics from recognized institutions and who have publications in outlets recognized to be respectable scientific journals. Certification depends on public acceptance. With the professions acceptance is controlled by legislation. The institutions of science are not formally and legally equivalent to those of the professions. Science is not technically self-governing because it has very limited power to police and to discipline its members and thus to take collective responsibility for failures. In most countries the records of suspect scientists cannot be subpoenaed by their own investigations, scientists cannot be compelled to attend hearings, and those guilty cannot be fined or *directly* disciplined. It is *indirectly* through employment, allotments of research monies, and ostracism that those convicted can be sanctioned. It is quite difficult to move such cases over into the regular legal system because the crimes – plagiarism, falsifying data, and their likes – hardly ever fit readily the definitions of the laws of theft or the laws of tort.

There is a very good reason why democratically minded people hesitate to recommend scrutinizing science too closely. It is widely held that scientific progress and freedom of thought go together. If a society, no matter how democratic, takes a proactive attitude to scientific organization, then there is a perceived danger that it will compromise freedom of thought and so the progress of science. Now this argument is flawed. There is a gap in it. To fill it we have to take a step back and ask, Where did the present scientific organization come from? The answer is: from earlier discussions. So far from being afraid of discussion we should be afraid that they were not as open and comprehensive as we should have made them.

Polanyi's model of science as run by an authoritarian leadership clashes with Popper's view of science as a Socratic seminar writ large. Neither view will do. Making science more authoritarian or more Socratic will not ameliorate the problems – whether of loose discipline or excess openness or alternatively of democratic deficit or closedness. The amorphous leadership of Polanyi is at best local, while the Socratic seminar of Popper is a spontaneous and virtually unstructured order, a congeries of cooperating individuals. The only structure I can discern in the latter is that Socrates is usually the one who invites participation and who usually dominates the discussion, and that there are temporal limits to each session. Science as collective activity cannot be modelled on such a structure for the simple and obvious reason that science contains no single dominant figure, that it is highly specialized and consists of many loosely organized groups, real and virtual, running simultaneously, and each simultaneous session contains parties and schools of thought, as always emerge in real debate. Polanyi's localized model of a workshop run by a master surrounded by apprentices, just as much as Popper's model of a loosely structured Socratic seminar, makes no place for politics in science and hence gives no account of their positive part in its functioning. In the spirit but not in the letter of their works one might argue that it is parties and schools of thought that are the institutional versions of scientific ideas. Cooperative critical discussion of ideas is fostered by these institutions, thus making them part and parcel of the collective functioning of scientific discussion. No institutions are perfect, as Popper notes, so these party formations bring problems as well as advantages. Yet there is little serious discussion of these matters. Isolated figures who try to raise them, like Joseph Agassi[7] and Steve Fuller,[8] are stonewalled.

As a collective activity, as a congeries of institutions, science is bound to have vital internal structures and no less vital external relations with other social structures. Both are problematic. The openness of the internal structures of science, *pace* Polanyi, is no mere internal concern: the interface of science with the wider society demands taking outside interests into account. A great deal of public money goes to research, and a democratic society demands public accountability wherever public money goes. Fraud in science captures the interest of the public, which also suggests that support of science is a form of public trust. A great deal of science is housed in universities, and these are public institutions subject to public scrutiny. Above all, science commands status and authority in society and makes great claims for the social benefits that it brings. This cannot but attract general forms of supervision. To think of this as outside interference is defensive and rests on the uncritical assumption that science can benefit by decreasing certain critical inputs.

The message of both Polanyi and Popper seems to be that political power should stay back and let science pursue its mission. However important this is, it will not do. There is a lot more to it than that. All institutions are arenas of political power, and they connect to political institutions in the wider society in diverse ways. Means have to be found to combine this fact with public supervision and genuine openness to free critical inquiry. Our societies have legislated the forms of governance for the professions of medicine, law, engineering, and accounting. It is not easy to weigh the pros and the cons of whether scientific research should be organized more like the free professions, with self-government plus properly created legal standing. So clearly this is a matter for democratic decision.

One gain from professionalization would be the possibility of codifying the difference between the authority of science as an institution (compare 'the majesty of the law') and the conjectural nature of scientific ideas (compare the flaws of current legislation). Above all such professionalization would impose responsibility and accountability to a degree that is only informal at present in science.

In short, the institutionalizing of incentives to maximum criticism, a proactive investigation of the democratic deficit in scientific institutions, and a greater emphasis on the responsibility to inform the public of the findings of science as the conjectures they are – these could do much good. And of course they should be open for free discussion and criticism.

NOTES

1 I would not deny that science has been appropriated as a substitute religion. If Merton is right about institutionalized scepticism being part of the ethos of science, then this is a strange religion. It is my observation that those who make a religion of science are strongly tempted to treat it as authoritative and thus to reproduce the dogmatism of traditional religions.

2 Polanyi could reply that Pauling and Schrödinger were neither cranks nor dabblers but distinguished scientists in their own right. Such a reply would lead to two (or more) grades of cranks and dabblers and the threat of infinite regress.

3 Cf. Crick's comment on the idea that scientists construct science: 'I am not completely convinced by this argument, at least in this case. Rather than believe that Watson and Crick made the DNA structure, I would rather stress that the structure made Watson and Crick. After all, I was almost totally unknown at the time, and Watson was regarded, in most circles, as too bright to be really sound. But what I think is overlooked in such arguments is the intrinsic beauty of the DNA double helix. It is the molecule that has the style, quite as much as the scientists' (Crick 1988: 76).

4 Fuller 2000b. He concentrates on Conant and Bush. I have added Lawrence, Oppenheimer, and Seaborg.

5 I leave aside the issue of national prestige.

6 Certainly the structure of DNA was not a problem assigned to either one. It belonged, as Watson wrote, to Maurice Wilkins. They had to ask the head of the laboratory, Sir Lawrence Bragg, for permission to build their models. Watson describes misunderstanding a talk of Rosalind Franklin's, catalogues his ignorance of relevant basic science, and tells also how his fellowship was revoked because of his unauthorized change of laboratory. It is interesting that both Wilkins and Crick were physicists who read and were greatly influ-

enced to shift to biology by Schrödinger's *What Is Life?* As for Wilkins, he had mixed feelings when he discovered that Watson and Crick were interested in and working on DNA. His autobiography shows a clear commitment to what Merton called the 'communism' of the scientific ethos. While it was his problem, he wonders whether he should have made greater efforts to engineer a London/Cambridge collaboration, so that cooperation would have short-circuited rivalry. His difficult relations with his head of lab, Sir John Randall, and his co-researcher, Rosalind Franklin, would have had to be different (Wilkins 2003). In general, the interdisciplinary training of all these Nobel laureates renders moot the whole question of who was an insider and who was not.

7 Agassi 1984, 1986a, 1986b; see also 1975, 1981, 1985, and 2003.

8 Fuller 1997, 2000a, 2000b, 2004, and the forty-two critics, mostly hostile, assembled in Gattei 2003.

REFERENCES

Agassi, J. 1975. *Science in Flux*. Dordrecht: Reidel.

– 1981. *Science and Society*. Dordrecht: Reidel.

– 1984. The Social Base of Scientific Theory and Practice. In *Rationality in Science and Politics*, edited by Gunnar Andersson, 15–28. Dordrecht: Reidel.

– 1985. *Technology*. Dordrecht: Reidel.

– 1986a. Scientific Leadership. In *Changing Conceptions of Leadership*, edited by Carl F. Graumann and Serge Moscovici, 223–39. New York: Springer.

– 1986b. The Politics of Science. *Journal of Applied Philosophy* 3, no. 1: 35–48.

– 2003. *Science and Culture*. Dordrecht: Kluwer.

Crick, F. 1988. *What Mad Pursuit*. New York: Knopf.

Fuller, S. 1997. *Science*. Minneapolis: University of Minnesota Press.

– 2000a. *The Governance of Science*. Buckingham: The Open University Press.

– 2000b. *Thomas Kuhn. A Philosophical History for Our Time*. Cambridge, MA: Harvard University Press.

– 2004. The Case of Fuller vs. Kuhn. *Social Epistemology* 18, no. 1: 3–49.

Gattei, S., ed. 2003. The Kuhn Controversy. *Social Epistemology* 17, no. 2–3: 85–324.

International Herald Tribune. 2004. Britain Fills Gap Left by U.S. in Stem Cell Research, by Elizabeth Rosenthal, 23 August. (Also in the *New York Times* of the following day.)

Kitcher, P. 2001. *Science, Truth, and Democracy*. Oxford: Oxford University Press.

Martin, Brian, and Evelleen Richards. 1995. Scientific Knowledge, Controversy, and Public Decision-Making. In *Handbook of Science and Technology Studies*, edited by Sheila Jasanoff, Gerald E. Markle, James C. Peterson, and Trevor Pinch, 506–26. Newbury Park, CA: Sage.

Medvedev, Z. 1969. *The Rise and Fall of T.D. Lysenko*. New York: Columbia University Press.

Mehta, M. 1998. Risk and Decision-Making: A Theoretical Approach to Public Participation in Techno-Scientific Conflict Situation. *Technology and Society* 20: 87–98.

Merton, R.K. 1938. Science and the Social Order. *Philosophy of Science* 5: 321–37.

Mill, John Stuart. 1859. *On Liberty*. London: Longman.

Oppenheimer, J. Robert. 1954. *Science and the Common Understanding*. Oxford: Oxford University Press.

Polanyi, M. 1969. *Knowing and Being*. Edited by M. Greene. Chicago: University of Chicago Press.

Popper, K.R. 1945. *The Open Society and Its Enemies*. London: George Routledge and Sons.

– 1992. *In Search of a Better World*. London: Routledge.

– 1994. *The Myth of the Framework*. London: Routledge.

Schrödinger, E. 1944. *What Is Life?* Cambridge: Cambridge University Press.

Watson, J.D. 1968. *The Double Helix*. New York: Atheneum.

Wilkins, M. 2003. *The Third Man of the Double Helix*. Oxford: Oxford University Press.

8 *New Atlantis* Reconsidered

LEON HAROLD CRAIG

The relationship between liberal democracy and modern science, heretofore symbiotic and thus relatively harmonious, shows clear signs of becoming increasingly troubled in the twenty-first century, affecting both entities in the relationship. The public's interest in promoting certain sectors of scientific research while restricting and even forbidding other sorts of research; the lure of fame and fortune implicit in the economy's ever-increasing dependence on commercializing technology; the growing electoral salience of public policy issues that involve scientific knowledge and theorizing, some of it highly speculative (as evidenced by an absence of unanimity among putative experts); the liability of such issues and expertise being hijacked for other partisan causes (e.g., exploiting environmental concerns as a cover for opposing free-market economics) – these and other developments threaten the intellectual autonomy and probity of science. On the other hand, the international anarchy that was always implicit in modern science's inherently universal and autonomous character is now a practical reality, thanks mainly to the powerful electronic technology spawned by modern science. As a result, any particular state's capacity to restrict or otherwise manage the growth, dissemination, and exploitation of scientific knowledge – to exercise sovereignty in these matters – or of any effective alliance of states to do so, is practically impossible.

Laws and the public purse strings still matter, of course, but they do not add up to strict control. It has long been the case that, however 'civilized' international commerce is conducted, economic competition among states dictates that certain lines of research be pursued by each, lest one 'fall behind technologically' at the cost of economic hardship, which readily translates into political discontent. More recently the stakes were raised. Given the dangerous capacities of modern science, the mere existence of actual and hypothetical 'rogue states' force on other states an active research agenda regarding unpalatable topics that liberal democracies in particular would prefer not to pursue and yet dare not risk neglecting to their mortal peril. For example, it would be politically irresponsible not to be secretly conducting 'illegal' research in biological and chemical weapons, if only for the defensive purpose of learning how to protect against them. But while these external factors compromised a state's freedom to set its own research priorities, heretofore it could still in principle manage the scientific research conducted within its own borders. That, however, may now be changing. Given the research potential of powerful personal computers, in conjunction with the internet and other means of international communication – this unprecedented electronic permeability of political borders – it is practically impossible to monitor, much less control, networks of like-minded individuals who wish to engage in collaborative research (sharing data, problems, analyses, and findings), irrespective of any and all nations' policies. And the very exclusivity of scientific competence, the glamour of scientific discovery, and the mystique of scientific progress – to say nothing of more material considerations – provide a ready self-justification for talented individuals, even if otherwise morally responsible, to exempt themselves from abiding by rules they regard as framed in ignorance of, if not outright hostility to, scientific inquiry and the growth of knowledge.[1]

Such being the situation, it invites – demands, I believe – a meditation aimed at imagining innovations whereby to resolve, or at least ameliorate, the very considerable dangers that lurk therein. While it is conceivable that further technology designed for the purpose might provide means of coping with certain aspects of the burgeoning problems, it is primarily political innovations that will be needed: in laws, institutions, policies, and even principles. I suspect that something more radical than piecemeal innovations within the parameters of liberal democracy may be required: a new form of regime. 'Unthinkable' as no doubt this is apt to strike most people, such a prospect must be contemplated, and not merely because of the problems posed by the need to manage science or somehow ensure that if it does not always serve the public good, at least it does no major harm. But also because these concerns intertwine with two other strands of problems, one internal and the other external, that in very different ways challenge the viability of liberal democracy, at least in its current form. This malign convergence of what are, for analytic purposes, three distinct kinds of assaults on the integrity of this regime renders unlikely the hope that any coherent set of reforms would suffice to salvage it. Should one conclude that such is the case, then the first step would be to determine the features of liberal democracy that would be most desirable to incorporate somehow in its replacement. But in making this assessment, it is imperative that one view political life in this regime objectively, how it really functions today, as opposed to its idealization. This is important, both for understanding why it has the internal problems that tend to undermine its viability, and for making a fair comparison with any possible alternative.

It bears remembering that Western civilization is not coterminous with liberal democracy. In fact, most of its greatest *cultural* achievements (as opposed to the bulk of scientific progress) antecede the emergence of even the earliest stages of the liberal democratic regime – as did the scientific revolution. No form of regime is an end in itself; all are means to some ulterior end, and *that* is what we need to be clear about. It is not the purpose of this chapter to present 'The Case for Transcending Liberal Democracy,' much less argue for a particular replacement. But it does seem to me that the perceived need to manage science must be viewed in a broader context of problems.

The Diseases Within

Let me begin with a reminder that, once upon a time, the surpassing superiority of democracy, in any of its actual or imagined incarnations, was not regarded as a self-evident truth. Indeed, quite the contrary – most thoughtful and prudent people regarded it as subject to some very damning criticisms: that it placed ultimate power in the hands of the mainly ignorant and thoughtless; that, as a consequence, public policies tended to be short-sighted, erratic, and inconsistent; that its political equality necessarily implied the equality of all desires, however base (for if one ranks desires, one *ipso facto* ranks people according to their dominant desires); that this failure to encourage the noble desires and discourage the base effectively favours the spread of base desires, incites an ever more licentious hedonism, and abets immoderation in general, especially among the youth; that the doctrine of equality itself licenses the indulgence of certain ugly and socially corrosive passions, notably spite and envy and resentment; and that democracy is vulnerable to every kind of demagoguery. Perhaps the modern experience with liberal and representative forms of democracy has proven some or all of these criticisms to be mistaken. I do not believe so, but that is not my primary point. Rather, it is that the architects of liberal democracy (e.g., Locke, Spinoza, Montesquieu, the American founders, et al.) actually had to make a convincing case for both the practical viability and moral superiority of this (then) new form of regime, and do so in the face of considerable scepticism on the part of proponents of other forms. Hence it would be useful to revisit those original debates and consider afresh the rationale for liberal democracy as well as the case made against it.

Practicalities dictate my disclaiming any intention of doing that here, but in noting some of our more serious internal problems, I am implicitly suggesting that liberal democracy is not performing as envisioned by its founders. Whether this is because of flaws in their original conception or because of the contemporary failure to comply with all of their principles and provisions (either through intentionally rejecting some or simply through neglect) – this, too, I shall not here address, except incidentally. Instead, I shall turn directly to what seem to me to be the most serious internal problems.

First of all, liberal democracy is not actually very democratic. If the point of sovereign power resting with 'the people' (the *dēmos*) is that government will thereby be in accordance with their wishes, especially regarding the things that most directly affect their lives – and a strong case can be made for the value of their judgment on those matters (unlike, say, on foreign affairs) – the actions of liberal regimes today are as apt to be contrary to majority views as consonant. Whether it is with respect to administering the criminal justice system, the content of public education, the management of sexual relations, or the regulation and funding of scientific research, laws and policies are more apt to reflect the wishes of various well-organized minority interests than 'the will' of even a sizable majority of 'the people.' Not for nothing have wise men over the ages warned of the dangers organized factions pose to the health of polities, and of democratic republics especially.

But the problem of public policy being distorted by small, well-organized sectors of the electorate (and even of aliens) who are focused exclusively on a very narrow set of concerns, be they economic or ideological, seems practically insoluble for liberal democracies, which enshrine the rights of political association and of petitioning one's government. In our time, the techniques of promoting special interests have become highly sophisticated, ever outstripping legislative efforts to control them. And scientific methods and technologies, including especially those that make possible the proliferating forms of both mass and selective communication media, play no small part in creating a political environment wherein such disproportionate influence can be exerted.

The net result is that the basic premise of democracy – that periodic elections will ultimately ensure that the preferences of the majority of citizens determine public policies – is undermined. Moreover, the growing public perception that such is the case, that the will of the majority is routinely subverted by 'special interests,' generates political cynicism, thereby eroding support for the regime. Similar remarks pertain to the increasing politicization of the court system, whereby narrow minority interests exploit the regime's commitment to the rule of law in order to circumvent the normal political process entirely, such that the rule of law becomes the rule of lawyers. This, in turn, tends to corrupt lawyers (and not a few who employ them), resulting in further attrition of the citizenry's respect for the law and for those they perceive to be professional exploiters of it. I most emphatically do not mean to suggest that I believe 'returning power to the people' is the solution to these and the other problems I discuss below. But it is important in assessing hypothetical alternative regimes, including those that would seem insufficiently responsive to popular will, that one bear in mind how little this is the case in contemporary liberal democracies.

A second set of problems results from divergent interpretations, hence expectations, of the liberal regime's commitment to equality. There can be no question but that its originators, and all of the thousands and millions who agreed with them, understood it to mean 'equality of opportunity'; that with equality before the law, and with there being no unnecessary laws, there would be no artificial obstacles to each individual's using his abilities and resources to pursue what seems to him the best life of which he is capable, or is willing to exert himself to achieve. And let the chips fall where they may. It was understood that a major advantage of this arrangement was its allowing for the free emergence of *scientific* talent from all ranks of society. But with people having different conceptions of the good life, not all of which are equally attainable, with chance still a significant factor in human affairs, and with there being innumerable other variables in circumstances and natures affecting actual outcomes, there is no reason to expect that everyone will be equally successful in whatever they choose to pursue. But so far as practical, people will, generally speaking, get to live the life they themselves have chosen and earned, and in that sense the life they deserve. And who could ask for anything more by way of a just political arrangement?

Over time, however, a very different notion of 'just equality' has been gaining ground, what is usually referred to as 'equality of result' (invariably as measured by material standard of living), a principle that pretty much turns the former idea on its head. Both views traced to the bottom, it becomes clear not merely that are they incompatible but that partisans of each must regard the other view as profoundly unjust.[2] Now, it can be shown in several different ways that, quite apart from its suspect justness, no society can be made to work well, or even work at all, on the basis of this latter idea.[3] However, there will always be people who prefer to believe that everyone is equally deserving, not merely of opportunity, but of happiness, and plenty more who see that they can personally benefit from the selective endorsement of that view. They provide a ready audience for any political opportunist able to articulate what they wish to hear. This alternative conception being rhetorically effective, it can be used to impugn the justness of any and all existing arrangements, insisting that they be 'reformed' in its direction. To the extent that this is done, contradictions are introduced into the functioning of the modern liberal regime, with all the frustrations, incriminations, and bitterness that contradictory views of justice are bound to generate.

Third, and perhaps most disturbing, is the accelerating decay of what is called popular culture.[4] Many of liberal democracies' other current troubles – from rampant drug abuse and irresponsible violence, especially among increasingly unruly children and youths, to the decay of basic morality, especially sexual morality, but including the civility, politeness, and respectful behaviour that make such a difference to the social comfort and general tone of life – these and other political problems are greatly exacerbated, if not simply caused, by the abject state of popular culture. For most people's, particularly young people's, ideas of how to behave, their understandings of how their world 'really' works, and their judgments about what is important (their so-called values) are profoundly shaped by the primary constituents of this culture: by the various modes of visual entertainment (not just the farces and melodramas of films and television, but also professional sports and psuedo-sports, news and psuedo-news, video games, internet browsing, etc.); by music (mostly of a quality and content that beggars description); and by all-pervasive commercial advertising (appealing to, hence encouraging and ratifying, the very hedonism and acquisitiveness that decent morality seeks to restrain).

The extent of the influence of these cultural components, steadily expanding because of their increasing availability through the proliferation of inexpensive new technical modes and devices, is almost entirely the result of modern science. For these phenomena presume not only the technology but the *luxury* that science has bestowed on liberal democracies, including a previously inconceivable amount of disposable wealth in the hands of children. What is more fundamental, however, is the dogmatic acceptance of the profoundly erroneous 'fact-value' distinction that has accompanied the saturation of everyday life by 'scientism.' According to the popular misunderstanding of modern science, it is the very paradigm of genuine, 'objective' knowledge, and it is so precisely because it is 'value-free' (as if science makes any sense apart from valuing truth, probity, clarity, etc.). But by the false light of this mistaken conception, principles of right and wrong, good and bad, noble and base, decent and obscene are regarded as incorrigibly 'subjective,' hence mere matters of opinion, and certainly as inadequate grounds for ruling any contrary 'values' as simply wrong, much less intolerable.

The issue is not primarily that of 'high-brow' versus 'low-brow' – though it is worth recalling there was once a time and place where ordinary citizens packed theatres to be entertained by the likes of Sophocles; and not so very long ago, unlettered common folk paid their precious pence to see performances of Shakespeare. The main point, however, is that popular entertainment does not have to be high-brow in order to be morally and politically wholesome, even edifying; and conversely, offerings aimed at a sophisticated cultural elite can be decadent in the extreme. But who other than a natural fool could seriously believe that a steady diet of witty insolence, slick pornography, celebrations of hedonism, and spectacular violence does not have a cumulative effect on human souls, especially the most impressionable souls: those of youths, life's eager imitators.

As for why the soul-suffusing effects of popular culture are not merely spreading outward (courtesy of innovative scientific technology), but spiralling downwards with a rapidity that would have defied belief a bare generation or two ago, it is a result of the conjunction of freedom of the marketplace with what is today often called (apparently without irony) 'freedom of artistic expression,' a wildly implausible expansion of liberal democracy's original commitment to the freedom of speech essential to an informed electorate. Applied commercially, this freedom of artistic expression has meant the freedom to make mountains of money by pandering to the lower impulses of human nature. Whether people's taste, as a result of ever more explicit portrayals of sexual ingenuity, glamorous criminality, wholesale killings, and spectacular destruction of property, has been corrupted beyond recovery is debatable. But given people's fondness for the pleasures to which they have grown accustomed, it is difficult to imagine any practical way that decency can be restored to popular culture by democratic means.

To be sure, these problems of cultural miseducation are not strictly the logical consequences of the commitment to liberty, not the political liberty indispensable to the practical functioning of representative democracy, nor such personal liberty as is requisite to the pursuit of happiness. But the practical question is, are they the *psycho*-logical consequences? That is, given the normal inclinations of human nature in the absence of intentional countervailing efforts, do people's expectations of liberty have a natural tendency over time to expand in degenerative ways? Does the commitment to rationally defensible liberties naturally tend to devolve so as to include indefensible ones, unless active measures are taken to resist this happening? Moreover, does not a regime's explicit dedication to liberty, unqualified as it invariably is in all popular praise, predispose a sizable portion of its citizens to reject any and all measures aimed at restraining or channelling it? Or is the Platonic Sokrates right about the long-term consequences of democracy's honouring a permissive conception of freedom as the highest good – that there is a relentless tendency for 'freedom to spread to everything,' with the souls of its citizens becoming so 'tender' that they chafe at any and every restriction on their liberty (*Republic* 562d–563e)? As a result, respect for moderation of any kind in regard to anything (including, most significantly, in expectations of what a just regime is responsible for providing) becomes despised as merely the rhetoric of puritans and the refuge of reactionaries.

The Devil Without

However, the evolving problems threatening the liberal democratic regime from within, whether inherent or adventitious, have of late been further complicated and hugely supplemented by a most sinister assault from without. 'International terrorism,' we have learned to call its frightening face, but that face has a calculating head that rests upon and is sustained by an enormous body. For although those who are prepared to engage in homicidal and suicidal terrorist activity may number in only the thousands, those who openly celebrate terrorist 'successes' number in the millions and those who privately condone terrorism directed at 'infidels' in the tens of millions.

It has become an established truism that the infamous deeds of 11 September 2001 'changed everything.' For on that day, the civilized world at large was suddenly jolted awake to a threat that, seen in retrospect, had been steadily growing for some three decades or more. This appalling climax (so far) of escalating violence and violent aspirations should, like nothing in the preceding century, awaken sympathy for Hobbes's view of the potential political problems posed by dogmatic religious faith: how readily the credulous are manipulated, even to the point of self-sacrifice, by those who have somehow acquired religious authority in their eyes ('So easie are men to be drawn to believe anything from such men as have gotten credit with them; and can with gentleness and dexterity, take hold of their fear, and ignorance' – Leviathan 12:19). However, if ignorance and the fears to which it gives rise were the sole source of these problems, they would in principle be eradicable by education, including so far as necessary a rationally enlightened religious education, but in any case an education grounded in modern science. Yet, as the events of that black day proved, the problems derive not simply from the possibility that 'crafty ambitious persons abuse the simple people' (ibid.) – that is, from the exploitation of ignorant rustics by clever, cynical pursuers of worldly power. For among those who commandeered the planes used in the suicidal attacks on the World Trade Center were college-educated men, amply acquainted with modern liberal society, including the endless wonders generated by the scientific technology to which it is devoted, but who rejected it root and branch, exploiting its freedom and technological vulnerability to attack it with multiple effectiveness. They represent a conjunction that heretofore was considered almost inconceivable: intimate familiarity with the advantages of the liberal, open, thoroughly modern lifestyle; technical sophistication; but a mentality we associate with the dark ages, which we presumed would be expunged simply through exposure to the wonders of modernity.

In the final analysis, one must conclude that their actions were profoundly, atrociously irrational, in which case they reveal the limits of the power of reason over the behaviour of certain people, including some supposedly well-educated. Even so, one may concede that the conception of reason whence the preceding judgment would be rendered might not be altogether satisfactory, its calculations neglecting if not excluding certain important concerns and constituents of human nature that are more adequately, or at least more readily, addressed by 'religion.' That is, addressed from within an understanding of this world as properly subordinate to another, promulgated by whomever, however, and on whatever basis, religion exercises intellectual and moral authority in the name of something higher than (mere) human desire, will, or reason. From within such a religious perspective, all valid reasoning necessarily rests upon the (supposed) revelations of a transcendent, supernatural power. But whether one regards suicidal terrorists as besotted acolytes of fanatics and nihilists, or inspired heroes of their faith, their very existence today suggests that a thorough reconsideration of the liberal democratic prescription is very much in order. And all the more urgently needed, one may add, insofar as some approximation of this regime is, practically speaking, the only kind regarded as legitimate by the vast majority, not only of the citizenry of such regimes, but of almost all people alive today who consider themselves politically enlightened.[5] Minimizing if not ignoring the deep, perhaps intractable problems that increasingly trouble this form of society, they too blithely presume that its further propagation is in every way both desirable and just, and the ultimate solution to the problem of international terrorism.

Many otherwise easy-going Westerners, products of societies long since secularized, regard with scepticism any public recognition of religious tradition by their fellows and view with downright suspicion expression of religious faith in the context of political activity. For them, the very idea that religion could pose a serious political problem is – or was – scarcely credible. After all, had not centuries of religious conflict (thankfully long since past) conclusively shown its patent irrationality? Indeed, growing recognition of the utter futility of attempting to coerce 'faith' was instrumental in both the fashioning and the ultimate political success of the liberal democratic prescription. The privatization of the sacred, which entails its practical subordination to the secular, is fundamental to the liberal formula for civil peace. And this peace, in turn, facilitates progress and prosperity – especially progress in the accumulation of scientific knowledge, and prosperity available through the commercialization of science-based technology. As for all theology-based understandings of the world, their intellectual authority in the

West has shrunk to a mere shadow under the relentless advance of a modern scientific understanding of all things. For modern science, its validity ratified by the stupendous technological power derived from its methods and mentality being brought to bear on sensual nature, has solidly established its claim not only to being *a* source of genuine knowledge, but also to setting the standard whereby all knowledge claims are judged, from cosmology to history. This includes, importantly, the history of religion. Suffice it to say, by the standards of science-influenced historiography and sociology, the origin and spread of the various religions will be in human, all-too-human terms; divine revelation and prophecy, miracles and holy spirits will have no explanatory role.

This 'scientism' being the dominant mindset through which most denizens of the liberal democracies view reality and so live our lives, it is with some bewilderment that we confront a surgent, or resurgent (and increasingly militant), religious fundamentalism, mainly Islamic in inspiration, that is implacably hostile to both liberal democracy and the untrammelled modern science that thrives therein. With respect to the latter, the objection is not merely to the atheism implicit in its metaphysical presuppositions, but to the entire battery of consequences that attend scientific enlightenment. With respect to the former, the clerical leadership and their friends assess the propriety of liberal democratic principles – quite sensibly – in light of the way of life that results. And whether for good reasons or bad, they do not like what they see.

For they see crass commercialization in every facet of life (religion included, judging from televised evangelism); and they see what goes with it, rampant 'materialism,' people preoccupied with money and with what it will buy, to the neglect if not exclusion of spiritual concerns; degenerate manners, especially in the youth, and erosion of the civility and sense of community that so profoundly affects the tone of people's everyday existence; the increasing deterioration of family life, proper locus of moral education, of respect for tradition, and of one's sense of belonging; and the deepening moral decadence that attends the natural devolution of hedonism – especially as it impacts sexual relations (the particular obsession of Islamic fundamentalists, whence derives the extreme repression of women that we find so grotesque). Even allowing for the narrowness and selectivity with which they view liberal democratic society, as if there were nothing of importance to be said in its favour; and ignoring the blatant hypocrisy of some of their criticisms, to say nothing of the gross injustices they would inflict on a majority of mankind, it seems to me that their objections to the liberal prescription are not without some validity. How ironic, then, that we seek to solve the problems of international terrorism by exporting liberal democracy, complete with scientific enlightenment, on the presumption that almost everybody would prefer it, but simply cannot manage to bring it about on their own. In our humane intention to make the world a safer and otherwise better place by encouraging everyone to become like us, we ignore legitimate objections to their doing so that are based on a credible critique of the way of life that ultimately results: that it is, in a word, decadent.

Nothing more clearly reveals the near-absolute dependency of contemporary liberal democracy on modern science and its technology, and the vulnerability that results, than do terrorist attacks, actual and imaginable. And the prospect of their being perpetrated by individuals who prove 'in deed' that they are not governed by the passion upon which the architects of liberal democracy premised their prescription – the 'universal' desire to live (whence derives the 'right to life' and to the liberty to live it as one pleases, pursuing happiness) – complicates by a whole order of magnitude the prevention of such attacks. As those architects well appreciated, no promise of earthly happiness through civil obedience to Leviathan can compete with a belief in the prospect of eternal bliss as reward for pleasing God. Hence, these political architects wrote in subtle and insinuating ways to undermine confidence in that prospect, which is to say, in the religious authority upon which such confidence is ultimately based. But, mainly, they placed their faith in the advancement of science: that as it progressed, its 'revelations' of the natural working of the world would eat away at the credibility of religious teachings, hence of religious authority. Judging from the recent historical experience of Western societies, they were right: this is pretty much what happened.

But this having become so glaringly evident to those who exercise religious leadership in the non-liberal (especially Islamic) societies of the world, as well as to the political leaders who rely upon and thus support them (whether or not true believers themselves), they are united in their determination not to let this happen to them. Whether their opposition is fuelled mainly by a genuine concern for the spiritual impoverishment of their people's lives through a mechanistic, materialistic, amoral worldview supplanting their intelligence-governed cosmology, or simply by the desire to sustain their own worldly power through manipulating people's fears and hopes concerning the eternal fate of their souls – whatever their actual motivation, untrammelled modern science has little appeal for them. So, complementing their antagonism to liberal democracy (ostensibly because of what they perceive to be the moral degeneracy that inevitably attends it) is a deep suspicion of modern science in general – liberalism's bedfellow as it were – and an open hostility to the popular dissemination of the science-based understanding of the world. And, to repeat, it is not as if there is no validity to what they believe, that all the truth is on our side, and that their resistance bespeaks nothing but ignorance and suspicion, envy, and resentment.

It is appropriate to focus attention on the intellectual leadership (mainly clerical) of the societies (mainly Islamic) whence come the individuals (fools or martyrs, as one prefers) who are willing to blow themselves up in striking a blow against the infidels. For as has become abundantly clear, the active terrorist networks are dependent for support and recruits on populations that have been nurtured on hate and bigotry since time out of mind. The prospects for their reform are not brilliant, as they imbibe an irrational, implacable hostility towards modernity with their mother's milk. To be sure, many of the teachers who do the nurturing are themselves 'ignorant as dirt,' but not all are. These latter can justify their exploiting the credulous dependence of everyone else for the sake of protecting their religion-based way of life from the corrosive effects of scientism and liberal democracy (which they can, and do, easily and plausibly conflate with colonialism and imperialism).

Be that as it may, the fomenters of jihad have no compunction in urging the use of modern technology against its crafters, thereby neutralizing somewhat the technological advantage their 'enemy' enjoys in this so-called clash of cultures. And they use it twice over, as it were, for they do not attack with scimitars but with high-tech weapons designed and manufactured in the societies they assault. And when not killing people indiscriminately for sheer terror effect, they aim at causing maximum social, economic, and/or political disruption by attacking one or another of the countless vulnerabilities of a high-tech society. Thus the 9/11 atrocity was such a stroke of evil genius: commandeer several high-tech marvels (jumbo jets) full of high-tech fuel, careen them into some other high-tech marvels (super skyscrapers) situated in the economic and cultural centre of the pre-eminent high-tech nation, kill thousands of people, damage dozens of other similar buildings in the vicinity, disable the commercial heart of that nation's leading city, precipitate a stock market plunge in which hundreds of billions of dollars of fiscal assets vaporize, and compel that nation to commit hundreds more billions to protecting itself from another such attack, on any of dozens of other vulnerable targets. We lose electricity, and our practical lives grind to an abrupt halt – literally, for almost nothing works without it, beginning with traffic lights. And when the lights go out, we are reminded that Hobbes's state of nature is not that far away after all: the city can become an urban jungle.

Such threats bring us straight back to the problems of managing science in a liberal democracy. Does not the need to protect the scientific research we sponsor from misappropriation by the enemies who would use it to destroy our society require some changes in how we 'do' science: Tighter screening of who has access to advanced research? Some governmental control over the circulation of certain scientific publications? Indeed, possibly even major changes in who is allowed access to educational programs in the sciences? As for those who respond, 'No more so than during the Cold War; we coped well enough with Soviet espionage,' they must bear in mind that the Soviets and those who served them were not suicidal, but quite the contrary: they were by comparison paragons of (worldly) rationality. However, controlling access to potentially dangerous scientific research (and can we be sure that we will always recognize that which is?) would not be simply a matter of excluding certain aliens. For due to the generous immigration and naturalization policies that characterize liberal democracies, and to their official or unofficial endorsement of a 'multiculturalism' that licenses ethnic groups to establish their own schools, the regime is endangered not only by enemies that infiltrate from without but by ones freely nurtured within.

And lest anyone think that the problem of international terrorism is bound to be only a short-time threat, that it is sure to 'burn itself out' when it is proven to be utterly futile (hence irrational), bear in mind: (1) it has been gestating for a century or more (at least since the aftermath of the First World War); (2) the accumulation of perceived grievances and hostility has been passed through several generations and saturates the minds of tens of millions of sympathizers; (3) they do not perceive terrorism as futile: while it has not made their lives one bit better, it has caused major discomfort to those they regard as their enemies and oppressors, and they take a spiteful satisfaction in that; and (4) most people are not rational.

There has arisen in the past few decades another external factor that has tended, on balance, to weaken liberal democracies. As of now, it is nothing like a problem on the scale of international terrorism, but it bears mentioning, not least because it can and has compromised the capacity of these regimes to combat the graver danger. I have in mind the proliferation in number and growth in influence of so-called non-governmental organizations (NGOs). Though virtually all of these organizations are the spawn of liberal democracies, they are formed expressly in order that like-minded individuals from wherever can act collectively upon, but outside of, the sovereign state system. Some do so in the name of science, or in the 'more equitable' sharing of its benefits; others either tacitly or expressly oppose various kinds of scientific research or its technological application (e.g., that it 'undermines indigenous cultures'). But whatever the case, such power as they wield is, quite literally, irresponsible, for they do not have to deal with the full consequences of their activities, nor in practice does the leadership that acts in the name of a typical NGO have to worry much about answering to anybody (including the rank and file that provides its financial base and foot soldiers). They are so many loose cannons battering away at both the protective sheathing and the structural supports of the liberal democratic ship of state.[6]

The Genie at Large

The third strand of problems facing the contemporary liberal democratic regime pertains more or less directly to science: how to generate sound public policies with respect to the increasing number of controversial issues that attend 'the generation, application and diffusion of scientific knowledge' itself (such as genetic research); or with respect to issues, the assessment and resolution of which inherently require special scientific expertise (such as possible global warming) – all the more perplexing when putative experts disagree among themselves. The problem, briefly put, is that of combining scientific expertise with political acumen – not 'in the abstract,' but within parameters acceptable to a liberal democracy. Seeing one's way to a solution requires a synoptic integration of two kinds of knowledge that are rarely found together in one head; and on those rare occasions when it might be, it would not be recognizable as such by all those called upon to approve it. For is that not precluded by the very hypothesis of its rarity: that it takes some considerable wisdom to recognize wisdom? If such is the case, is there a collective substitute, a way that quite diverse intellectual resources can be pooled and substantive conclusions reached that would command the whole-hearted acceptance of virtually all contributors? That there is such a possibility is the premise of this book (whose contributors, I need hardly remind you, represent an intellectual elite – certainly not a 'representative cross-section' of the general citizenry). Assessment of the prospect might begin with a consideration of the two main kinds of knowledge and knowers.

So, consider first what is required on the 'scientific side.' Obviously, it would entail a solid general understanding of the practicalities of conducting scientific research, but with due regard for the diversity *within* science; the research methods and resources of galactic astronomy are quite different from those of microbiology, similarly those of plate tectonics vis-à-vis primate ethology, or nuclear physics vis-à-vis meteorology. Moreover, it has long been acknowledged among practising scientists that there is not a single 'scientific mentality,' and in particular that 'theoreticians' tend to differ markedly from 'experimentalists.' In short, one has to be wary of the idea that this or that particular scientist, or even an assortment thereof, can be treated as representative of 'scientists in general.' But, in addition, we require from the scientific side sufficient familiarity with frontier research in various fields to alert us to the very thing that gives rise to our perceived need to manage science in the public interest: the existence of the potential sociopolitical problems implicit in such research. This presumes not merely the ability to imagine various possible technological applications of it, but some modicum of political imagination also. One quote from Phillip Kitcher's *Science, Truth, and Democracy* will suffice to illustrate my point:

> Within a decade, biotechnology companies will be offering hundreds, if not thousands, of predictive genetic tests. Given the character of the practice of medicine in much of the affluent world, it is highly likely that a significant number of people will confront information that is psychologically devastating, or be excluded from a job on genetic grounds, or be denied insurance through genetic discrimination, or face an acute dilemma about continuing a pregnancy. (2001: 5)

Note the self-evident: one needs some appreciation of the possible impact of such technical powers on people's lives in order to see potential problems that may require political attention through laws and regulatory agencies.

Of course, scientists being themselves intelligent citizens, most can be counted upon to imagine some of the more obvious social implications of some of the more obvious technological applications, and thus call them to the attention of any relevant authorities or of the general public – presuming, that is, that their concern for the public interest outweighs any personal interest they may have to the contrary. Even were that presumption a thousand times safer than it is, clearly this reliance upon scientists' sense of civic responsibility is not adequate for protecting the long-term interests of society at large from the dangers inherent in the present anarchical state of modern science. Fully grasping why not requires an understanding of political reality that can come only from an adequate intelligence studiously applied to the topic over a long time. For it is not primarily a matter of knowing how the present institutional structure works, nor of being familiar with present modes of behaviour. It includes that, of course, but far more important is knowledge of human nature in its full significant diversity, so as (for example) to foresee the likely long-term consequences of various actual or hypothetical technological possibilities, so as to assess what is 'politically possible' in a liberal democracy, and 'politically possible' *simpliciter*; and so as to understand 'political necessity' in a liberal democracy, and 'political necessity' *simpliciter*.

As I admitted at the outset, I am deeply sceptical about contemporary liberal democracy's capacity to deal successfully with the conjunction of forces that are assaulting it from within as well as without. For it is not simply a matter of envisioning measures whereby to manage science while doing damage to neither science nor society, nor of instituting reforms that could restore decency to popular culture, restore moderation to people's expectations, and at the same time restore some greater degree of democratic control over the everyday concerns of people. There must also be a politically realistic way to get from where we are today to where we imagine we would like to be. The same problem applies even more forcefully, of course, with respect to any proposed replacement regime, since in addition one would have to overcome its people's deep-rooted prejudice that liberal democracy, for all of its deficiencies, is the only legitimate and desirable arrangement of political life. The truth may be that, contrary to the congenial assumption we have inherited from the Enlightenment, there is no politically feasible way to surmount the challenges we face. The world offers us no guarantee that all our problems are resolvable on terms we would regard as acceptable.

With this much said, I shall limit the balance of my remarks to some preliminary considerations pertinent to any proposals for resolving, or at least ameliorating, the problems inherent in the interdependence of modern science and liberal democratic regimes. They rest on the following assumptions. First, the case has been made by others for there being sufficiently serious issues involving the relations between modern science and liberal democracy to warrant the recognition of a distinct problem terrain called 'public science.' Second, I have made the case that the appropriation of scientific technology by terrorists to attack liberal democracies where they are most vulnerable (their high-tech dependencies) both heightens the urgency and compounds the problems of managing science. Third, the practical measures realistically available in liberal democracies for dealing with foreseeable but distant problems and dangers is severely limited by their politicians' necessary preoccupation with the electoral cycle in order to continue in power, a 'political necessity' which inclines them to favour their own short-term interest over the long-term public good. It bears emphasizing, however, that there are two quite different desiderata that putative solutions must answer to: (a) ensuring that scientific research and its applications on the whole benefit the supporting society, and work no serious harm; (b) preventing the appropriation of scientific technology for malign use by terrorists who are incited and supported by societies that are effectively led by the theocratic enemies of both liberal democracy and modern science.

How, then, to incorporate needed scientific expertise in public policy decision making without unduly privileging the tiny minority of experts at the expense of democratic majoritarian rule? Typically, liberal democracies favour procedural answers, being loath to trust in anyone's disinterested prudence (since the premise of the regime's entire rationale is that people are motivated primarily by self-interest). So, is there a liberal-democratic way to marry scientific knowledge and political responsibility? Some politically realistic way, that is, not some utopian fantasy featuring idealized scientists and statesmen?

Begin by considering the reality with respect to 'the community of scientists.' First of all, as I and others have noted, it is far from a community in the sense of united in perspective and by commonality of interest. It is riven with competition for resources and recognition, both among and within disciplines, as well as divided by intellectual controversies and personal animosities. Even where there is consensus among putative experts on some question, it may as readily be an indication of 'shared paradigm' orthodoxy as it is of reliable knowledge. On the other hand, regarding issues where such consensus is lacking, we non-experts are scarcely better off than were there no scientific experts to consult. This is but the first problem with any proposal that leans toward the 'letting the scientists (first) decide among themselves' approach. And yet, as I noted before, there is an extent to which some reliance on the scientists themselves is unavoidable – and not just on their scientific knowledge and opinions, but also on their political understanding. For we rely mainly on the various scientists to bring to public attention both the potential benefits and the potential hazards implicit in the research they are pursuing (or are otherwise aware of). And one need only state this fact in order to recognize the 'asymmetric reliability' inherent in such reliance: we non-scientists can be a lot more confident in scientific specialists impressing upon us the beneficial side of their work (and of scientific research in general). Moreover, their appreciation of the broader political implications is not apt to be reliable at all.

Is this corrigible by education? And, more generally, could 'science' remain quasi-autonomous – as doubtless most scientists would prefer (they of course want public financing, but would prefer no strings attached) – and yet more effectively 'police itself' were scientific education substantially changed – were it, for example, given a more 'humanistic' foundation? I am sceptical. First of all, people who are naturally attracted to specialize in some science – and particularly those who prosper in it – are not naturally very sympathetic to the humanistic studies (after all, that is why they instead pursued science). There are notable exceptions, of course, but they are indeed 'exceptions that prove the rule' (i.e., prove the existence of a *norm* against which these few stand out). In any event, requiring more 'liberal arts' of science majors, while it might be good for them, does not really touch the problem, since it presumes that understanding political life does not really take that much effort. And that view itself is only too typical of the scientific mentality: they (like most people) have no clue of what is involved in gaining a barely adequate understanding of political life, much less a profound understanding, entailing as it does a comprehensive view of human life in its natural setting.

At the risk of caricaturing the typical scientist's perspective on politics, let me briefly sketch its essentials. Everything about his training, reflective of the ever-narrowing specialization that is intrinsic to modern science, renders him suspicious of 'generalists.' After all, specialization is the price of acquiring the kind of detailed understanding of nature ('real knowledge') that is required in order to manipulate it to serve human purposes. For this technological commitment – 'the relief of man's estate,' in Bacon's formulation; or, as Descartes promised, 'to make us masters and possessors of nature' – is what distinguishes modern science from its Aristotelian predecessor, both ratifying its findings and generating public support for its further progress. Hence, general understandings are, by scientific standards, necessarily superficial. And what is 'a comprehensive view' if not a kind of general understanding? Moreover, the typical scientist, who necessarily approaches his work in an organized, logical, dispassionate manner, looks with some disdain on the messy, irrational, emotional, often violent realm of politics (and probably with some disdain on those who spend a lot of time studying it, with almost nothing solid in the way of 'real knowledge' to show for their efforts). He is apt to think that, were the political realm put in charge of people such as himself, it would soon be set right, rationally organized, efficiently run, and everyone's life improved.

When I hear scientists express their political opinions, they often strike me as political ignoramuses. And this is puzzling, since they are obviously intelligent. So why do their political views so often sound naive, unrealistic, if not downright stupid to someone such as myself who has spent his life studying politics? Apart from the fact that they have not invested the time and effort it takes to see very deeply into political reality, such that 'thinking politically' has become second nature, they seem to me to suffer from three limitations.

First, they do not really appreciate the psychic range and political significance of human diversity. Though I presume most scientists would deny it, their view of human nature is tacitly based on their (probably pretty superficial) understanding of themselves. That is, they presume everyone is basically like them, only less rational to a greater or lesser extent (most likely not smart enough to be scientists), and that this is sufficient to account for the observable diversity among people. They may make an exception for artistic talent, but when asked to explain extraordinary political ambition (a Caesar, say, or a Napoleon, a Churchill), they are apt to invoke some reductionist psychology.

Second, and related, most scientists do not seem to realize the power of passion in most people's lives, and of political passions in particular. However passionate they may be about their own, that work does not require them to deal with passion as a factor in what they study; indeed, they may be attracted to science partly for this very reason. Thus, whether contemplating the 'cycle of violence' between Israelis and Palestinians, the ever-festering animosities that 'balkanize' the Balkans, or India's and Pakistan's hostile intransigence respecting Kashmir (Why would anyone care *who* controls Kashmir? What's in Kashmir worth fighting about?) – to say nothing of suicidal terrorism – deep down they, like most people not involved, just don't get it.

Third, the practice of their chosen profession does not require them to confront evil (nor, typically, does the environment in which they work); they do not have to take seriously the possibility that there is such a thing as evil – manifested, for example, in the sheer will to do harm, to kill and destroy and disrupt as ends in themselves, being perverse expressions of power. Thus, they are not prepared to conceive of political life as the domain of both good and evil, the very ideas of which they are apt to dismiss as 'pre-scientific concepts' (and most certainly non-scientific). If they are moved to political activity, they approach the problems of political life in the spirit of a social engineer, for whom there is no such thing as evil, just bad management. Hence, they are apt to underestimate the dangers of possible technological applications of scientific research, as if it were simply a matter of misuse for private gain. The collective consequence of these intellectual tendencies is that scientists typically have an excessively 'rationalistic' view of political life, and thus of how to approach the problems thereof.

The point is, whatever procedures and structures one adopts for bringing scientists into whatever processes and venues of political deliberation, they will bring not only their expertise and selfish interests with them but also the limitations characteristic of their perspective on people and politics.

Turn now to the politicians' side of the requisite collaboration. Most are, of course, scientific ignoramuses. Moreover, just as it is politically important not only that justice be done, but that it be *seen* to be done (according to the onlookers idea of justice, that is), so, too, with the allocation of resources for scientific research: it must not only be according to the public's long-term best interest, but perceived as such by that public. The voting public, to be more precise. Thus what Lewontin (2002) reports – that of the billions upon billions of dollars the U.S. government has spent on 'academic research and development' in the past fifty years 'over half ... went to the National Institutes of Health' (28) – is hardly surprising. People believe in health, and by that they mean primarily bodily health (certainly not moral health). The human genome project was sold to the U.S. Congress by scientists, mainly on exaggerated claims for its potential health benefits (e.g., in eliminating genetically transmitted diseases and disabilities), whereas Congress turned down funding for the proposed superconducting supercollider, whose proponents could offer no publicly appealing rationale that would justify its enormous expense. Whether either decision was truly in the long-term public interest is far from clear. Not all the imaginable uses to which knowledge of human genetic structure might be put, for example, are benign.

Both Bacon and Descartes saw clearly from the outset the most effective way to garner popular support for their effort to revolutionize science: promise enormous gains in health, increased longevity, enhancement of intellect, perhaps even the overcoming of mortality. The science they founded has certainly delivered on the first two but in doing so has massively changed people's basic posture towards life, and their very conception of political justice insofar as they have come to regard equal access to state-of-the-art health care as a natural right. Given the ever-escalating costs of high-tech medicine, it does not take a genius to foresee that this will breed even greater political problems in the future than it has thus far (which are plenty). But what politician who cared about his or her political fate (i.e., remaining in office) would dare suggest that we severely curtail medical research and only fund that which can plausibly promise to lower treatment costs? Moreover, what medical research scientist would recommend doing so?

As for the other two Baconian-Cartesian come-ons, modern science has not and cannot rule them out; hence, they serve as paradigm illustrations of the perplexing problems that scientific progress can present to political life. Given the means to genetically engineer greater intelligence, and that virtually everyone acknowledges more intelligence is better than less, what publicly convincing grounds are there for legislating against it? Doubtless there would be politically safe arguments for not publicly funding it (since there are always competing claims for the money), but on what grounds can it be prohibited in a regime that prides itself on the freedom it offers its citizens? That this freedom is exercised by so many who believe dogmatically that all 'value judgments' are subjective, thus incorrigibly 'relative' – and who vote accordingly – simply compounds the difficulty of dealing foresightfully with any long-term dangers that may lurk in this technical power. Scientific progress created the issue, but scientific expertise can contribute nothing to resolving whatever sociopolitical problems might result. The moral of this example is pertinent to so many: it is politically very difficult in a liberal democracy (meaning, for the politicians who govern it at the periodic sufferance of the voters) to foreclose developing technological applications of scientific knowledge once they are extant. Best not let the genie out of the bottle in the first place. But how is that to be decided and done, given the polity's reliance on competent scientists to so advise them, meaning the very ones who are competent to undertake the research? After all, pursuing knowledge, not shunning it, is what defines them.

Similar remarks apply when expert scientific opinion warns of a long-term threat to the common good (degradation of the environment, say) and recommends some corrective or remedial action that entails more or less immediate adverse consequences for some significant sector of the public. The scientific case can be absolutely conclusive but not carry the day with politicians who must face the injured constituency. It *should* carry the day, but it will not. Now, is there anything that can be done to rectify this irrationality – within the framework of liberal democracy, that is? Scientists complain that their advice is often sought but seldom accepted as the basis for policy decisions. Perhaps it seldom should be, limited as I believe their political knowledge typically is. But doubtless sometimes it should be, but nonetheless is not. Is there some way to determine which is which – when scientific expertise should be accepted as simply authoritative, when given extra but not decisive weight, when treated as but one relevant factor, and when regarded as merely another opinion?

A moment's reflection on this question returns a pessimistic verdict. For one realizes that it is just another face of the basic dilemma. For if there were some reliable way to make this determination, requiring as it does a synthesis of scientific knowledge and political understanding, one would have already solved the puzzle of how to ensure effective collaboration between representatives of science and the representatives of liberal democracy.

Virtually from its inception in the mind of Francis Bacon, visionary author of *The New Atlantis*, there have been thoughtful people who understood the frightening potential of modern science to become Dr Frankenstein's monster. It is only in our time, however, that the potential is on the verge of becoming actual. But, as I have attempted to show, the modern predicament caused by liberal democracy's Faustian bargain with science has facets that go well beyond the obvious and challenge the very viability of the only regime most of its citizens can imagine supporting. This, it seems to me, is the most urgent problem for political philosophy in our time. It is not one that is apt to be solved by modern science.

NOTES

1 Oviatt (2000: 1) asserts: 'Multimodal research represents "Science without Borders" because it requires combining expertise from different component technologies, academic disciplines, and cultural/international perspectives. It also is rapidly erasing borders as it promotes the increased accessibility of computing for diverse and non-specialist users, and for field and mobile usage environments. This paper reviews two studies that highlight recent advances within the field. It also draws parallels between the multimodal areas of speech/pen and speech/lip movement research. Finally, it indicates new research challenges that will require additional bold "border crossings" in the near future.'

2 Rousseau sees this clearly. As he expresses it in the final note he appended to his *Discourse on Inequality*, 'Distributive justice itself would be at odds with the rigorous equality of the state of Nature, even if it were practical in civil society; and as all members of the State owe it services proportionate to their talents and forces, Citizens ought, in return, to be distinguished and favoured in proportion to their services. It is in this sense that a passage in Isocrates has to be understood, in which he praises the first Athenians for having correctly discerned the more advantageous of the two sorts of equality, one of which consists in allotting the same advantages to all Citizens indifferently, and the other in distributing them according to each one's merit. These skilled politicians, adds the orator, by banishing the unjust equality which draws no distinction between wicked and good men, inviolably adhered to the equality that rewards and punishes everyone according to his merit' (221).

It should be noted that matters are merely muddied by the introduction of a third variant that is sometimes argued for: 'equality of condition,' referring to the totality of circumstances in which 'opportunity' is exercised – insisting that equality of opportunity is a sham, if not simply meaningless, unless people's 'initial conditions' are equal. Whether or not it is the intention of those who argue for this interpretation of 'just equality,' the consequence of its acceptance is an indirect endorsement of 'equality of result' in one of two ways. Since even a superficial analysis reveals that complete equality of condition is practically impossible – and that only political horrors result from over-zealous efforts to achieve such equality – the effect is the endorsement of equality of result by default of the only credible alternative. Or, by treating inequality of result as *prima facie* proof of inequality of condition, one licenses the manipulation of conditions so as to produce the only outcome that (given the premise) would ratify their equality: equality of result.

3 To see why, one might begin with Aristotle's critique of its impracticality in

Book Two of his *Politics*, which takes due cognizance of the fact that gross disproportions in the distribution of wealth and property raise political problems that justify measures for their amelioration in the names of civility, charity, humanity, and peaceful stability.

4 Mis-called, actually, at least according to the original meaning of 'culture': cultivating something – be it plant, animal, or human – in such a manner as will perfect its specific nature.

5 As Leo Strauss observed in his *Natural Right and History* (1953), Hobbes, as 'the creator of political hedonism' upon which is based his political prescription, has produced 'a doctrine which has revolutionized human life everywhere on a scale never approached by any other teaching' (169).

6 This is most readily evident in the proliferation of NGOs modelling themselves on Doctors Without Borders, thereby tacitly declaring themselves 'above' national partisanship. It is a development welcomed by those who foresee with favour the decline of the nation-state (regarding it as anachronistic in this globalized world, as an irrational source of violent conflict, etc.), and urge its supplantation by supranational governing organizations. But this is a mode of thinking that should constitute a new worry for the partisans of liberal democracy. As both theoretical analysis and historical experience attest, the more the jurisdiction of governing agency expands, the more rules and regulations multiply – along with bureaucracies to administer them. This not only adds to the already enormous expense of government, it renders it that much more remote from the influence of individual citizens; bureaucratic rule tends toward totalitarian rule, not towards democracy.

REFERENCES

Hobbes, T. *Leviathan*. 1991. Edited by R. Tuck. Cambridge: Cambridge University Press.

Kitcher, P. 2001. *Science, Truth, and Democracy*. Oxford University Press.

Lewontin, R.C. 2002. The Politics of Science. *New York Review of Books*, 9 May, 28.

Oviatt, S. 2000. *Multimodal Interface Science: A Science without Borders*. Proceedings of the Sixth International Conference on Spoken Language Processing (ICSLP). http://www.cse.ogi.edu/CHCC/publications/multimodal_interface_research_science_without_borders_oviatt.pdf (accessed 26 April 2007).

Rousseau, J.J. *Discourse on Inequality*, 1997.. Edited and translated by V. Gourevitch. Cambridge: Cambridge University Press.

Strauss, L. 1953. *Natural Right and History*. Chicago: University of Chicago Press.

9 Expertise, Common Sense, and the Atkins Diet

STEVEN SHAPIN

Different kinds of scientific expertise bear different kinds of relation to common sense and to lay concerns. Particle physics, for example, is hard to understand, and much of it is counter-commonsensical. But, for the most part, that is not a problem for the laity, just because it is not an expert practice that they care much about. Experts may regret this state of affairs; they may recommend new initiatives in 'the public understanding of science,' but, for present purposes, it is more pertinent to understand why there is such a lack of concern. In such cases, the public let experts, so to speak, 'get on with it,' and, if and when the subject gets enshrined in the curriculum and there is a school examination to be passed, then the kids will have to learn the stuff. And if and when there is a concrete technology that emerges from such work, then the public may give it some attention themselves or they may assume that their democratically elected representatives should have the matter in hand. There just are not the sort of intermediate structures that allow the public to engage with such expertise in any other ways.

However, there are other technical practices whose scientific content may be hard to understand, and whose intellectual content may even conflict with lay models, but about which the laity care quite a lot. Any technical practice which purports to have a bearing on the doings of everyday life may engage lay concerns in this way – expertise on how to have good sex, how to raise the kids, and how to launch a start-up company are all in this position. In the 1890s, Leo Tolstoy (1937: 186) criticized modern science for not having an answer to the question 'What should we do?' and in 1918 Max Weber (1991) celebrated modern science for just the same reason: 'Science is meaningless because it gives no answer to our question, the only question important for us: "What shall we do and how shall we live?"' (143).[1] But, in fact, quite a lot of current scientific practices are in the business of supplying answers to that kind of question.

I want to talk about one version of this sort of modern expertise, one that illustrates different strands of technical expertise, enormous lay interest in the general field, enthusiastic lay take-up of some kinds of expertise, and vigorous lay rejection of other sorts. The case is dietetics and such cognate practices as nutrition science – anyway, expertise about what you ought to eat, and not eat, to avoid a range of diseases and to live a long and healthy life. Dietetics is now, and always has been, an expert practice that has the possibility of uniquely tight engagement with quotidian life, lay knowledge, and the texture of morality (Shapin 2003a). If you are lucky, you get to eat three times a day, seven days a week, so there are at least twenty-one occasions per week when the fork approaches the mouth. Each one of these is a possible occasion for the intercalation of expert knowledge between fork and mouth. In our sort of society, it is just not possible for food to be bought, prepared, and ingested unaccompanied by thoughts flowing from bodies of technical expertise: the relationship between dietary cholesterol and heart disease, fibre and colon cancer; the boundary between the amount of wine that is good for you and the amount that is not; and, of course, calorie content in relation to weight gain and loss. Every meal now comes polysaturated with technical expertise, and, while you can, at cost, decide to ignore it on *this* occasion, even ignoring expertise is a decision you probably have to take. In the main, however, the decision seems to be which type of expert commentary gets pride of place at *this* meal.

Dietetics is about exceedingly complex phenomena, and attempts to understand complex phenomena typically generate endemic expert disagreement. The British Astronomer-Royal Sir Martin Rees (2005) recently drew an instructive contrast between the security of scientific knowledge of simple things – the properties of celestial objects, for example – and the fragility of our knowledge of complicated things like human nutrition: 'Scientists still can't agree on what food is good for us. There is a real sense in which dietetics is harder than cosmology.' Questions about which expertise to invite to the table can be, and now often are, vexed decisions for the average eater. Almost needless to say, modern technical expertise about what you ought to eat is seriously divided.[2] The heterogeneity of expertise on a very large number of occasions of public concern is a notable feature of our culture (as argued, for example, in Collins and Pinch 1998a and 1998b, Shapin 1992, and Barnes 1982). Walter Willett (2001a), a Harvard expert, freely admits that 'research about diet and nutrition seems to contradict itself with aggravating regularity' (27). Most laypeople probably know that expert dietetic claims are various and change rapidly, though it would be good to have more securely grounded understandings of what people do and do not know. Sometimes the diversity of conflicting expert voices encourages the laity to suspect that genuine expertise is not available in the area – as in psychiatric expert testimony on legal responsibility – but that is not a necessary inference for all areas of internally conflicted expertise. Dietetics seems to be a domain characterized by both very great expert heterogeneity and marked lay faith that authentic expertise must be available *somewhere* in the welter of conflicting claims. If one strand of dietetic expertise is treated with scepticism, then residual faith in the existence of expertise seems to be transferred to another claimant. The laity assert their freedom to pick and choose which expertise is credible, while giving few signs that they find the whole domain of dietary expertise wanting.

This is one further indicator of the way in which the body of expertise has become entrenched in our culture and what a long way we have come from the late sixteenth century, when Montaigne (1965: 827) thumbed his nose at the very idea of such expertise and its authority over his daily patterns of eating and drinking: 'The arts that promise to keep our body and our soul in health promise us much; but at the same time there are none that keep their promise less.' Expert physicians disagreed among themselves, and, so, if you did not like the dietary advice that one offered, you could always pitch one doctor's favoured rules against another's: 'If your doctor does not think it good for you to sleep, to drink wine, or to eat such-and-such a food, don't worry: I'll find you another who will not agree with him' (833). For that reason alone, you might as well do what you thought best: you were, and ought to be, your own dietary physician (Shapin 2003a).

The USDA Food Pyramid (figure 9.1), initially published in 1992 with wide academic consultation, and much industry feedback, still seems to represent the condensation of academic expert knowledge among the nutrition scientists. Marion Nestle (2002: 66–8), one of the leading academic nutrition scientists, assures us that two-thirds of the American public know the original USDA Pyramid; that there is 'almost universal' expert consensus about its recommendations; and that genuine expert counsel has long been and will remain stable: it is a high-carbo-

Figure 9.1 USDA Food Guide Pyramid

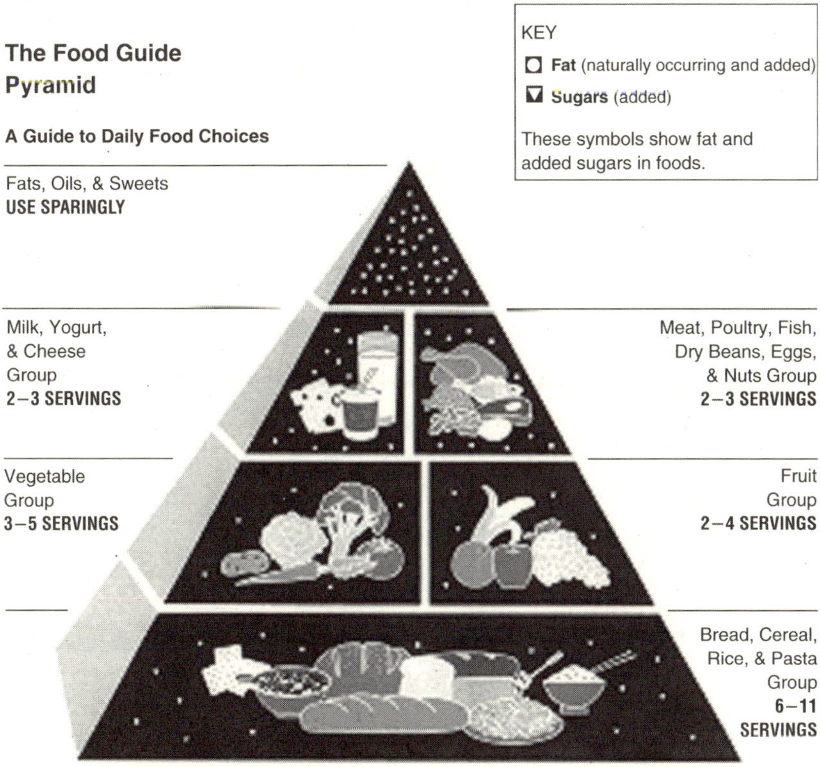

hydrate, high-fibre, low-protein, low-fat diet – pretty close to 'the Mediterranean diet' associated with the 1950s and 1960s work of the pioneering nutrition scientist Ancel Keys (Keys and Keys 1959).[3]

However, when Nestle published her book *Food Politics* in 2002, the Pyramid was already considered ripe for revision. In February 2000, the USDA staged a public 'Great Nutrition Debate' among nutrition experts as a preliminary to revising the Pyramid and in September 2003 announced a program for that revision, partly as a response to anxieties that two-thirds of Americans were now classified as overweight or obese (Helmich 2002; Martin 2003).[4] During the writing of the original draft of this paper, it was understood that the revised Pyramid was to be announced early in 2005, and the odds were that, while it would still be a relatively low-fat, low-protein diet, it would be substantially dif-

Figure 9.2 MyPyramid

ferent from the existing form. Important distinctions might be made between 'good' and 'bad' carbohydrates, 'good' and 'bad' fats – for example, whole wheat bread is pretty good, white bread very bad; olive oil is good, Crisco very bad (Buckley 2004: 4). As it turned out, the revision released in April 2005 – and labelled as 'MyPyramid' – was widely viewed as an incoherent compromise, the USDA avoiding the responsibility of giving blanket dietary advice to the population as a whole and certainly not counselling eating 'less' of anything (figure 9.2). 'One pyramid does not fit all of us,' said the Agriculture Secretary Mike Johanns, 'so we created 12 different ones' (Burros 2005: A21).[5]

Among academic experts, Dr Walter C. Willett, of the Department of Nutrition at the Harvard School of Public Health, was leading the charge: 'The [old] Food Pyramid is tremendously flawed. It says all fats are bad; all complex carbohydrates are good; all protein sources offer the same nutrition; and dairy should be eaten in high amounts. None of this is accurate' (Willett 2001b). The probable causes of this knowing

Figure 9.3 Willett's Proposed Food Pyramid

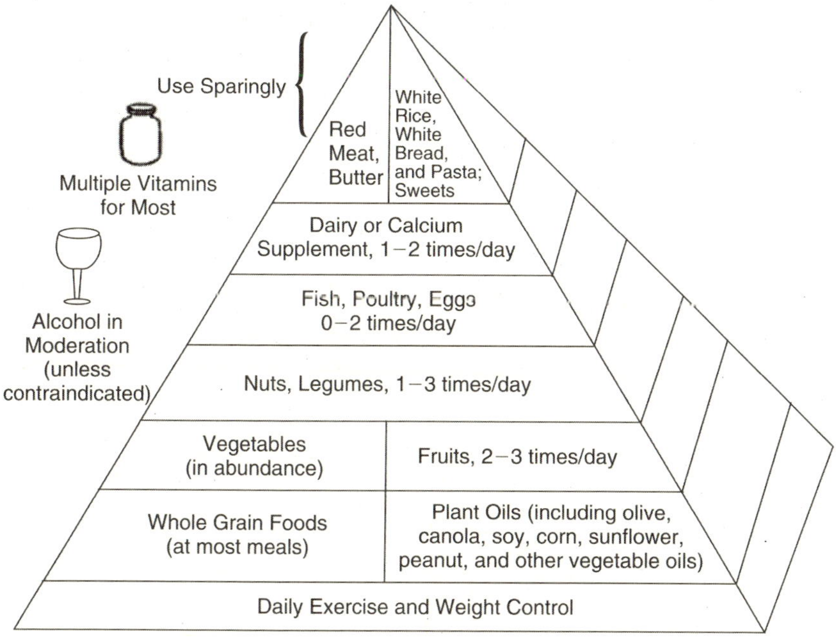

© Reproduced from *Eat, Drink, and Be Healthy* by Dr. Walter C. Willett (New York: Simon and Shuster 2001).

dissemination of bad advice were, first, that 'many nutritionists decided it would be too difficult to educate the public about [the] subtleties' involved in discriminating among carbohydrates and fats and, second, that the USDA Pyramid was a nutritional icon substantially shaped by the political and economic interests of the food industry. Nor was there any convincing evidence that individuals adhering most rigorously to the counsel of the USDA Pyramid had improved health outcomes attributable to their diet: true, they did have 'a lower risk of major chronic disease,' but, Willett found, these same individuals also smoked less and, in general, led 'healthier lifestyles' (Willett and Stampfer 2003: 20). Instead, Willett proposed a conceptually jumbled 'Healthy Eating Pyramid' (figure 9.3) with 'exercise and weight control' at the base, complex carbohydrates and mono- or polyunsaturated vegetable oils just above, and moderate alcohol and multi-vitamins hanging in space off to the side (24).

Figure 9.4 The Atkins Food Pyramid

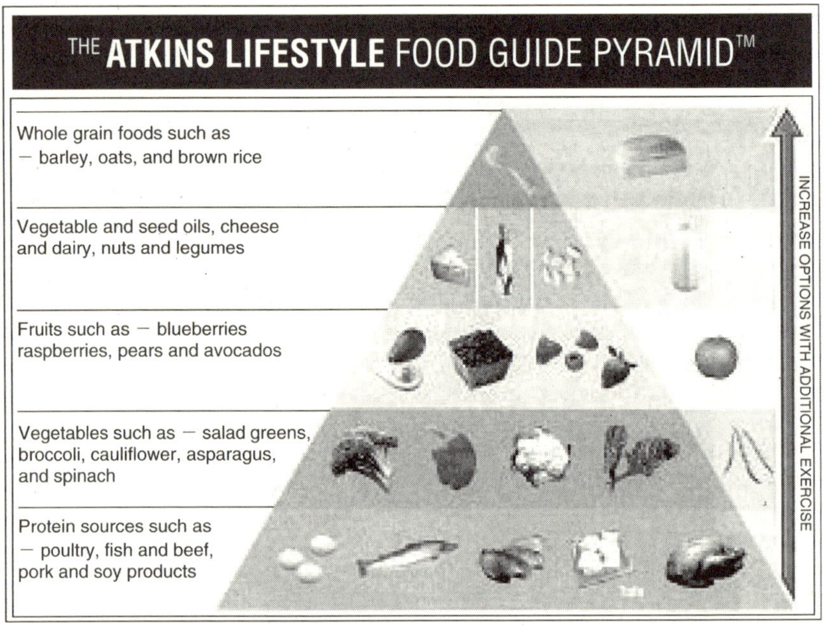

And then there are the experts who actually sell books – lots of books – the ones who are clearly winning in the marketplace of dietary credibility. And here the crucial names were Atkins and Agatston – both of whom turned the USDA Pyramid upside down and commended lots of protein and fat and tight control of the carbs (Atkins 1972, 1992, 2003; Agatston 2003) (figure 9.4). Since Dr Robert C. Atkins's *Diet Revolution* first appeared in 1972, over 16 million copies of its various versions have been sold worldwide; 30 million Americans have said they have given it a go; and it was almost as successfully globalized as McDonald's, that paradigm of American globalization (Bentley 2004).[6] Atkins was very big in Britain, in Scandinavia, and, amazingly, in Italy, where there was a certain amount of panic among the pasta makers. In its U.S. heartland, few *academic* experts were won over, but their lives were being made more and more uncomfortable as their constituencies increasingly voted with their gut instincts. *The New York Times Magazine* pronounced Atkins vindicated, declaring that USDA Pyramid-type academic expertise may all along have been 'a Big Fat Lie' (Taubes 2002).[7]

And so the question is put: what are the grounds of dietetic credibility? Why did Atkins, and many others of the low-carb popular writers, do so well in the public credibility market, while academic expertise is struggling to make its voices heard? That is a complex question, and all I want to do is to suggest some features of the credibility economy that figure here and that have got some interesting connections with how we tend to think about science and common sense (or lay knowledge) more generally. There are all sorts of considerations I am going to set aside for these purposes, and which are doubtless very important: for example, the fact that Americans like to eat meat and that it is relatively cheap; the possible route that low-carb regimes offer for the masculinization of dieting; a dialectical oscillation in diet fads – low-carb, high-protein fashions alternating with the high-carb, low-protein orthodoxies of the recent past; a rhetoric of luxury, metabolic causation of obesity, and the disavowal of will power that resonates beautifully with current American fashion and academic theory.[8]

First, there is the matter of who is an expert and, therefore, entitled to public authority. That is not a simple question either. Presumably, attending to any given writer implies the recognition of expertise: who would put their health in the hands of someone who they reckoned did not know what they were talking about or who did not possess special and superior knowledge? What is, however, crystal clear is that membership in an academic institution, or in a credentialed expert community, is not, in this specific case, much of a public warrant of expertise and may even count as a liability. One of the most pervasive tropes one finds in the popular dietetic literature is a studied disavowal of credentialed expertise. So while Atkins was a physician, he quickly assured his vast readership that he was a cardiologist and not a diet doctor, drawn ineluctably to his calling by simple, but vast, experience of what actually worked and did not work: 'Looking back at this period in my life I realize what a fortunate thing it was that my training has been in cardiology and not in metabolism. If I had been trained in nutrition and metabolism, I'd be parroting the same classical misconceptions that so many of my colleagues still hang on to. Being free of these misconceptions, allowed my mind then – and since – to approach the observable facts without prejudice' (Atkins 1972: 22).

Dr Agatston of the South Beach Diet does just the same, insisting that his institutionally acquired expertise was even more distant from dietetics. He was a physician specializing in CT imaging technology. Right up front, Agatston boasts that 'I'm not a diet doctor.' Discontented with the obvious inefficacy of the American Heart Association–approved low-fat regimen, he was forced to face facts and to develop a diet that actually worked (Agatston 2003: 6–7). So too Dr Tarnower of the late 1970s Scarsdale Diet: 'Nobody could be more delighted and surprised at the extraordinary success of the Diet than I, since I am not a "diet doctor"' (Tarnower and Baker 1979: 2). Not being a doctor at all can even be a recommendation: Judy Mazel (1981) of the early 1980s Beverly Hills Diet so stated, getting her physician friends to testify to their amazement and delight at her success. A book by two registered dietitians details their initial struggle to arrive at a radical weight-control Truth that nevertheless stood starkly opposed to their profession's key tenets: 'How could we, as nutritionists (registered dieticians), trained to look at the connections between nutrition and health, sanction a way of eating that seemed to reject the very foundation of our [profession's] knowledge and philosophy?' (Tribole and Resch 1995: xv). Of course, there is a lot of metabolic science in Atkins's books – though not so much in Agatston and many others: all the stuff about ketosis, hyperinsulinism, and glycemic indices that is well on its way to being vernacularized in American lay talk about the human body and its aliment. But, although Atkins reckoned it was important that you be told the science of why his diet works, his protestation that he is not a 'diet doctor' gives adequate indication that this is not his personal expertise, so there is no more reason to believe him on the theory of ketosis than many of his expert opponents. Expertise can be an ambiguous category in dietetics: what count as the insignia of expertise in one register may be identified in another as disqualifications.

The grounds on which credentialed expertise is so vigorously disavowed bear inspection: the popular writers identify that kind of expertise as interested, invested in the theories of orthodoxy, and, accordingly, biased, while they want themselves seen as free-booting, anti-establishment independents. They speak a Truth inconvenient to the Vested Interests and cut free of those expert theoretical investments that afflict the academy and the medical profession. That is a sentiment well worth considering for those who think that the academy and the professions still represent the voice of integrity in the wider culture. But in this case studied rejection of credentialed expertise points to something else going on. It testifies to substantial differences in how expert knowledge is represented and brought to bear on questions about 'what should be done.'

One telling contrast between the writings of not just Atkins and the academic nutritional establishment, but between virtually all the best-selling writers and the establishment, concerns the question of a science of the particular versus a science of the universal.[9] Look up dictionary definitions of 'science,' and you will see an overwhelming disposition to define science as the search for, and accumulation of, knowledge of universal scope.[10] That is why Simplicio lost the argument with Salviati in Galileo's *Dialogue*: when Simplicio noted that Galilean idealizations worked well in the abstract but rarely if ever fit concrete instances, Salviati recognized a peasant when he heard one. Salviati/Galileo (1967) wanted to know how bodies moved through ideal, frictionless media; Simplicio wanted to know how a particular body moved: 'These mathematical subtleties do very well in the abstract, but they do not work out when applied to sensible and physical matters' (203). In fact, there are many present-day sciences of the particular, though you would not know it from the definitions: all those people trying to work out when the next Cascade volcano will go off, or when the next earthquake in San Francisco will occur, are practising sciences of the particular. I would guess – though I can provide no firm statistics to support the guess – that the majority of people categorized in census data as 'scientists' work within sciences of the particular. Many expert practices can be viewed as hybrid: civil engineering, it is said, 'depends upon' physics but every bridge is different; medicine 'depends upon' physiology in the same way, but every sick body is different. And you cannot 'idealize' away either the unique geology of San Francisco or the unique physiological state of Professor Shapin's liver.

You can eliminate Simplicio's voice from the body of expertise, but there is a price to be paid. That price may not be consequential public disapproval – the American public have almost always been willing to pay the substantial bill for theoretical physics, and, for all the talk of 'public hostility to science,' scientists continue to be held in exceptionally high esteem and science research budgets continue to go up – but it is likely to be public 'unconcern.' This is not a big problem for physics, but it is a very big problem for those expert practices that purport to engage with the question 'What should de done?' Dietetics is certainly one of these practices. It would be quite wrong to say that academic nutritional experts are trying to shrug off the responsibility of giving practical advice. But they go about it in a significantly different way than Dr Atkins.

First, consider the philosophically crucial divide between the objective and the subjective. As a global generalization, I do not think that more than a tiny fraction of academic scientific writing on practical dietetics ever addresses what food and eating mean. Eating is treated as an instrumental act: here is what you want to do to avoid diabetes, heart attacks, colon cancer, Alzheimer's, death. The assumption is that the recipients of advice are like the subjects in rational decision theory, maximizing their utilities, and expert advice just tells people – to the best of the expert's ability – how to achieve this maximization most effectively. This is how Harvard's Walter Willett (2001a) starts his exercise in extending academic nutritional expertise to a wider audience: 'We eat to live. It is a simple, obvious truth. We need food for the basics of everyday life – to pump blood, move muscles, think thoughts. But we can also eat to live well and live longer. By making the right choices, you will help yourself avoid some of the things we think of as the inevitable penalties of getting older' (15). As he says later in the book, with a self-conscious gesture at a Wall Street *Weltanschauung*, 'that is not a bad return on investment' (180).[11] I do not doubt that people want to live as long and as healthily as possible, but this is not the only thing that people do when they eat. They eat because they feel bad, or because they feel good. They eat to celebrate or to mourn, to establish claims to status, to affirm or deny group identity. They eat to be sociable, and sometimes they eat like other people eat in order to be sociable. You would not know any of this if you read the works of almost all academic scientific experts on nutrition. (Actually, I tell a slight lie: these writers do have a view on why people eat irrationally, and that is because the rational voice has been made inaudible in the culture by advertising and marketing. So people are still taken as rational maximizers, but they have been given the wrong information on which to do their rational calculations.) In the best-selling literature, however, you cannot miss engagement with the emotional and social meanings of eating. It should come as no surprise, then, that the laity might recognize in this literature a sense of pertinent aboutness that they may not see in academic writing.

Second, academic dietetic expertise treats human bodies in general and as populations. If you weigh this amount, or if you drink this amount, your chances of contracting or avoiding this or that dread disease is *n* per cent, or your risk of it is elevated or decreased *n* per cent. This can be useful information, most especially for policymakers, and let us grant for the moment that it is reliable, even though the domain of such academic statements remains notoriously conflicted. The problem, however, is that individual people tend to be interested in what is likely to happen to *them*. The answer for any specific individual will either be colon cancer or no colon cancer, die before age 75 or after age 75. What any individual wants to know is what it is going to be like for them, an answer to what one historian of medicine has called 'the patient's fundamental question – why me?' (Marks 1997: 171). As the cognitive psychologist Gerd Gigerenzer (2003) has brilliantly pointed out, we are organisms that seem to have evolved to take in natural frequencies (say, three out of ten people) rather than probabilities (say, 30 per cent), and there is a consequent problem of misunderstanding risk between experts and laity. But the issue to which I draw attention is that people – the recipients of expert counsel – are interested in themselves as unique particulars, not as members of populations, and I see no way to say they are wrong to do so. They *are* particulars, and it is plausible to think that they approach statements about populations as statements about particulars. Expertise about populations just cannot give that answer to any individual, whether expressed in natural frequencies or probabilities, but it does not follow that individuals may not want it anyway and look for expert writing that gives signs that it considers such a desire legitimate and intelligible. The difference should not be thought of as one between knowledge and ignorance, between cognitive virtue and cognitive error, but one between different kinds of concerns. You cannot use better logic or more evidence to refute a different kind of concern.

The popular literature acknowledges this circumstance. The style is overwhelmingly casuistical. That is, while there may be much metabolic science of general scope in any given text, the books are quite generally structured around particular cases.[12] Indeed, they often are organized as a kind of counterpoint between the general and the populational, on the one hand, and stories about, say, Mrs Finkelstein, a forty-nine-year-old woman weighing 212 pounds, on the other. Case follows case, testimonial follows testimonial. Dr Atkins barely gets his story underway before we are introduced to Stanley Moskowitz, 'a vigorous 64-year-old sculptor who had survived three heart attacks,' who was overweight, who was addicted to junk food, and who, under Atkins's tutelage, learned the delights of low-carb dieting. Then Mary Anne Evans gets to 'tell her own story,' a forty-two-year-old, 209-pound woman who had abandoned all hope before she achieved low-carb redemption (Atkins 1992: 10–14). The somewhat less folksy and rhetorically more restrained Dr Agatston (2003) interleaves expert nutritional reasoning with typographically distinct stories about individual cases: Karen G. shares with us the effects of divorce on eating patterns (14–16); Ellen P. explains that she decided to lose weight because her 'oldest daughter's bat mitzvah was coming up' (23–4); Daniel S. confesses to back-sliding pasta consumption and serves as an illustration that South Beach does not presuppose absolute purity of commitment (30–1).

You could describe this casuistical style as a deliberate attempt at credibility engineering, and that would almost certainly be correct, but it is also possible to interpret it as a response to the concerns which deliver the reader to the text. That is the sort of thing they expect to hear. True, Mrs Finkelstein is not you, but she is, and is here presented as, a unique individual, not a population, and the metaphorical extension from aspects of her predicament to yours is available for you to make if you happen to feel her case to be sufficiently like your own. That is one way in which the popular writings are in a direct lineage of early modern medical traditions, which were focused on the idiosyncrasies of individual sick bodies whose workings the physician attempted intimately to understand (Jewson 1974). Enjoying special authority, both in early modern medicine and in the contemporary best-selling dietetic literature is 'the case of the author' – a confession of how the author stood before undertaking the proposed dietary discipline and how it turned out for him or her – invariably, of course, very well indeed. In the mid-eighteenth century, the most celebrated diet doctor in England, the enormously fat Dr George Cheyne, devoted thirty pages of his celebrated *The English Malady* to 'The Case of the Author,' and in 1972 the most famous diet doctor of *his* day, Dr Robert Atkins, told his readers how, as a medical student, he 'had the reputation of being the biggest chow-hound in the hospital' (Shapin 2003b; Guerrini 2000). So deep was he in denial of his own condition that it was not until he was in his thirties that he realized he had 'three chins.' If you find dieting hard, Dr Atkins (1972: 21) assures you that so too did he: he understands your difficulties. So far as I can now see – though future reading may prove me wrong – there are no cases in the academic literature.[13]

An early modern maxim – used as much by analogy outside of medicine as within it – was 'Physician, cure thyself.' You might well want to take the expert advice of healthy and slim physicians, just because they gave embodied warrants of sincerity as well as efficacy, and one thing you might well want to know about those attempting to give you expert advice is whether they took it themselves, and to what effect.[14] That is what an interviewer asked Walter Willett, and, having told readers next to nothing about his own regimen in his expert writings, Willett responded to his interviewer directly: Interviewer: 'What do you typically eat in any given day?' Willett: 'For breakfast, I'll usually have something like cooked whole grains with fruit. Lunch may be leftover grains from breakfast as part of a salad, with vegetables, and nuts. Dinner is usually salad or vegetables with tofu, chicken or fish.' Interviewer: 'What about secret indulgences?' Willett: 'Sometimes I'll have some flavorful cheese or a bit of chocolate' (Willett 2001b).[15] Having offered his own case to launch his 1972 text, when Atkins died in 2003, much attention focused on the circumstances of his death. From within the academic scientific point of view, it is hard to see why that should matter: probabilities being what they are, why shouldn't you 'do as I say, not as I do,' and why shouldn't someone following his own excellent advice nevertheless suffer a heart attack at age 71 and wind up dead on a Manhattan pavement at age 72? Assuming that Atkins followed his own diet, and assuming that he reduced his chance of a heart attack by, say, 50 per cent, the fact that he nevertheless suffered a heart attack should not affect the credibility of his diet: that's probabilities for you. But in this case it was rather harder to gloss away a New York death certificate which shows that at the time of his death Dr Robert Atkins weighed 258 lbs and was, according to his own criteria, and those of the Centers for Disease Control, 'clinically obese' (McLaughlin 2004 and Kleinfield 2004). So consider the contrast between regimes of credibility in which such a circumstance makes no difference and one in which it makes all the difference. Out of such a consideration, I suggest, might come an interesting new way of parsing the sciences and of describing the different conditions of their credibility. Less ambitiously, I have offered a sketch of some reasons – other than the simple distribution of unpalatable truth and appealing falsity – why the popular writers *are* popular and why the academic experts are not.

NOTES

1 For contrasts in the conditions of credibility confronted by different kinds of expert practices, see Shapin (1995).
2 A fine study of divided and politicized nutritional expertise and expert advice to the American government during the 1980s is Hilgartner (2000).
3 According to Dr Keith-Thomas Ayoob (of the Albert Einstein College of Medicine), testifying on behalf of the American Dietetic Association in the USDA's Great Nutrition Debate of February 2000, 'a recent Gallup poll

found that 86 percent of dieters like the USDA food guide pyramid. They thought it was a common, no nonsense approach to managing weight in the long term, even though 40 percent of those people had at some point tried a high protein/low carbohydrate diet.' http://www.usda.gov/cnpp/Seminars/GND/Proceedings.txt (accessed 24 August 2004).

4 For a transcript of the Great Nutrition Debate, see http://www.usda.gov/cnpp/Seminars/GND/Proceedings.txt (accessed 24 August 2004).

5 Critics declared that the Pyramid did not accurately represent advice contained in the 2005 *Dietary Guidelines for Americans*, published in January 2005, and soon MyPyramid was becoming an object of satire and political criticism – some opponents referring to it as 'McPyramid' because of its alleged compromises with the food industry. See, e.g., Severson (2005).

6 As of August 2005, Arthur Agatston's *South Beach* diet books have sold 10 million copies world-wide (Warner 2005a).

7 Amy Bentley (2004: 37) notes that Atkins's book 'maintained a modicum of popularity' through the 1970s and 1980s – a bit of an understatement – but caught a swiftly rising American tide of low-carb sentiment in the late 1990s and enjoyed an enormous surge of popularity, probably peaking in the summer of 2002, when Taubes's *New York Times Magazine* article was published. As late as September 2004, carb-dependent companies were facing enormous financial pressures from the low-carb craze: Interstate Bakeries (the makers of Wonder Bread and Twinkies) filed for Chapter 11 bankruptcy, citing pressure from the Atkins Diet. But from the middle of 2004, journalistic pieces began to appear announcing that the low-carb phenomenon had either passed its peak or had reached a plateau (e.g., Warner 2004). A year-end review announced that 'it was probably inevitable that the year that started with a boom in low-carb diets would end in a bust in low-carb diets' (Zernicke 2004). The piece described mounting inventories of unsold low-carb foods in manufacturers' warehouses and speculatively attributed the demise of the Atkins diet to a combination of health warnings, ineffectiveness, resulting bad breath, and the fact that it was just too boring. By spring 2005, there was solid evidence that the low-carb fad had passed its peak, and newspaper pieces appeared with such headlines as 'Ding Dong, the Craze Is Dead' (Yonan 2005). That article claimed that the number of Americans on a low-carb diet had declined from a February 2004 peak of 9.1 per cent to 2.4 per cent in December 2004, before recovering to 4.4 per cent in March 2005. And on 31 July 2005 Atkins Nutritionals, Inc. itself filed for Chapter 11 bankruptcy protection, a last-ditch effort to save the company, but a clear sign that the end was nigh. Accordingly, a British television news agency announced that the Atkins 'craze' was 'officially' over ('The Atkins

Diet Crashes' 2005). The company at the same time announced its intention to refocus on nutrition bars and drinks, becoming a smaller firm not specifically orientated towards weight loss but catering 'more broadly for consumers who are concerned about health and wellness' (Warner 2005b: A12; LeSure 2005: A5). At the time, a food industry consulting firm concluded, 'Now there is almost no market for low-carb food products' (Bob Goldin, of Technomic Inc., quoted in Hirsch 2005: A1). Early in 2006, the company emerged from bankruptcy with a new management team – much of which came from the French food and bottled-water group Danone – aiming to compete in the highly competitive nutrition bar and shakes market (Grant 2006: 13).

8 The best source for the history of dietary fashion is Schwartz (1986). Modern low-carb fashion is related to changing sensibilities about personal responsibility in Shapin (2004). Bentley (2004) offers a concise summary of the current state-of-the-low-carb art, while making the valuable suggestion that 'Atkins allows men to feel comfortable admitting in public that they are dieting' and observing that 'companies such as Atkins Nutritionals ... appear to be courting the male dieter with resolve' (35).

9 I gesture here at Socrates' rejection of the very idea of a science of the particular, as, for example, in the *Phaedrus* (249c): 'For a man must have intelligence of universals, and be able to proceed from the many particulars of sense to one conception of reason; – this is the recollection of those things which our soul once saw while following God – when regardless of that which we now call being she raised her head up towards the true being. And therefore the mind of the philosopher alone has wings.' Jacques Lacan referred to psychoanalysis as 'the science of the particular' (Lacan 1954).

10 *Webster's 3rd Unabridged*, definition 3a: 'accumulated and accepted knowledge that has been systematized and formulated with reference to the discovery of general truths or the operation of general laws.' *Webster's 10th Unabridged*, definition 3a: 'knowledge or a system of knowledge covering general truths or the operation of general laws.' *Oxford English Dictionary*, definition 4a: 'A branch of study which is concerned either with a connected body of demonstrated truths or with observed facts systematically classified and more or less colligated by being brought under general laws.'

11 I focus attention on Dr Willett's work because he is a well-known nutritional expert, writing from the centre of academic authority, who has been dedicated to addressing a wider public and who has attempted to meet that public on what he takes to be its own terms. This is a well-written, well-organized, and passionate book, and if I seem to criticize some of its assumptions and expository conventions, it is because it is one of the best

representatives of a genre of academic expert writing. Needless to say, I pretend to no technical expertise in dietetics and can have no opinion on scientific claims and practical outcomes.

12 For the history of casuistry in moral reasoning, see Jonsen and Toulmin (1988).

13 I except the clinical journal literature, where discussions of clinical cases are the norm.

14 A study of René Descartes's medical counsel from this point of view is Shapin (2000); see also Shapin (2003b). Of course, there are reasons you might wish to take the advice of obese physicians – because they might be thought to understand your predicament – and this was indeed the case with many of Cheyne's patients.

15 Willett (2001a: 11–13) tells readers briefly about the author's Michigan dairy farming family, his membership in the 4-H club, and his victory in a National Junior Vegetable Growers Association competition, but I marked only two brief mentions of his own dietary habits – first, an assurance that he and his Harvard colleagues have 'substantially modified our activity levels and diet' (13) as a result of their expert findings, and, later, in the recipe section, some remarks about his 'Busy Day Menu': some left-over Kashi with bits and pieces added: 'I put all this in my backpack, dash out the door with my bike, and I'm in my office at the Harvard Medical area in fifteen minutes ... In the evening, if I'm lucky, my wife, Gail, may have made one of the entreés from this book, or one of countless other healthy creations, for dinner' (213–14).

REFERENCES

Agatston, A. 2003. *The South Beach Diet: The Delicious, Doctor-Designed, Foolproof Plan for Fast and Healthy Weight Loss.* New York: Random House.

Atkins, R.C. 1972. *Dr Atkins' Diet Revolution: The High Calorie Way to Stay Thin Forever.* New York: David McKay.

– 1992. *Dr Atkins' New Diet Revolution.* New York: Avon Books.

– 2003. *Atkins for Life: The Complete Controlled Carb Program for Permanent Weight Loss and Good Health.* New York: St Martin's Press.

The Atkins Diet Crashes. 2005. ITV.com, 1 August. http://www.itv.com/news/world_1987245.html (accessed 5 September 2005).

Barnes, B. 1982. *About Science.* Oxford: Basil Blackwell.

Bentley, A. 2004. The Other Atkins Revolution: Atkins and the Shifting Culture of Dieting. *Gastronomica: The Journal of Food and Culture* 4, no. 3: 34–45.

Buckley, N. 2004. Battle Rages over How to Build the New Food Pyramid. *Financial Times*, 19 August.

Burros, M. 2005. U.S. Introduces a Revised Food Pyramid. *The New York Times*, 20 April.

Collins H., and T. Pinch. 1998a [1993]. *The Golem: What You Should Know about Science*, 2nd ed. Cambridge: Cambridge University Press.

Collins, H., and T. Pinch. 1998b. *The Golem at Large: What You Should Know about Technology*. Cambridge: Cambridge University Press, 1998.

Galileo Galilei. 1967 [1632]. *Dialogue Concerning Two Chief World Systems*, 2nd ed. Translated by Stillman Drake. Berkeley: University of California Press.

Gigerenzer, G. 2003. *Calculated Risks: How to Know When Numbers Deceive You*. New York: Simon and Schuster. Published in the United Kingdom as *Reckoning with Risk: Learning to Live with Uncertainty*. London: Penguin.

Grant, Jeremy. 2006. Atkins' Quest to Put the Lean Years Behind It. *Financial Times*, 13 April.

Guerrini, A. 2000. *Obesity and Depression in the Enlightenment: The Life and Times of George Cheyne*. Norman: University of Oklahoma Press.

Helmich, N. 2002. Scales Tip in Favor of a New Food Pyramid. *USA Today*, 4 November. http://www.usatoday.com/news/health/2002-11-04-food-pyramid_x.htm (accessed 28 September 2004).

Hilgartner, S. 2000. *Science on Stage: Expert Advice as Public Drama*. Stanford, CA: Stanford University Press.

Hirsch, Jerry. 2005. The Nation: Carb Supporters Rejoice as Atkins Goes Belly Up. *Los Angeles Times*, 2 August.

Jewson, N.D. 1974. Medical Knowledge and the Patronage System in Eighteenth-Century England. *Sociology* 8: 369–85.

Jonsen, Albert R., and Stephen Toulmin. 1988. *The Abuse of Casuistry: A History of Moral Reasoning*. Berkeley: University of California Press.

Keys, A., and M. Keys. 1959. *Eat Well and Stay Well*. Garden City, NY: Doubleday.

Kleinfield, N.R. 2004. Just What Killed the Diet Doctor, and What Keeps the Issue Alive? *The New York Times*, 11 February.

Lacan, Jacques. 1954. *Le Séminaire de Jacques Lacan*. Seminar I, 20 January. http://perso.numericable.fr/~desylvie/sem01/03Sem1.htm (accessed 12 October 2004).

LeSure, Elizabeth. 2005. Low-Carb Company Atkins Files for Bankruptcy. *Boston Globe*, 1 August.

Marks, H.M. 1997. *The Progress of Experiment: Science and Therapeutic Reform in the United States, 1900–1990*. Cambridge: Cambridge University Press.

Martin, P. 2003. Shaping Our Diet. *Spiked*, 28 October. http://www.spiked-online.com/Printable/00000006DF9C.htm (accessed 28 September 2004).

Mazel, J. 1981. *The Beverly Hills Diet.* New York: Macmillan.

McLaughlin, K. 2004. Report Details Dr Atkins' Health Problems. *Wall Street Journal*, 10 February.

Montaigne, M.E. de. 1965 [1580, 1588]. Of Experience. In *The Complete Essays of Montaigne*, translated by Donald M. Frame, 815–57. Stanford, CA: Stanford University Press.

Nestle, M. 2002. *Food Politics: How the Food Industry Influences Nutrition and Health.* Berkeley: University of California Press.

Rees, M. 2005. Think Big, Like Einstein. *Connected.Telegraph*, 26 January. http://www.telegraph.co.uk/connected/main.jhtml?xml=/connected/2005/01/26/ecrein26.xml&sSheet=/connected/2005/01/26/ixconnrite.html (accessed 23 April 2007).

Schwartz, Hillel. 1986. *Never Satisfied: A Cultural History of Diets, Fantasies and Fat.* New York: Doubleday Anchor.

Severson, Kim. 2005. The Government's Pyramid Scheme. *The New York Times*, 24 April.

Shapin, S. 1992. Why the Public Ought to Understand Science-in-the Making. *Public Understanding of Science* 1 (1992): 27–30.

– 1995. Cordelia's Love: Credibility and the Social Studies of Science. *Perspectives on Science* 3: 255–75.

– 2000. Descartes the Doctor: Rationalism and Its Therapies. *The British Journal for the History of Science* 33: 131–54

– 2002. Barbecue of the Vanities. *London Review of Books*, 22 August. http://books.guardian.co.uk/lrb/articles/0,6109,777310,00.html (accessed 23 April 2007).

– 2003a. How to Eat Like a Gentleman: Dietetics and Ethics in Early Modern England. In *Right Living: An Anglo-American Tradition of Self-Help Medicine and Hygiene*, edited by Charles E. Rosenberg, 21–58. Baltimore, MD: Johns Hopkins University Press.

– 2003b. Trusting George Cheyne: Scientific Expertise, Common Sense, and Moral Authority in Early Eighteenth-Century Dietetic Medicine. *Bulletin of the History of Medicine* 77: 263–97.

– 2004. The Great Neurotic Art. *London Review of Books*, 4 August, 14–16. http://www.lrb.co.uk/v26/n15/shap01_.html (accessed 23 April 2007).

Tarnower, H., and S.S. Baker. 1979. *The Complete Scarsdale Medical Diet, Plus Dr Tarnower's Lifetime Keep-Slim Program.* New York: Bantam.

Taubes, G. 2002. What If It's All Been a Big, Fat Lie? *New York Times Magazine*, 7 July. http://query.nytimes.com/gst/fullpage.html?sec=health&res=9F04E2D61F3EF934A35754C0A9649C8B63 (accessed 23 April 2007).

Tribole, E., and E. Resch. 1995. *Intuitive Eating: A Revolutionary Program That Works*. New York: St Martin's Press.

Tolstoy, L. 1937 [1899]. Modern Science. In *Recollections and Essays*, translated by Aylmer Maude, 176–87. London: Oxford University Press.

Warner, Melanie. 2004. Is the Low-Carb Boom Over? *The New York Times*, 5 December.

– 2005a. Atkins Diet Saga Now at Chapter 11. *The New York Times*, 3 August.

– 2005b. Atkins Files Bankruptcy Petition. *The New York Times*, 1 August.

Weber, M. 1991 [1918]. Science as a Vocation. In *From Max Weber: Essays in Sociology*, edited by H.H. Gerth and C. Wright Mills, 129–56. London: Routledge.

Willett, W.C. 2001a. *Eat, Drink, and Be Healthy: The Harvard Medical School Guide to Healthy Eating*. New York: Free Press.

– 2001b. Nutrition Book Author Willett Rebuilds USDA Food Pyramid. *Harvard Public Health NOW*, 24 August. http://www.hsph.harvard.edu/now/aug24/ (accessed 27 April 2007).

Willett, W.C., and M.J. Stampfer. 2003. Rebuilding the Food Pyramid. *Scientific American Exclusive Online Issue* (January): 19–24. http://www.sciam.com/article.cfm?articleID=0007C5B6-7152-1DF6-9733809EC588EEDF (accessed 23 April 2007).

Yonan, Joe. 2005. Ding Dong, the Craze Is Dead. *Boston Globe*, 30 March.

Zernicke, Kate. 2004. Fatkins. *The New York Times*, 26 December.

10 The Role of the Public Academic Scientist in the Twenty-first Century: Who Is Protecting the Public Interest?

ALAN McHUGHEN

For much of the twentieth century, scientists working in public universities and research institutes provided a pool of experts capable of offering specialized expertise on scientific and technical issues and advice on how developments in those areas might affect the public good. Such scientists were, in general, held in high esteem as protectors of the public interest in scientific and technical matters and were sought out as expert resources for public policy formulation. Scientists at public universities and other institutions conducted scientific experiments and the results were disseminated at conferences and in technical literature, where the results became 'public domain.'

In the final years of the century, however, as research grants from taxpayer-funded sources diminished, academic scientists were increasingly encouraged, if not required, to seek research funding from private sources. Taxpayers were increasingly unwilling to foot the bill for vague, curiosity-driven research, which was often seen as supporting grown men (and few women) playing in a lab with no true purpose or apparent accountability. The alternative source of support, private funds, typically came with various strings attached; no more curiosity-driven research, and no more 'public domain' depositions of research findings. Instead the research was driven by the private research agenda, which generally sought an applied, and intellectual property protected, endpoint.

A second factor, at least in the United States, was the Bayh-Dole Act of 1980. In this act, universities and other institutions conducting research projects supported by federal funds became entitled to patent the findings and license the inventions exclusively (Thursby and Thursby 2003). The result of Bayh-Dole was a dramatic 238 per cent increase in patent applications from universities between 1991 and 2000 and a whopping 520 per cent increase in royalty stream in the same period (AUTM 2003). Was Bayh-Dole distracting U.S. public scientists from working on public-good research and diverting them instead to pursue more lucrative objectives? Perhaps more importantly, were patents and other intellectual property restrictions impeding the application of research findings for the public good? Roger Beachy cites the hindered development and deployment of beta carotene–enhanced 'Golden Rice' to vitamin A–deficient diets in impoverished nations due (among other things) to '16 important patents and 72 potential IP barriers' (Beachy 2003). Not all of these were from public institutions, of course, but the fact remains that research and development in plant breeding, particularly, has suffered from the increased presence of IP barriers. In Canada, the Plant Breeders' Rights Act of 1990[1] has chilled the traditional practice of plant breeders to exchange seeds and breeding lines. Similar legislation around the world has made plant breeding a morass of paperwork and legal constraints: MOUs, FTOs, MTAs and others form an alphabet soup of acronyms restricting usage of seeds and genotypes that at one time would be freely provided.

As research funding from private sources comprised an increasing proportion of total funds available to the scientist, and as public scientists increasingly sought patent and other intellectual property protection instead of public domain depositions for their findings, the image of the 'unbiased' public scientist as champion of the public good began to tarnish. This development provoked many questions, among them:

1. Can science retain independence and objectivity in the face of demands to meet commercial and public policy objectives?
2. In what ways is scientific discourse privileged in the formation of public policy?
3. How can scientific knowledge and scientific methodology be made compatible with the interdisciplinarity and integration required in public policy discourse and formation?
4. Can public scientists, whose salary is paid from taxpayer funds but whose research is supported privately, still be considered credible champions of the public good? If not, who is providing the highly technical expertise and undivided loyalty to the public?
5. Are public scientists still seen as resources and advocates of the public good? What are the consequences of losing public scientists as a public utility?

Is Scientific Fact Politically Malleable?

Humans can observe and even use the laws of nature (for example, to generate hydroelectric power) but cannot modify or rescind them – consider King Canute's attempt to turn the tides[2] or the apocryphal story of the Indiana legislature simplifying *Pi* to 3.2.[3] But we need not invoke historical allegories: we have current examples of electorates who have decided they do not like nature's facts and try to alter them to a political end. These modern attempts to impose political values on science invariably turn out wrong and establish the populace as scientific illiterates – or worse. For example, California's Mendocino County passed a ballot initiative in 2004 to ban GMOs and attempted to redefine DNA – the physical manifestation of all biological genetic material – as a protein.[4] During the campaign, a number of assertions appeared in local outlets, in letters to the editor, in regular columns by journalists in local papers, as well as in public forums. Among the claims, all unjustified and unsubstantiated yet mostly unchallenged, were the following: 'Federal agencies give almost no oversight to genetically altered organisms'; 'Gene-spliced salmon have deformed organs.' 'Unstable genes cause mass failure of pesticide-resistant crops'; 'Cows treated with genetically altered growth hormone develop multiple diseases'; 'Infant allergies to soy formula have risen by 50 per cent since widespread use of genetically engineered soy'; '24 people died from eating Starlink corn'; 'A German farmer fed GMO to his cattle; 12 died'; 'GMO canola is toxic; mutated into superweeds'; 'Most GMOs are sterile and have a suicide gene.' Fortunately (or unfortunately, for the majority of voters of Mendocino County), scientific fact is not subject to democracy or the whim of popular opinion.

Scientific illiteracy is not the exclusive domain of the People's Republic of Northern California. In the Eurobarometer survey from 1996, for example, only 45 per cent of Europeans agreed that cloning makes genetically identical offspring, less than half knew that eating a genetically engineered fruit does not alter the person's own genes, and only a third disagreed with the statement 'Ordinary tomatoes do not contain genes while genetically modified tomatoes do' (Durant, Bauer, and Gaskell 1998). Even in western Europe, many suffer under scientifically incorrect beliefs and misunderstandings, a likely cause for the assumptions that ultimately drive European public policy on biotechnology.

Examples like these prove the necessity of having trusted public scientists inform public debate involving scientific issues. This does not mean advocacy for one side or the other, but factual, technically accurate information – even if it is as simple as pointing out that DNA is not now, and cannot ever, even under democratic legislation, be a protein, or that ordinary tomatoes do indeed contain genes.

Unfortunately, there are people with PhDs who present themselves as 'scientists' and argue in a very unscientific manner. Such people can improperly influence the public by asserting science when they actually espouse a political stand. When another PhD holder enters the debate and advocates the opposite position, the voters become confused and increasingly doubtful of the non-partisan objectivity of science and scientists.

But science should not be 'he said/she said,' should it? The scientific method of answering questions is alien to anyone with a particular agenda. A true scientist, public or private, approaches a question with no *a priori* desired conclusion. A scientifically sound answer is satisfying enough, regardless of whether that sound answer supports one side or the other.

I would argue that a non-scientific approach would start with a conclusion, search for evidence to support it (cherry picking), discredit alternative views, and often lack context. In contrast, the scientific approach would lead individuals to collect and analyse all available evidence before (perhaps) reaching a conclusion, actively seek alternative interpretations, and involve active critical thinking skills and critical assessment of results.

Any reputable scientist, and all reputable scientific societies, will follow the latter format. Nevertheless, all communities suffer their fringe, and scientists are not immune. The scientific community has its own professional societies, and these effectively screen out the extremists and dilute the envelope-pushing fringe. This is why a report by, for example the American Medical Association or the National Academy of Sciences, carries far more influence than any report from an individual scientist.

There is a popular concern about the safety of foods derived from genetically engineered crops. Some partisan groups have at the outset declared their opposition to such foods, claiming that they present an inherent safety hazard but without offering any supporting evidence. In contrast, and exemplified by the National Academy of Sciences (NAS) and Institute of Medicine (IOM), a panel was struck with the objective to research the safety of genetically engineered foods, to collect and analyse all relevant information, and to make recommendations if warranted by the evidence. There was no prior intent to reach any particular conclusion, the table was open for all interested parties to contribute evidence, and the committee evaluated all submitted data before deliberating on the findings (NAS/IOM 2004).

The committee was drawn almost entirely from public institutions, in order to minimize criticisms of bias, and the NAS goes to great lengths to ensure balance on committees. Indeed, there is a public comment process to allow those with undue bias to be identified and removed from the committee. As a result, the findings and recommendations from such studies enjoy a high degree of respect among the public and the scientific and regulatory community, even if they contradict 'preferred,' if not entrenched, political positions. Other scientific societies have similar, if less rigorous, operating rules to ensure the public good is protected both from conflict of interest and from incompetence. As a result, when policy analyses are issued by NAS, IOM, or other scientific societies, the degree of public trust and credibility remains high. This is fine for purely scientific studies, but what about for policy issues?

Government regulators face increasing scrutiny from NGOs across the political spectrum. It is common to hear concern voiced about, say, the FDA being influenced by drug companies. The assumption is that any regulator should be working purely for the public good and making regulatory decisions based exclusively on that premise. In recent years, NGOs have complained that supposedly 'pure' civil servants may be tainted, perhaps because they previously worked for the drug company whose products the regulator is now assessing. Is there a conflict of interest?

As explored in McHughen (2000), those people best able to evaluate safety and efficacy of a new product or technology are usually the ones with relevant industry connections, either through previous training or employment; they are exactly the ones raising the greatest concerns over divided loyalties. Employing only those who have no such industrial taint or background ensures that the public is being served by the people with lesser expertise. Scientists who have never been paid by industry may superficially seem to be the least likely to suffer conflict of interest, but they are also the least likely to have the specialized skills and experience to protect the public interest by asking the right questions and properly analysing the evidence.

A nasty public bloodletting in the medical scientific community arose in Toronto, in the case of Dr Nancy Olivieri, who was researching a particular drug at the University of Toronto and the Hospital for Sick Children. The research was funded by the drug developer Apotex. When her experiments revealed possible nasty side effects, her academic duty was to report the findings through the standard academic channel: peer review publication. However, Dr Olivieri had signed a confidentiality agreement with the company, who promptly threatened to sue her if she released the findings without their approval, and, considering how the news would affect commercial prospects for the drug, they were unlikely to approve such publicity. Defiantly, Dr Olivieri published (in the prestigious *New England Journal of Medicine*) (Olivieri et al. 1998) and was damned, demoted, and later fired outright. The drug company struck a committee to review Dr Olivieri's results; the committee disagreed with her interpretations. At the same time, the company was dangling a multimillion-dollar donation for a new medical building at the University of Toronto. To mitigate the bad press, the Hospital for Sick Children established a review board, headed by a prominent and respected academic, Arnold Naimark, former president of the University of Manitoba. Unfortunately, Dr Naimark was also 'tainted,' having accepted a donation for his university from the same company some years earlier. Eventually, an independent panel of international experts of unimpeachable record was struck; their review concluded that Dr Olivieri was fulfilling her role as a public scientist, that she had been denied her academic freedom, and that she should be reinstated (Jeffrey 1997; Turk 1999). But larger questions remain. Apart from the specific merits of this case, we must consider the policy implications of public scientists conducting research funded by private interests.

While it appears that Dr Olivieri was vindicated in her dispute (Thompson, Baird, and Downie 2001), the drug in question was approved for release in Europe over Dr Olivieri's objections (Apotex 2003). To casual observers like me, it appears that Dr Olivieri had indeed noted potential problems with the drug and properly reported her data but was vilified by the sponsoring company. At the same time, the company, citing the need to maintain confidentiality for competitive reasons and waving the confidentiality agreement signed by Dr Olivieri, also understandably did not appreciate any bad publicity about their potentially lucrative new product. In the final analysis, the drug is on the market and presumably helping patients in spite of the occasional difficulties documented by Dr Olivieri. Most drugs have limitations, contraindications, side effects, and other undesirable features, yet still provide relief to suffering patients. Clearly, Dr Olivieri suffered divided loyalties: to warn the public of the potential hazards as she saw them, on the one hand, and to honour her contractual commitment to confidentiality, on the other.

The difficulty in subjectively judging such whistleblowers is that the revealed hazards may emanate from a single source or perspective, but when placed into the context of the 'big picture' the product may still provide value to society (not to mention the company). In this case, Dr Olivieri may well have legitimately found a problem with the drug, but when the hazard was ultimately considered in the context of value, it was insufficient to kill the product. This situation differs, of course, from that in which whistleblowers get their facts or interpretations wrong and cause unnecessary anxiety and fear.

A Question of Trust

Science and technology are increasingly impacting modern life – from computers, cell phones, and entertainment, to genetic applications in medicine and even food. Not only is science becoming more pervasive, it is becoming more complex – witness the relative simplicity of the traditional telephone versus the modern electronics in today's cell phones, or the difference between traditional Mendelian genetics in crop breeding versus modern molecular DNA modifications to our foods. Even if ordinary consumers were supplied with full and complete information about these technologies and their risks, they lack the training (not to mention time) to evaluate and decide for themselves whether to support or reject a new product. Instead, most people are prepared to defer those risk assessments to others, to trust the experts (Luhmann 1979).

Traditional, public-supported academic scientists were trusted implicitly – when they had anything at all to say in public. Consider the stereotypical scientist: brilliant and kind hearted but absent-minded and politically naive, easily duped into supporting an evil agenda. With the shift to private support for public academics, the taxpayer may have saved money by ceasing to support academic scientists, but the cost was the loss – or at least the perceived loss – of a cadre of trusted, unbiased experts in technological fields who worked exclusively for the public good.

Discrediting the scientists' arguments, the legitimate form of dispute discourse, rarely succeeds these days, as that requires some technical knowledge; so the tactic now is to bypass the content and go straight to discrediting the scientist. These ad hominem attacks are both more effective and more exciting to onlookers than a dispassionate analysis of the data. As a result of this increasingly widespread practice, academic scientists in general are losing public trust and credibility.

If an academic scientist appears in a public forum to express an opinion or even to relay factual, non-partisan information, those on the other side of the discussion attempt to discredit the scientist, usually by charging a conflict based on industry funding. Recently, University of California at Berkeley, one of the premier public universities anywhere, was painted by a broad and discrediting brush. Because Berkeley had accepted industry funding from Novartis, the public should disregard anything any Berkeley faculty member said or wrote (Dr Peggy G. Lemaux, University of California, Berkeley, pers. comm. 2004). So even those Berkeley faculty having no gain from the industry connection are unjustly caught in the trap and suffered a loss of public trust and respect.

What about other scientists charged with protecting the public interest? Scientists working as regulators in government agencies are in general technically well trained and have an obligation to use their scientific skills to protect the society and the environment in discharging public policy. But is there public trust? Interestingly, this depends on the agency and the nation. In Europe, public trust in government agencies is extremely low, usually attributed to the technical mistakes and loss of public confidence over a series of food safety scandals, including the 'mad cow' (BSE) outbreak, *Salmonella*-contaminated poultry products, *E.coli* in meat, and other nasty and natural disasters ordinarily controllable by competent regulation and regulators. In a Eurobarometer survey on safety of biotechnology food products, only 4 per cent of Europeans said government agencies were trustworthy (Eurobarometer 2005). In contrast, 90 per cent of Americans trust USDA, followed closely by FDA at 84 per cent (Gaskell et al. 1999). Both of these U.S. agencies have an enviable track record of protecting the American public from the kinds of safety fears suffered by European consumers.

Nevertheless, in spite of the degree of public trust and the paucity of mistakes from USDA and FDA, charges are frequent that U.S. government scientists are untrustworthy because of private sector influence over government appointments and decision making. Ideally, scientific data should be argued based on scientific criteria, but too often the reality is an attack on the scientist or institution. Even the most loyal and capable scientific champion of the public good can be discredited and libelled without evidence, simply by accusations of some tenuous and unsubstantiated 'industry connection.'

Conclusion

Whom should we trust to evaluate public policy that involves highly technical, scientifically specialized issues? We discount any industry-based scientist because, the argument goes – ordinarily without any supporting evidence – industry has no hesitation in manipulating data to support their products. And public academic scientists are also suspect because, even if a particular professor accepts no industry funds, his or her university almost certainly does, and the university is not going to bite the hand that feeds it. Scientists working in regulatory agencies should champion the public good, but even here, it is argued that industry has too much influence over regulatory decisions; that such decisions are not entirely objective and unbiased – witness the frequent flow of personnel from industry to the agencies. This reasoning is appealing to many, although evidence of improper industry influence or favour is rare.

But now we have eliminated virtually everyone with scientific competence to evaluate a technology or product with the public good in mind. With human society increasingly dependent on science and technology to provide food, shelter, labour, entertainment, and transport, whom can we trust to champion our safety and welfare?

Science and scientists have come a long way from the historical days of patronage support and, more recently, exclusive popular (taxpayer) support for research programs. Taxpayers still provide salary support for scientists in public universities and research institutes, but the expense to conduct experiments – equipment, supplies, technical staff, and so forth – is largely borne by private sources. It is this influence from private sources that leads to concern over the unbiased objectivity of the research and the researcher, rightly or wrongly tarnishing the image of the public scientist as a loyal champion of the public good. While I am not aware of any reputable scientist who has succumbed to external pressure to influence research findings, the perception of the opportunity for misconduct is real and must be taken into account.

Unfortunately, this is the price society pays when public research is supported by private funds. In the current political climate in Western democracies, taxpayers are unlikely to revert to the earlier system in which the public paid the entire scientific research bill. If anything, it appears taxpayers are contributing less to academic research, thus forcing public scientists more into dependency on private support for research projects. The public purse may feel money saved, but the public should be aware that the saving comes at the price of a perceived, if not real, loss of an expert champion of the public good.

NOTES

1 Available online at http://laws.justice.gc.ca/en/P-14.6/.
2 See, for example, http://www.viking.no/e/people/e-knud.htm.
3 Debated but failed second reading in Senate, 1897. See http://db.uwaterloo.ca/~alopez-o/math-faq/mathtext/node18.html.
4 See http://www.gmofreemendo.com/moreh.html#proposed for the exact wording of Measure H.

REFERENCES

Apotex. 2003. European Court Rules against Blood Doctor, Nancy Olivieri, in Her Challenge of the Marketing Authorization of Ferriprox® (Deferiprone). Decision Ensures Patient Access to Life-Saving Treatment. Press release, 18 December.

Association of University Technology Managers (AUTM). 2003. AUTM Licensing Survey, FY 1991 and FY 2001. http://www.autm.net/surveys/01/01summarypublicversion.pdf.

Beachy, R. 2003. IP Policies and Serving the Public. *Science* 299: 473.

Durant, J., M.W. Bauer, and G. Gaskell, 1998. *Biotechnology in the Public Sphere: A European Sourcebook*. London: Science Museum.

Gaskell, G., M.W. Bauer, J. Durant, and N.C. Allum. 1999. Worlds Apart? The Reception of Genetically Modified Foods in Europe and the U.S. *Science* 285: 386.

Gaskell, G., S. Stares, A Allansdottir, N. Allum, C. Corchero, C. Fischler, J. Hampel, J. Jackson, N. Kronberger, N. Mejlgaard, G. Revuelta, C. Schreiner, H. Torgersen, and W. Wagner. 2006. *Europeans and Biotechnology in 2005: Patterns and Trends*. Special Eurobarometer 244b: Wave 64.3. http://ec.europa.eu/public_opinion/archives/ebs/ebs_244b_en.pdf.

Jeffrey, S. 1997. Doctor May Be in Legal Battle after Reporting Negative Findings of Drug Company Study. *The Medical Post*, 21 January.

Luhmann, N. 1979 *Trust and Power*. Chichester, UK: Wiley.

McHughen, A. 2000. The Regulations of GM Foods: Who Represents the Public Interest? *International Journal* (The Canadian Institute for International Affairs) (Summer): 624–32.

NAS/IOM. 2004. *Safety of Genetically Engineered Foods: Approaches to Assessing Unintended Health Effects*. Washington, DC: National Academies Press.

Olivieri, N.F., G.M. Brittenham, C.E. McLaren, D.M. Templeton, R.G. Cameron, R.A. McClelland, et al. 1998. Long-Term Safety and Effectiveness of Iron-chelation Therapy with Deferiprone for Thalassemia Major. *New England Journal of Medicine* 339, no. 7: 417–23.

Thompson, J., P. Baird, and J. Downie. 2001. *Report of the Committee of Inquiry on the Case Involving Dr. Nancy Olivieri, the Hospital for Sick Children, the University of Toronto, and Apotex, Inc.* Ottawa: Canadian Association of University Teachers.

Thursby, J.G., and M.C. Thursby. 2003. Enhanced: University Licensing and the Bayh-Dole Act. *Science* 301, no. 5636 (22 August): 1052.

Turk, J. 1999. The Greatest Academic Scandal of Our Era. *Interview.* University of Calgary Faculty Association newsletter, May.

11 The Science Literacy Gap: Enabling Society to Critically Evaluate New Scientific Developments

ERIC S. SACHS

The pursuit of science and knowledge is generally regarded as essential for innovation and technology advancement. Scientific progress has served society's needs well throughout history, sometimes with associated risks and health or environmental consequences, but broadly speaking the impacts have been positive (Mokyr 2002). But with advancement comes change. Change fosters fear. Fear generates headlines. Facts become blurred by beliefs, and beliefs become powered by politics. Fear exploited by ideological groups is a proven weapon against innovation. Science today is a battleground, and society desperately needs to become better equipped to understand and make judgments about new technologies. But how does this happen? Where does society turn for information?

Information sources are abundant – television, radio, newspapers, magazines, state and government officials, universities, and the internet all offer information. But how good is the information? Is it factual? Does it provide a complete context? Are the sources impartial and credible? These are legitimate questions, and, for many, the answers are troubling.

The fact is that advancements in science are proceeding at a rapid pace, and the public is not equipped to discern fact from fiction in the media. Opponents use an arsenal of weaponry, including incomplete or misleading information, false and unsupported statements, opinions masquerading as facts, and facts taken out of context – together they tell a tale of gloom and doom, igniting fears and exposing doubts. Journalists compound this problem by focusing on alleged risks and presenting as fact unsubstantiated opinions by special interest groups. But the problem cannot be laid only at the feet of journalists – the problem is exacerbated by a lack of science literacy among the public and a lack of engagement and input from the scientific community. Science literacy in society, in the schools, in the kitchen, must become a priority.

The solution to the science literacy problem is for more scientists to serve society by engaging the media and playing a principal role in increasing the public's understanding of science and technology. Scientists must accept more responsibility to serve public interests, to respond to false allegations with facts and knowledge, and to communicate in a manner that is understandable and informative. They must acknowledge legitimate technology risks and explain how scientists and regulatory authorities assess these risks prior to technology use. The net result will be a more informed public with improved critical-thinking skills that will enable sound decision making that is based on knowledge and reason rather than ignorance and fear.

Informing the Public about Science and Technology

In the face of uncertainty and potential risk, each of us examines the available information and uses our knowledge and experience to determine what is acceptable or unacceptable. Although society at large accepts risks everyday, the acceptability threshold for specific risks varies greatly from individual to individual. Europeans generally reject genetically engineered foods but accept nuclear power (Eurobarometer 2005). People fear flying in airplanes but drive cars. Some even prefer organic food and follow every meal with a cigarette. In spite of these inconsistencies, individuals make risk judgments and look at available information to form opinions and draw conclusions (Phillips, Smyth, and Kerr 2006).

Unfortunately, for most of society, scientific knowledge has expanded rapidly, and the discoveries of today were the science fiction of our youth. Most people lack a strong scientific foundation and can easily fall prey to unscientific information that is based on attitudes and beliefs rather than on scientific analysis and testing.[1] Journalists compound this problem by 'framing' stories, focusing on alleged risks, and presenting as fact unsubstantiated opinions by special interest groups.[2] Journalists turn a newsworthy event into a story and frame the content according to their own understandings, ideologies, styles, and practical limitations such as deadlines and space (Marks and Kalaitzandonakes 2001). Judging information and determining whether risks are exaggerated or real, given a poor understanding of science and framing by journalists, makes drawing scientifically literate conclusions difficult. Realistically, society will continue to look to the media for information and will evaluate and make judgments about new scientific developments and the associated risks and benefits based upon readily available information sources.

The Role of Journalists

Much of the information that is readily accessible and comprehensible is available through the media – newspapers, magazines, television, radio, and the Web. Journalists are trained to communicate information but they do not pass the information through a scientific lens; instead they strive to treat all information – opinions, beliefs, misinformation, and facts – with a degree of neutrality, presenting the public with various points of view and allowing them to draw conclusions. This process enables journalistic freedom but places the burden of responsible and critical analysis on society (Einsiedel 2006).

Scientists and journalists have a shared responsibility to help society to critically examine new technologies and make informed decisions. Hartz and Chappell (1997) state: 'The public – including the media – has grown less and less familiar with the basic tenets of science and technology, even as achievement in both areas has become more and more essential to modern life' (xii). Recognizing how most people judge science and technology risks and benefits by relying on a variety of media sources, it is imperative that scientists engage media and effectively communicate relevant and critical information to inform sound decision making. Only by stepping out of the research setting into society's classroom can scientists fulfill their mission to expand the knowledge base, increase public understanding of science and technology, allay unsubstantiated concerns and perceived risks, and enable society to critically evaluate new scientific developments. As scientists play a larger role in engaging journalists, there is a great opportunity to improve the quality of information delivered to the public and to better prepare the public to make scientifically literate decisions.

The Role of Scientists

While there is evidence that framing by the news media can shape public opinion, such allegations cannot and should not be laid only at the feet of journalists; the problem is exacerbated by a lack of science literacy among journalists and the public and, in particular, a lack of input from the scientific community. Scientists have a responsibility to inform society about the world of science. Advancements in science are proceeding at a rapid pace, and the broad scientific community needs to reach out and equip journalists to better serve the science and technology literacy needs of society.

Scientists at universities are generally regarded as objective, reliable sources of information on technology, including safety assessment, regulation, and potential societal impacts (Einsiedel, Finlay, and Arko 2000). While the public views the scientific community as credible and trustworthy, only a relatively few scientists participate in public discourse. It is ironic that the vast majority of scientists in universities and research institutes are focused on understanding the physical and biological world, breaking though scientific barriers, and making discoveries to develop and expand the use of technology but are uninvolved or unmotivated to communicate science to society. Public researchers are frustrated by the lack of public funding for scientific research but seem numb to the fact that, in part, shrinking funds are a direct result of an expanding science literacy gap in society and a growing misunderstanding of science and technology (Hartz and Chappell 1997).

Given the important role for scientists to prepare the public to make rational and informed decisions about new technologies, why are so many scientists staying on the sidelines? There are at least five reasons why scientists refrain from speaking to journalists: (1) scientists generally are afraid of being misquoted or embarrassed; (2) scientists prefer not to take sides or to become aligned with commercial or private interests; (3) scientists often are uncomfortable interpreting data, especially when the data are not from their area of expertise; (4) the nature of the scientific method requires that scientists recognize uncertainties and constraints on interpretation of findings, leading to equivocal statements that often weaken an otherwise strong scientific case; and (5) public education and interaction with media are not regarded as scholarly endeavours and do not lead to tenure and promotion. While all of these may be valid and legitimate reasons for staying on the sidelines, they lead to an outcome that does not serve society's interests. Change is essential – and the road to change often is difficult.

Engaging more scientists to address the science literacy needs of society will require fundamental changes in university policies, particularly relating to recognition and reward systems. In general, criteria for tenure advancement focus on scholarly research, publication, and relevant contributions to science and knowledge, not media interviews and education of non-science audiences. While this is an important hurdle, it is not insurmountable, so long as high-stature scientists take the societal issues seriously and make it a priority to challenge university decision makers. There also is a need to encourage university scientists to explore and address the barriers that limit broader engagement of science issues in society. Some scientists are deeply concerned with these issues, and discussions among senior scientists, both those engaged and not engaged, should help to elucidate key barriers and provide insights for ways to increase scientists' interface with the public.

The Case of Agricultural Biotechnology

The arena of agricultural biotechnology is fertile ground for controversy and regularly receives the attention of journalists. The stories generally focus on potential health or environmental risks, such as potential for allergenicity, effects on animal health, harm to monarch butterflies, gene flow, threats to biodiversity in centres of origin, insect and weed resistance, and impacts on pesticide use, to name only a few. Stories also target the adequacy of regulatory policies and decision making, questions about product benefits, alleged impacts on organic production systems, derived product labelling, and so on. These stories are framed around potential adverse effects and are relevant to public interests.

At the same time, journalists have focused less on the positive effects of agricultural biotechnology and the broad support of scientific organizations and international scientific bodies. There are literally hundreds of peer-reviewed publications that support the safety and positive economic and environmental impacts of biotech crops. Only rarely do these publications in scientific journals draw attention from media. There also is a host of organizations[3] and scientific bodies[4] that have examined potential and alleged risks of agricultural biotechnology, as well as the regulatory procedures for assessing safety. However, journalists often overlook authoritative findings of a 'reasonable certainty of no harm' or determinations of 'small and manageable risks.' Journalists argue this information lacks a 'news hook' or is of little interest to the public. Just as potential or alleged risks are relevant topics in the media, so are findings of safety and benefits.

In the case of agricultural biotechnology, the media have seized our attention but unfortunately have tended to focus more on the perceived risks than the demonstrated benefits. Marks and Kalaitzandonakes (2001) found that the UK media linked benefit information with catastrophic and memorable negative events (e.g., BSE, dioxin, listeria or *E. coli* food contamination). Regrettably, this can look like one step forward and two steps back. Media coverage in the United States was generally more positive, emphasizing potential benefits over risks. Marks (2001) points out that media coverage of agricultural biotechnology has had 'subtle, indirect, and cumulative' (153) effects over time and that the public tends to overestimate risks that receive a lot of media attention. She argues that 'open, clear, and transparent communication with the public about the risks and the benefits of agrobiotechnology seems to be the best way to proceed' (154). In other words, the job of journalists is to write about events and issues that are important to society, including relevant, factual information about the potential risks and benefits of agricultural biotechnology, and to clearly state the views of experts and interested stakeholders. In short, the role of journalists is to aptly tell us what to think about, not what to think (Marks and Kalaitzandonakes 2001).

Scientists, in both the public and private sectors, are active on several fronts in the agricultural biotechnology arena. They participate in basic research and technology development, evaluate the potential for technology risks and benefits, recommend science-based regulatory safety assessment methods and policies, oversee and conduct experimental product testing, participate in post-commercial testing and monitoring, validate current paradigms and practices through data gathering and analysis, and communicate all of this information in scientific forums, government and stakeholder workshops, publications, and through the media. There is little doubt that the scientific community is fully engaged in discovery, development, evaluation, regulation, validation, monitoring, and publication of new technologies, but there is a paucity of scientists that are actively engaged in communicating about new technologies to society through the media.

Despite a large number of scientists who are actively engaged in university research programs directed toward agricultural biotechnology, universities could better serve society by encouraging more researchers to communicate their scientific research to local communities. Such a paradigm shift not only would help to address the science literacy gap in society but also would demonstrate the important role that universities are playing in technology development and assessment in agricultural and food biotechnology, food security and hunger, nutrition and obesity, synthetic and plant-made pharmaceuticals, bioethics and social justice, as well as other topics of relevance to society.

Summary

Scientific exploration will continue to provide new solutions and new technologies to address current or future societal challenges. At the same time, knowledge and understanding must extend beyond the scientific realm to society at large, or else society may ignore or reject important technological innovations as a result of ignorance or poor-quality information. Scientists and journalists have a shared responsibility to help society to critically examine new technologies and make informed decisions. It is essential that both groups collaborate in explaining technological advancements to society, making it possible for the public to appreciate the benefits and risks of new or evolving technologies and to critically evaluate the available information, separating fact from fiction and truth from opinion, in order to draw legitimate and knowledge-based conclusions.

NOTES

1 The International Food Information Council (IFIC) (2006) asserts that many consumers are unaware of the nature of new technologies related to food.

2 The problem lies not really in the normal reporting of scientific developments but perhaps in the framing of those articles. Bubela and Caulfield (2002) found that only 11 per cent of a set of newspaper articles about genomics were categorized as having moderately to highly exaggerated claims; the majority were categorized as having no claims (63 per cent) or only slightly exaggerated claims (26 per cent). One related concern is that while the text of most stories was relatively balanced, the headlines for the stories often created a somewhat more hyped perception.

3 For example, American Medical Association, British Medical Association, Institute of Food Technologists, American Dietetic Association, Council for Agricultural Science and Technology, American Society of Plant Biologists, Federation of Animal Science Societies, The United Kingdom Royal Society.

4 For example, United Nations Food and Agriculture Organization, World Health Organization, Organization for Economic Co-operation and Development, International Life Sciences Institute.

REFERENCES

Bubela, T., and T. Caulfield. 2002. Do the Print Media 'Hype' Genetic Research? A Comparison of Newspaper Stories and Peer-Reviewed Research Papers. *Canadian Medical Association Journal* 170, no. 9: 1399.

Einsiedel, E. 2006. Telling Technology Tales: The Media and the Evolution of Biotechnology. In *Crossing Over: Genomics in the Public Arena*, edited by E. Einsiedel and F. Timmermans, 134–54. Calgary: University of Calgary Press.

Einsiedel, E., K. Finlay, and J. Arko. 2000. *Meeting the Public's Need for Information on Biotechnology.* Commissioned paper for CBAC. www.cbac-cccb.ca/documents/en/PublicNeed_Info_Einsiedel.pdf (accessed 26 June 2002).

Eurobarometer. 2006. *Europeans and Biotechnology in 2005: Patterns and Trends* (July). Eurobarometer 64.3, March. http://ec.europa.eu/Research/press/2006/pdf/pr1906_eb_64_3_final_report_May2006_en.pdf (accessed 16 April 2007).

Hartz, J., and R. Chappell. 1997. *Worlds Apart – How the Distance between Science and Journalism Threatens America's Future.* Nashville, TN: First Amendment Center. http://www.freedomforum.org/templates/document.asp?documentID=13649 (accessed 16 April 2007).

International Food Information Council (IFIC). 2006. *Food Biotechnology: A Study of U.S. Consumer Attitudinal Trends, 2006 Report* (November). http://www.ific.org/research/upload/2006%20Biotech%20Consumer%20Research%20Report.pdf (accessed 16 April 2007).

Marks, L. 2001. Communicating about Agrobiotechnology. *AgBioForum* 4, nos. 3 and 4: 152–4.

Marks, L.A., and N. Kalaitzandonakes. 2001. Mass Media Communications about Agrobiotechnology. *AgBioForum* 4, nos. 3 and 4: 199–208.

Mokyr, J. 2002. Innovation in an Historical Perspective: Tales of Technology and Evolution. In *Technological Innovation and Economic Performance*, edited by B. Steil, G. Victor, and R. Nelson, 23–46. Princeton: Princeton University Press.

Phillips, P., S. Smyth, and W. Kerr, eds. 2006. *Governing Risk in the 21st Century: Lessons from the World of Biotechnology.* New York: Nova Science Publishers.

SECTION THREE

Solutions to the Problems: Institutional

12 Science and Policymaking: The Legitimation Conundrum

GRACE SKOGSTAD AND SARAH HARTLEY

It is important to consider the contribution of science to public policy in liberal democracies. Does science have a privileged role in public policy? If it does, does such a role permit science to retain its 'independence and objectivity'? Both questions invite us to reflect on the appropriate relationship between science and democracy and the circumstances under which science can contribute to legitimate governing in liberal democracies. What image of science and what role for scientific knowledge in the public policy process will enhance legitimate policymaking?

There is no single answer to this question. Rather, societies divide within and across one another in their views of the authority of science and their beliefs regarding what constitutes an appropriate role for science in public policy processes. Canadians' collective views of the authority of science and its contribution to legitimate policymaking depart in some important respects from those of many Europeans, at least when it comes to the role of science in risk regulation. And yet, in a world of integrated markets, and the rapid diffusion of norms and values across national borders, the willingness of other political communities to grant authority to science, to recognize it as a legitimate basis of policymaking, bears on the status of science-based policymaking in Canadian governance. It also affects the capacity of Canadians to secure economic and trade goals through science-based international regulatory governance.

The comments which follow here are directed to the role and legitimacy of science in agri-food regulatory policies, and more particularly, those designed to regulate new technologies. The argument advanced is that the traditional authority that scientific knowledge has enjoyed as a basis of legitimate policymaking is eroding. Three developments in particular are currently undermining its authority. The first development is a political cultural shift in liberal democracies which has seen a decline in deference to authority and the elevation in importance of democratic norms of public participation, public deliberation, and government accountability. The second development is the emergence of new understandings of science that stress its limitations as a basis of decision making. The failure of regulatory science to protect human health in Europe, the privatization of science, and the development of technologies and products that pose unforeseen dangers and sometimes ethical dilemmas, all raise doubts about a straightforward equation of science with progress and undermine trust in science-based regulation. The third development is a counter-reaction to the institutionalization of science in international trade agreements, where science has been granted the authority of a neutral arbiter of the legality of domestic policies to safeguard the health of citizens. Collectively, these developments have diminished science's status as a legitimate basis of both domestic and international regulatory governance. This conundrum puts the onus on democratic governments like those in Canada to find ways to render more compatible democratic norms and science-based policymaking.

The argument that science as a source of legitimate policymaking is now under stress in liberal democracies, including Canada, is developed as follows. The first section discusses standards of legitimate policymaking in liberal democracies, outlining the inherent tension between science-based policymaking and democracy. Although science-based policymaking undermines procedural norms of public participation and deliberation, it usually contributes to more effective policy outcomes and, by extension, to the legitimacy of liberal democracies. The second section examines the Canadian tradition of science-based policymaking. It advances the claim that Canada's dominant political culture of liberalism and Canadian citizens' deference to authority, combined with our political institutional framework of executive-dominated government, lent legitimacy historically to expert, including scientific, authority. The dominance of science-based policymaking can be seen in agri-food regulatory policies, as can the consequent effect of narrowing the scope of these debates. Section three discusses how the legitimacy of science in policymaking has been destabilized and undermined by political cultural shifts that elevate democratic norms and create new understandings of science as a basis of policymaking. Section four considers the implications of these developments for Canada, with respect to both domestic decision-making processes and reasonable expectations of science-based international regulatory governance. It probes the possibility of strengthening the legitimacy of science-based policymaking both in Canada and at the multilateral level, drawing on the experience of the United Kingdom, where the challenge to the image of neutral science is considerably greater than in Canada.

Science, Democracy, and Legitimate Policymaking

Public policies must pass the test of legitimacy in order to be stable and willingly consented to by the public. Policies that have legitimacy are viewed as proper and appropriate, the outcome of decision-making procedures that conform to social norms and standards of acceptability. In liberal democracies, the legitimacy of public policies rests overwhelming on the belief that they originate in a democratic process. Although understandings of democracy differ across political communities, the common thread is that the people or their elected representatives are in control of decision making. In representative democracies, popular control requires that political representatives be accountable for their decisions, answering for them and taking responsibility when things go wrong. Accountability is, in turn, promoted by transparency of governing operations; opening the 'black box' of decision making enables citizens to track who makes decisions and on what basis. Citizens are thereby better positioned to hold to account those who rule on their behalf. In some democracies, norms that foster public deliberation and participation in policy debates are given added emphasis. Adherence to these democratic procedures confers procedural legitimacy on public policies and contributes to citizens' sense of obligation to respect them.

Science can also be granted authority when it is perceived to contribute to good governing in liberal democracies.[1] Specialized knowledge of the kind that professional experts and scientists possess is requisite to supplying valued public goods like safe food and a sustainable environment. Elsewhere, it improves the substantive quality of public policies. In technically complex policy domains, the 'everyday' knowledge of the public is not an effective replacement for the specialist knowledge of experts. Nor is that of the elected representative, whose expertise of technical matters is unlikely to exceed that of certain members of the public. Without the input of scientific or other expert knowledge, valued public goods will not be procured as effectively or at all. And in the absence of effective policy outcomes and policy failures, governments, no matter how democratic their functioning, will find their legitimacy questioned.

If science adds to the legitimacy of governments by improving the effectiveness of their policy outcomes and performance, it nonetheless is an anti-democratic source of legitimacy. Science as a basis of policy-making undermines democratic norms in some important respects. Democratic procedures by definition respect the preferences of citizens in policy outputs and provide opportunities for citizens to make those preferences known to decision makers. Both conditions are jeopardized when the judgments of scientifically trained experts trump citizen preferences, when problems are so narrowly defined that only those armed with scientific knowledge can participate, and/or when such a narrow problem definition excludes from the debate other non-technical dimensions of a policy issue. The likelihood of policy debates being constricted in this fashion is greater, scholars argue, when science is depicted as capable of yielding universal and neutral knowledge.

This image of scientific knowledge as universal and objective is, many would suggest, an ideal and unrealistic image of science. It portrays the scientific enterprise in the Mertonian fashion: one in which scientists assess competing claims on the basis of pre-established universal criteria, are accountable through a process of peer review to fellow experts, and approach problems with a detached scepticism that questions 'the taken for granted of society.' This positivist view does not deny that uncertainty arises in empirical science, but it assumes that when scientists adhere to explicit guidelines about appropriate inferences in the face of uncertainty, scientific methods can provide 'objective' information; that is, knowledge that maintains consistency in core assumptions across cases and provides transparent and reasonable conclusions (Walker 1998: 267).

On this view of science as neutral and universal, it is argued that scientists are uniquely positioned to provide governments with advice that is free of the partisan or political bias displayed by virtually all others in the political arena. Experts' shared knowledge and judgments are non-political, and therefore relying on them will improve the quality of public policies. Establishing a boundary line between the scientific and the political in this way does, however, undermine democratic norms of public participation and deliberation. Those with scientific knowledge are within the boundaries of the policy community; those lacking that knowledge are outside it. Defining policy problems in this way also risks shunting off to the sidelines the very policy concerns – the distributional effects of policies, their social implications – that are of major concern to citizens.

Supporters of science-based regulation, with its professed depoliticization of policy debates, argue that these democratic limitations can be offset by constraints on the latitude of action of delegated agents. Regulatory governance, even when it is predicated on objective scientific expertise, can still provide opportunities for public participation. As well, norms of democratic accountability and transparency can be built in to require regulatory agents to give reasons for their decisions and to have those decisions subject to review (Majone 1999; Shapiro 1997).

Science and Canadian Regulatory Policy

Writing in the late 1960s, George Grant (1969) argued that the predominance of liberalism, which he defined to be 'a set of beliefs which proceed from the central assumption that man's essence is his freedom and therefore that what chiefly concerns man in this life is to shape the world as we want it' (114), explains North Americans' faith in technology. Grant described Canada and the United States as 'technological societies' in which technology and liberalism support each other as 'the governing faith of the society' (28). Technology is viewed as a principal means of economic and social development; science and technology are equated with human progress.[2]

Brunk, Haworth, and Lee (1991: 2) also look to the predominance of liberalism in North America to explain the heavy reliance on science in regulatory policymaking. They argue that the priority placed on freedom in liberal democracies requires that any limits on the freedoms of individuals and firms must be backed by a 'clearly articulated rationale' which either reflects a societal consensus or is neutral. Only under these conditions could the infringement on freedom not be perceived as 'seriously biased in favour of the values of one interest group against the values of others' (2). This requirement for a clear rationale favours 'an image of science as a neutral arbiter whose authority is unquestioned' (2).

If the dominance of liberalism predisposes Canadians to an image of science as neutral, Canada's political institutional framework has an inherent hierarchical bias consistent with deference to authority. As David Smith (1995: 189) observes in his book *The Invisible Crown*, 'power emanates from above and flows down.' The fusion of the legislature and the executive, and the concentration of authority in the executive facilitates the delegation of regulatory authority from principals (the executive/legislature) to agents (public servants or even non-governmental bodies).

The Canadian institutional context is readily distinguished from the American. The U.S. institutional separation of powers makes Congress loath to delegate rule-making discretion to executive agencies, and the courts have been given strong powers to ensure that administrative agencies both carry out and do not exceed their mandate. As a result, courts have come to play a prominent role in U.S. regulatory governance, with court battles over the scope of regulation often turning on the quantitative and scientific data of contending parties. The result is frequently a contest of scientists, a phenomenon that works to undermine an image of science as objective and universal. By contrast, the image of neutral science is arguably easier to sustain in the Canadian parliamentary system, where knowledgeable officials upon whom governments rely for regulatory policymaking are comparatively insulated from public scrutiny and judicial review. Providing Canadian government officials have sufficient in-house expertise upon which to make their regulatory decisions, they can be assured of considerable independence of civil society actors as well.

Canadian governments have historically relied on the authority of neutral science to stem controversy around regulatory decisions. This turning to science has a democratic cost when science is used to cut off debate around non-scientific dimensions of policy and to preclude public participation in the policy debate. Canada offers the public far fewer institutionalized opportunities to participate in science-based regulatory policymaking than exist for the American public. Harrison and Hoberg (1994) argue that science-based regulation in Canada closes the regulatory process to public scrutiny but, in their view, does not close out cooperation with regulated actors. A couple of examples from the agri-food sector are instructive of Canadian government regulators' reliance on the authority of science. They offer different lessons for the ability of science-based regulation to be reconciled with democratic norms like opportunities for public participation and influence on policymaking.

The rbST Controversy

The first instructive example is the controversy that developed over an application for the licensing of recombinant bovine somatotropin (rbST). The request to approve the product, initially made in 1990, provoked controversy, not only among Canadian dairy farmers, consumers, and environmentalists, but also among government scientists in the Bureau of Veterinary Drugs of Health Canada. Some officials in the bureau publicly alleged Health Canada was under pressure from drug sponsors to approve the drug. Other scientists in the bureau, who had not been involved in the review to determine the safety of rbST to people who drank milk from dairy cows injected with the hormone, also spoke out publicly to declare there was insufficient evidence on which to base an earlier bureau decision that the drug posed no human health risks. Monsanto's application for licensing rbST was declined in 1995 on the grounds the hormone was unsafe to dairy cows injected with it. Nonetheless, controversy continued within Health Canada. In an effort to put it to rest, the government of Canada turned to two external expert panels to review the bureau's decision on human and animal safety. Their reports upheld the decision of Health Canada regulators not to approve rbST (Mills 2002: ch. 4; Turner 2001). In giving scientists external to government what it hoped would be the last word, the Canadian government put its faith in the authority of neutral science to quell the controversy. Whether it did so cannot be determined because the hormone was not approved. Had it been approved – despite apprehension of its social and economic consequences to the dairy industry and over mobilized consumers' opposition – the ability of science to resolve authoritatively a controversy would have been tested.

What lessons does the rbST debate have about how science-based regulation undermines democratic norms? Were non-scientific dimensions of a policy issue sidelined? And was policy influence restricted to those who wield such knowledge? Yes and No to both questions. Both dairy farmers and consumers mobilized against the licensing of rbST, concerned not only about the drug's effects on animal and human health, but also on the social and economic well-being of the dairy sector. The government refused to consider these socioeconomic effects, arguing these factors were not a regulatory criterion and 'could pre-empt decisions based on safety, and effectiveness' (as quoted in Mills 2002: 89). It also explicitly precluded the expert veterinary panel struck to consider the animal health effects of rbST from examining the economic issues raised by the drug. A parliamentary committee recommended that the government of Canada investigate these socioeconomic effects. If the regulatory decision, at least officially, ignored the broader social and economic factors raised by rbST, the agri-food policy community nonetheless had the opportunity to publicize its concerns. They did so before both House of Commons and Senate committees and a task force appointed by the government of Canada to investigate both the animal and health effects of rbST as well as its impact on the dairy industry.

Regulating Genetically Modified Crops and Foods

A second instructive example of government reliance on an image of neutral science to legitimize policy is found in the regulation of plant biotechnology. Like rbST, genetically modified (GM) crops and foods not only pose potential human and environmental safety issues, they also have important socioeconomic dimensions. The latter include the distribution of the benefits and risks of the technology, biotechnology developers' structural power over farmers, patenting rights, consumers' rights to avoid GM products, and the viability of markets for GM products. Canadian government officials sought information and advice from scientific networks and advisory committees in developing the GM policy and regulatory framework (Moore 2000). The framework they picked adhered to existing practice of using scientific expertise to assess the safety of GM products based on the risks they posed. The degree of risk GM products were scientifically determined to pose triggered their degree of regulation (Moore 2000).

At the outset and in its implementation, this regulatory framework limited the opportunity for and the scope of a public debate on plant biotechnology. There was no wide public or parliamentary debate surrounding the introduction of GM foods and crops because existing legislation was relied on to regulate these products. Subsequently, multistakeholder consultations and workshops allowed for some public input into regulations and policy guidelines. Although some of these consultations in the early and mid-1990s were ostensibly designed to elicit feedback on the social and ethical dimensions of plant biotechnology, critics say they failed to realize this objective (Abergel and Barrett 2002). Decisions to approve GM crops and foods in Canada provided little opportunity for public input or little transparency. There were no provisions for public comment throughout the process to assess the risks of GM crops and foods, and no opportunity to appeal regulatory decisions.

Even so, until the late 1990s, there was very little public debate or controversy in Canada about either this regulatory framework or GM foods and crops. When controversy arose then, it was largely as a spillover effect of the rejection of plant biotechnology in Europe.

In response to the growing controversy around plant biotechnology and its regulation since the late 1990s, the government of Canada turned to science to bolster policy legitimacy. It asked scientists external to the government of Canada, in the Royal Society of Canada, to provide advice on scientific issues of GM foods. As discussed further below, this strategy initially backfired, as the Expert Panel of the Royal Society produced a damning critique of the scientific integrity of the genetic engineering risk regulatory framework. In the wake of its recommendations, the government of Canada committed itself to an Action Plan in 2001, one element of which is to add external experts to the previously exclusively internal regulatory review process.

The government of Canada also took steps to improve the democratic credentials of plant biotechnology regulation. The Canadian Biotechnology Advisory Committee (CBAC) was created in 1999 to provide advice to a ministerial committee on the social, ethical, regulatory, economic, health, environmental, and scientific considerations of biotechnology policy. It is also mandated to enhance public awareness of biotechnology and facilitate 'an open, transparent national conversation on key issues' related to its development and application. Twenty-one experts external to government and drawn from diverse backgrounds serve on the CBAC. They devised a consultation document, in concert with various 'stakeholder' groups (industry, agricultural producers, a nutritionist, and an environmental group), that was the focus of discussion in a series of 'multi-stakeholder' workshops and internet discussions in 2001 on GM food. The success of this consultative exercise was impaired by the decision by the vast majority of environmental and public interest non-governmental organizations to boycott the consultations on the grounds that the CBAC's remit was too narrow and lacked independence from government.[3]

During the course of its consultations, the CBAC (2002) observed the lack of a mechanism by which to incorporate non-scientific – social and ethical – factors into discussion of plant biotechnology policy development. In an effort to correct this shortcoming, the CBAC urged the government of Canada to engage in a further dialogue with the Canadian public. At this point, there is little evidence that the government has done so. The controversy that envelops the introduction of GM wheat into Canada opened a window for a debate over the distribution of benefits and costs of plant biotechnology but did not provide any clear process to do so.

The examples of rbST and plant biotechnology regulation affirm the Canadian government's reliance on an image of neutral science both to justify regulatory policy outcomes and to sideline the legitimacy of non-scientific considerations entering into food safety and animal health regulations. However, several developments cast doubt on its continuing ability to turn to science as an authoritative basis of regulatory policymaking.

Science-Based Policymaking under Challenge

The legitimacy of a non-democratic source of political authority, like science, is not unconditional. Rather, it is contingent upon citizens' perception that science can contribute positively to valued public goods and that it is trustworthy. The capacity of science to retain the public trust it needs in order serve as a legitimate basis of policymaking has been undermined. One reason is the perception that science-based regulatory policies have failed to protect the public good in the late twentieth century. Another reason for a loss of legitimacy of science-based regulation is the shift in the political culture of Western democracies to put a higher value on democratic norms of public participation, deliberation, and government accountability. And finally, the authority of science has been undermined by its deployment as an instrument to promote economic globalization at the expense of a loss of government accountability. Some of these developments are more characteristic of other countries than they are of Canada. In each case, however, they have ricocheted across national boundaries to destabilize the legitimacy of science-based regulation.

Canadian Developments: Loss of In-House Capacity and Weakened Credibility

Among the factors that contribute to trust in science are its autonomy and its capacity to pass peer review among other scientists. In Canada, the authority of science has been weakened by three developments in particular. The first is a diminution of in-house scientific capacity. The second is a blurring of public–private lines in the production of scientific knowledge. And the third – arguably the most important given Canadians' respect for science – is scientists' own public questioning of the quality and autonomy of the science that underwrites regulatory policy.

In the realm of agri-food policy, the government of Canada traditionally had considerable in-house research and scientific capacity, so much so that its leadership and autonomy within the Canadian agricultural research effort were unquestioned. However, in the late twentieth century, fiscal restraints reduced financial support for public research and caused considerable slippage in that capacity.[4] At the same time, a shift to collaboration with the private sector in public–private research partnerships (as a way to cope with reduced funds for research) raised concerns about diminished government autonomy to define public research priorities independently of private interests. In its endorsement of public–private research partnerships, Canada was following the earlier path of the United States (Moore 2002).

Both developments – diminished in-government expertise and public–private partnerships – make the Canadian government more reliant on the science of industry and other governments (Doern and Reed 2000; McHughen 2000).[5] These deficiencies came to the Canadian public's attention with the publication of the Royal Society of Canada's (2001) report on Canada's biotechnology risk regulation framework. Its expert panel concluded that existing regulatory procedures failed to conform to scientific standards of 'an open, completely transparent enterprise in which any and all aspects of scientific research are open to full review by scientific peers' (214). It targeted provisions in existing regulations that protected developers' confidentiality and compromised the public transparency of scientific data and the scientific rationales on which regulatory decisions are based. It criticized the absence of independent review of data and inadequate consultation of the expert scientific community in the design and execution of testing regimes for GM organisms. The panel concluded that GM plants were being approved without sufficient research to appraise their environmental impacts and that the existing regulatory system did not warrant Canadians' confidence (132, 215).

The Royal Society report also pointed to two conflicts of interest. The first is the dual mandate that regulatory agencies have to promote the development of agricultural biotechnology as well as to regulate it. The second is within the scientific community, where 'entrepreneurial interests in resulting technologies and the increasing domination of the research agenda by private corporate interest' make it difficult for scientists to be disinterested (ix). Both conflicts of interest, in its view, 'compromise the integrity of regulatory science and decision making' (212).

The Royal Society's criticisms were echoed two years later in a report by the Canadian Biotechnology Advisory Committee (CBAC), the independent expert body established in 1999 to advise the government of Canada on a range of issues relating to biotechnology. Its 2002 report also called for greater scientific rigour and transparency in Canadian regulatory policies by, for example, more involvement of outside experts in risk assessments of GM products, clearer information on how GM products are regulated, and public disclosure of the data on which safety assessments are made (CBAC 2002). And, like the Royal Society Expert Panel, the CBAC urged clarification of the mandates of government agencies involved in plant biotechnology to avoid overlapping responsibilities for regulation and other functions like economic development and trade promotion.

These two reports have given reasons to doubt the integrity of the science that underwrites biotechnology regulation in Canada. Recognizing that danger, the Canadian government has indicated in recent statements and budgets its commitment to improving its in-house scientific capacity. How much the credibility of science as an authoritative source of Canadian regulatory policymaking has been eroded to date is difficult to say. In any event, it is decidedly less than in Europe, where the image of science as a source of neutral and objective knowledge has been discredited.

The European Union: New Understandings of Science-Based Policymaking

Europeans have experienced a crisis of confidence in science in the late twentieth century of a degree that can be equated to a 'risk society' mentality on the part of many Europeans. Ulrich Beck (1992, 1999) coined the appellation 'risk society' to describe societies in which the risks of scientific and technological innovations to individuals, society, and the environment are perceived to have escaped the control of society, its institutions, and national boundaries. Rather than being equated with progress, the industrial and technological developments that mark modernity are singled out as the main producers of risk and the 'bads' that we perceive to be threatening. This risk society, argues Beck, questions the continuing appropriateness of traditional concepts of authority based on hierarchy and neutral expertise. The science that has created the new products and technologies that are the source of manufactured and unforeseen dangers cannot be relied upon to protect us from these dangers.

Since the late 1990s, there has been evidence that a risk society mentality characterizes the European public's reaction to the technology of genetic engineering (Skogstad 2005). It was initiated in no small degree by regulatory policy failures that discredited both government regulators and the scientific experts on whom they relied to protect public health. (These policy failures were most prominently the BSE crisis, the discovery of dioxin in animal food in Belgium, and the tainted blood scandal in France.) These fiascoes caused a deterioration in Europeans' faith in science and its capacity to produce public benefits (Eurobarometer 2001: 26). The BSE crisis, in particular, undermined the idea of scientists as 'objective' and 'independent' (Joerges and Neyer 1997: 612). The widespread perception was that the EU committees of scientific (veterinarian) experts responsible for advising the commission on BSE had jeopardized human health by underplaying the nature and severity of the crisis (Chambers 1999).

One facet of a risk society, argues Beck, is a diminished faith in science. Survey data provide evidence that Europeans' faith in the ability of science to produce benefits has deteriorated over time. The preponderant view today is that science and technology are not 'a panacea' and should be subject to 'social control' (Eurobarometer 2001: 30, 37). These views contrast sharply with the predominant understandings of science in North America, where a benevolent view of science and technology still prevails (Gaskell et al. 2001: 101–2).[6]

A risk society mentality is consistent with, but not exclusive to, an emphasis on the indeterminate nature of scientific knowledge, particularly when it comes to appraising the risks of novel technologies and products derived from them. In this regulatory arena, scholars agree that science is inexact, uncertain, complex, and even ignorant (Brunk, Haworth, and Lee 1991; Liberatore 2001; MacRae 1976; Shrader-Frechette 1985). In addition to the uncertainties that could be reduced in principle with more empirical research, there are other unknowable uncertainties whose probability cannot be theoretically or empirically calculated because they are rooted in the technology's novelty, complexity, and variability in different contexts (Christoforou 2003a). As a result, scientists' assessment of risks necessarily entail methodological, conceptual, and normative assumptions that contradict claims of neutrality and objectivity. Scientific knowledge is therefore not value-free; the values and goals of scientific assessors, and to some degree those of their sponsors, implicitly infiltrate the assessment (Brunk, Haworth, and Lee 1991; Mazur 1980; Salter 1988; Shrader-Frechette 1985).

In Europe, new understandings of science's limitation as a source of neutral and value-free knowledge are linked to the ascendance and institutionalization of the precautionary principle. This principle, which has long roots in Germany in particular, was incorporated in the 1997 Treaty of Amsterdam as part of the EU's constitution. The precautionary principle deals with scientific uncertainty, where epistemological knowledge and regulatory consensus break down or diverge (Christoforou 2003b). It places an obligation on policymakers to take action to avoid harm to the environment or human health 'even when the harms sought to be prevented are not as yet precisely understood or subject to formal analysis' (Jasanoff 2003: 229).

A precautionary approach does not eschew the need for science as a basis of regulation (Commission of the European Communities 2000). However, as endorsed in European Union regulation, it regards scientific opinion as only a necessary, not a sufficient, basis for risk regulation because expert knowledge can provide 'only a partial characterization of possible hazards' (Jasanoff 2003a).[7] The European Commission Science and Society Action Plan states that 'science is often perceived as dealing with certainty and hard facts, whereas this is rarely the case, particularly at the frontier of research' (cited in De Marchi 2003: 173). Such a view vests authority in science not by virtue of its claims to neutral and elite knowledge, but rather on admission of its limitations. As De Marchi (2003: 149) phrases it: 'Expertise has legitimacy when it is exercised in ways that make visible its contingent, negotiated character.'

The gap between this precautionary European image of science and that which dominates in North America – where regulatory decisions are justified by 'appeals to objective knowledge' (Jasanoff 1995: 32) and science is perceived as capable of yielding universal and certain facts (Isaac 2002) – has been pointed out by several observers. Europe's endorsement of the precautionary principle has occurred even as the United States has backed away from a precautionary approach (Jasanoff 2003b; Vogel 2003). If the two polities have drifted apart in their views of the authority of science, they are also separated by their norms of democratic decision making. However, here the gap is not so acute.

Shifting Political Cultures: The Demand for More Democracy

There is considerable research to document shifts in the political culture of liberal democracies that include a decline in citizens' deference to authority (Pharr and Putnam 2000). In the European Union, economic and political integration has spawned a broad academic and societal debate around the 'democratic deficit' that occurs when decision-making authority is transferred from accountable national decision makers to unaccountable officials in supranational institutions. In order to continue with the integration project, it has been necessary to construct a more democratic political architecture. The elected European Parliament has gained increasing powers to veto legislation, decision makers at the EU level have constructed elaborate networks that draw actors representing civil society more closely into policy formulation processes, and national governments have implemented direct democracy mechanisms like referenda to give citizens a voice at each step toward ever closer integration (Bellamy and Castiglione 2000; Eising and Kohler-Koch 1999; Eriksen and Fossum 2000; Lord and Beetham 2001).

Canada has not been spared these cultural shifts. Survey data document that the hierarchical and often non-transparent processes of executive-based decision making have lost favour with Canadians. The public is demanding more accountability of governments between elections and wants more open and transparent decision-making processes. They also want a more active and influential role in government decisions that affect them (Bricker and Greenspon 2001; Nadeau 2002). The Canadian government is not ignorant of this demand. Indeed, government planning documents, including those of Health Canada's Health Products and Food Branch, which is responsible for health and safety risk assessments, state a commitment to public engagement, public participation, transparency, and accountability. As yet, however, citizens' expectations for more democratic decision making along these lines remain largely unfulfilled, having been confined to piecemeal initiatives (Skogstad 2003).

A Counter-Reaction to the International Institutionalization of Science

A third development which has destabilized science as a basis of authoritative regulatory policymaking is the very effort to vest it with such status in international law. International agreements in the 1990s turned to science as an arbiter between legitimate and illegitimate health and safety measures in the international trade regime. The Sanitary and Phytosanitary (SPS) Agreement of the World Trade Organization is the foremost example. It requires measures to protect human, animal, or plant health to be based on 'scientific principles' and a risk assessment that is supported by 'available scientific evidence.' It vests science with the legal authority to determine which SPS measures are legitimate and which are not. SPS measures which conform to international standards are assumed to meet this scientific standard. Those which do not must be supported by a scientific risk assessment.

The nature of the scientific evidence required to support SPS measures in the international trade regime was clarified to some extent by the WTO Appellate Body in the *Hormones* case. It rejected the argument of the European Union's reference to the precautionary principle (article 5.7 of the SPS Agreement) to defend its prohibition of imports of hormone-treated beef. The Appellate Body ruled that the precautionary principle did not override the need for a scientific risk assessment. At the same time, however, the WTO body ruled that the scientific evidence needed to justify a trade-restrictive SPS measure did not need to embody majority scientific views and did not need to be quantitative. It could represent 'divergent opinion coming from qualified and respected sources' and it could consist of evidence 'not susceptible to the empirical or laboratory methods commonly associated with the physical sciences' (World Trade Organization 1998: para. 187). The Appellate Body did not, however, define what constitutes sufficient scientific evidence to support WTO-legal SPS measures.

The *Hormones* ruling provoked a public backlash in the European Union. It generated a debate about the WTO risk assessment requirement, provoking questions about whether science can be a 'neutral arbiter' for triggering precautionary measures (see especially Quintillan 1999; Walker 2003). EU decision makers chose to accept suspension of symmetrical concessions rather than comply with the ruling (Skogstad 2001). It also gave the EU strong incentives to promote the precautionary principle in other international agreements, namely the Cartagena Protocol on Biosafety.

Signed in January 2000 and in effect since September 2003, the Biosafety Protocol applies to living modified organisms, that is, genetically modified seeds, feeds, and food. It endorses the precautionary principle, stating that a lack of scientific certainty shall not prevent countries from taking action to prohibit or restrict genetically modified organisms for the purpose of promoting environmental or health concerns. The Biosafety Protocol stands on an equal footing with the WTO agreements, but, as scholars like Phillips and Kerr (2000) have pointed out, it is potentially in conflict with them. The Biosafety Protocol allows socioeconomic criteria to be considered in determining the safety of imports while the WTO does not. Winham (2003: 151) argues that the conflict between the principles of scientific risk assessment in the international (WTO-centred) trade regime and the precautionary principle in the international environmental regime (formalized in the Biosafety Protocol) will promote international conflict, not cooperation.

In short, the effort to institutionalize science as an authoritative basis for international regulation of measures to protect human, animal, and plant health and safety has resulted in competing international regimes that ensure continuing conflict over the authority of science as a basis for risk regulation. This conflict has undoubtedly been exacerbated by the weak democratic credentials of the WTO and, more generally, the limited legitimacy of international law that does not originate in a democratic process (Joerges and Neyer 2003). Critics raise concerns that the international bodies that are recognized by the WTO to set the science-based standards of SPS measures are not immune to political influence. For instance, industry representatives dominate the non-voting members of countries' delegations to the Codex Alimentarius Commission (Braithwaite and Drahos 2000; Quintillan 1999). Recognizing Codex standards as legally enforceable has served to politicize its deliberations as well as the epistemic community on which it relies for advice (Braithwaite and Drahos 2000: 403).

Implications for Canada

If the foregoing analyses are correct, and the image of science as a neutral source of knowledge is eroding and its authority as a basis of legitimate policymaking slipping, then what are the implications for regulatory policymaking in Canada, and for risk regulation in particular? If science is to retain its appropriate place in regulation as an authoritative basis of knowledge, then there are two matters that need the attention of Canadian governments. The first is to get our own house in order in terms of providing a democratic infrastructure that exposes science-based regulation to more public scrutiny and that provides opportunities for non-scientific concerns to be aired. The second, which is arguably more contentious, is to recognize the legitimacy of other countries' legitimation standards for risk regulation, even to the point of providing more scope for non-scientific factors to shape food and environmental regulations where indeterminate risks are potentially significant.

There seems little option but to provide for greater democratic scrutiny of the science that underwrites public policies. When science-based regulation is necessary, any claims it makes to a privileged status as a source of truth and objectivity are bound to be controversial. As Sheila

Jasanoff (2003b: 160) observes, 'science invoked to support policy tends to unravel under the stresses of politics: those wishing to question a given scientific interpretation can generally find errors, hidden biases or subjective judgments that undercut their opponents' claims to truth and objectivity.' There are numerous examples (in the United States and Europe) to confirm Jasanoff's observations. Looking at the measures that other countries have taken to bolster the credibility of science-based regulation when it is under attack can be instructive in terms of suggesting steps Canada might take. As a parliamentary democracy, the UK experience provides some insight, even if we have not yet experienced anything like the crisis of confidence in science that it has endured in recent years.

The lessons of the UK experience around plant biotechnology would appear to be twofold. The first is that chambers of democratic representation can play an important role in broadening the debate around risk regulation and lending legitimacy to regulatory processes and outcomes. Committees in both the UK House of Commons and Lords alerted the government to the decline in deference to scientific authority and recommended ways to shore it up, for example, by acknowledging the uncertainty of science in risk decision making and increasing the transparency and peer review of scientific data. On the latter point, they were in accord with Canada's Royal Society, but for the British parliamentary committees, transparency meant public access to the facts and analysis behind regulatory decisions. The British government acted on many of the parliamentary committees' recommendations, including the advice for more pluralistic epistemological representation on its scientific and strategic advisory bodies. The second lesson of the UK experience would thus be the imperative for scientific advisory bodies to operate at arm's length from government and to include a fairly broad range of independent scientists not linked to the biotechnology industry. The UK scientific advisory body also provides public access to regulatory decisions and the data on which they are based.

What gave the UK parliamentary committees their influence over government policy was their appearance of being 'balanced' and politically neutral forums. If parliamentary committees are to play the same legitimation role in Canada, it seems evident that they will need to be restructured so as to erase their current image as forums of partisanship and clientele domination. The present functional distribution of policy matters to parliamentary committees – plant biotechnology issues to the Agriculture and Agri-Food Standing Committee of the House of

Commons, for example – undercuts capacity to represent a spectrum of views that extends beyond those of their self-defined clientele. In the particular matter of risk regulation, parliamentary committees whose mandate transcends several policy domains – science, agriculture, health, the environment – are likely necessary. Of course, these structural changes will not add to the democratic quality of policy debates unless the political influence of committees' reports on government policy is increased.

Providing a more democratic infrastructure for public deliberation and input around science-based regulation is necessary because of the inherent democratic limitations of international regulatory forums. International regulatory governance is a reality of our era. However, international institutions have weak legitimacy owing to their secrecy, limited opportunities for broad public participation, and their lack of direct accountability. Their legitimacy is therefore derivative to an important extent of the democratic states that created them (Keohane and Nye 2003). If science-based international bodies like Codex and the WTO are to be authoritative, then the rules they make and enforce will need to have passed democratic standards of legitimate policy processes and outcomes in member states.

What does this imperative of respect for democratic norms and public policy processes at the domestic level mean for the current conflict over the understanding of science on which international regulatory regimes should be based? It means that if international science-based regulation is to be widely accepted as legitimate, it will likely have to rest on a more sophisticated and sceptical understanding of scientific knowledge. Such a view of science would inevitably give countries more scope to devise public policies consistent with the preferences of their citizens, as these are made evident through forums of informed public deliberation.[8]

NOTES

1 Political authority and legitimacy are closely related. Political authority is the right to exercise binding commands – to wield political power. In democracies, the exercise of political power must be viewed as legitimate – desirable, proper, or appropriate – if it is to be willingly obeyed.

2 The relevance of the writings of Grant and Brunk et al. was drawn to our attention by Moore (2000).

3 The CBAC office was located in a government department (Industry Canada) which is a paid member of BIOTECanada, an advocacy group for biotechnology companies (Baxter 2000). CBAC was disbanded in May 2007.
4 Agriculture and Agri-Food Canada's research budget dropped from $367 million in 1985–6 to $252 million in 2001–2 (Moore 2002).
5 The federal and provincial governments committed themselves to greater funding of agri-food research, including that for food safety and environmental sustainability purposes, in the 2002 Agricultural Policy Framework. A government official reported to the House of Commons Standing Committee on Agriculture and Agri-Food on 12 June 2003 that government regulators are doing more in-house work to confirm independently the scientific data submitted by biotechnology developers.
6 Nevitte and Kanji (2004: 86–7), using survey data gathered in 2000, show rising levels of support for science and technology, with more than three out of five Canadians believing that more emphasis on the development of technology and scientific advances will benefit mankind.
7 Christoforou (2003b: 209) observes that the European Court of First Instance has ruled that 'scientific legitimacy is not a sufficient basis for the exercise of public authority.'
8 Howse (2000) offers some important insights into how democracy and science can be reconciled through 'rational democratic deliberation.'

REFERENCES

Abergel, E., and K. Barrett. 2002. Putting the Cart before the Horse: A Review of Biotechnology Policy in Canada. *Journal of Canadian Studies* 37, no. 3: 135–61.

Baxter, J. 2000. Government Paying to Lobby Itself: Industry Canada Is a Member – and a Client – of Group Organized to Influence Biotech Policy. *Ottawa Citizen*, 10 April.

Beck, U. 1992. *Risk Society.* London: Sage.

– 1999. *World Risk Society.* Cambridge: Polity Press.

Bellamy, R., and D. Castiglione. 2000. The Uses of Democracy: Reflections on the European Democratic Deficit. In *Democracy in the European Union: Integration Through Deliberation?* edited by E.O. Eriksen and J.E. Fossum, 65–84. London: Routledge.

Braithwaite, J., and P. Drahos. 2000. *Global Business Regulation.* Cambridge: Cambridge University Press.

Bricker, D., and E. Greenspon. 2001. *Searching for Certainty: Inside the New Canadian Mindset.* Toronto: Doubleday.

Brunk, C.G., L. Haworth, and B. Lee. 1991. *Value Assumptions in Risk Assessment: A Case Study of the Alachlor Controversy.* Waterloo: Wilfrid Laurier University Press.

CBAC (Canadian Biotechnology Advisory Committee). 2002. *Improving the Regulation of Genetically Modified Foods and Other Novel Foods in Canada.* http://cbac-cccb.ca/epic/site/cbac-cccb.nsf/en/ah00186e.html (accessed 17 April 2007).

Chambers, G.R. 1999. The BSE Crisis and the European Parliament. In *EU Committees: Social Regulation, Law and Politics,* edited by C. Joerges and E. Vos, 95–106. Oxford: Hart Publishing.

Christoforou, T. 2003a. The Precautionary Principle in European Community Law and Law and Science. In *Precaution, Environmental Science and Preventive Public Policy,* edited by J.A Tickner, 241–62. Washington, DC: Island Press.

– 2003b. The Precautionary Principle and Democratizing Expertise: A European Legal Perspective. In *Science and Public Policy* 30, no. 3: 205–11.

Commission of the European Communities. 2000. *Communication from the Commission on the Precautionary Principle.* COM(2000). Brussels. 2 February.

De Marchi, Bruna. 2003. Public Participation and Risk Governance. *Science and Public Policy* 30, no. 3: 171–6.

Doern, G.B., and T. Reed. 2000. Canada's Changing Science-Based Policy and Regulatory Regime: Issues and Framework. In *Risky Business: Canada's Changing Science-Based Policy and Regulatory Regime,* edited by G.B. Doern and T. Reed, 3–28. Toronto: University of Toronto Press.

Eising, R., and B. Kohler-Koch. 1999. Governance in the European Union: A Comparative Sssessment. In *The Transformation of Governance in the European Union,* edited by B. Kohler-Koch and R. Eising, 267–85. London: Routledge.

Eriksen, E.O., and J.E. Fossum. 2000. *Democracy in the European Union: Integration through Deliberation?* London: Routledge.

Eurobarometer. 2001. *Europeans, Science and Technology* 55: 2. Brussels: Director-General for Research.

Gaskell, G., E. Einseidel, S. Priest, T.A. Ten Eyck, N.C. Allum, and H. Torgerson. 2001. Troubled Waters: The Atlantic Divide on Biotechnology Policy. *Biotechnology 1996–2000,* edited by G. Gaskell and M.W. Bauer, 96–115. London: Science Museum.

Grant, G. 1969. *Technology and Empire.* Toronto: House of Anansi.

Harrison, K., and G. Hoberg. 1994. *Risk, Science, and Politics.* Montreal and Kingston: McGill-Queen's University Press.

Howse, R. 2000. Democracy, Science, and Free Trade: Risk Regulation on Trial at the World Trade Organization. *Michigan Law Review* 98, no. 7: 2329–57.

Isaac, G.E. 2002. *Agricultural Biotechnology and Transatlantic Trade: Regulatory Barriers to GM Crops*. London: CABI Publishing.

Jasanoff, S. 1995. *Science at the Bar: Law, Science and Technology in America*. Cambridge, MA: Harvard University Press.

– 2003a. A Living Legacy: The Precautionary Ideal in American Law. In *Precaution, Environmental Science and Preventive Public Policy*, edited by J. Tickner, 227–40. Washington, DC: Island Press.

– 2003b. (No?) Accounting for Expertise. *Science and Public Policy* 30: 157–62.

Joerges, C., and J. Neyer. 1997. Transforming Strategic Interaction into Deliberative Problem-Solving: European Comitology in the Foodstuffs Sector. *Journal of European Public Policy* 4, no. 4: 572–90.

– 2003. Politics, Risk Management, World Trade Organisation Governance and the Limits of Legalization. *Science and Public Policy* 30, no. 1: 219–25.

Keohane, R.O., and J.S. Nye, Jr. 2003. Redefining Accountability for Global Governance. In *Governance in a Global Economy*, edited by M. Kahler and D.A. Lake, 386–411. Princeton: Princeton University Press.

Libertore, A. 2001. *Democratising Expertise and Establishing Scientific Reference Systems*. Report of the Working Group. White Paper on Governance. Brussels: European Commission.

Lord, C., and D. Beetham. 2001. Legitimizing the EU: Is There a 'Post-Parliamentary Basis' for Its Legitimation? *Journal of Common Market Studies* 39, no. 3: 443–62.

MacRae, D. Jr. 1976. *The Social Function of Social Science*. New Haven: Yale University Press.

Majone, G. 1999. The Regulatory State and Its Legitimacy Problems. *West European Politics* 22, no. 1: 1–13.

Mazur, A. 1980. Societal and Scientific Causes of the Historical Development of Risk Assessment. In *Society, Technology and Risk Assessment* edited by J. Conrad, 151–7. London: Academic Press.

McHughen, A. 2000. The Regulation of GM Foods: Who Represents the Public Interest? *International Journal* 4: 624–32.

Mills, L.N. 2002. *Science and Social Context: The Regulation of Recombinant Bovine Growth Hormone in North America.* Montreal and Kingston: McGill-Queen's University Press.

Moore, E. 2000. Science, Internationalization, and Policy Networks: Regulating Genetically-Engineered Food Crops in Canada and the United States, 1973–1998. PhD dissertation, University of Toronto.

Moore, Elizabeth. 2002. The New Direction of Federal Agricultural Research in Canada: From Public Good to Private Gain? *Journal of Canadian Studies* 37, no. 3: 112–33.

Nadeau, R. 2002. Satisfaction with Democracy: The Canadian Paradox. In *Value Change and Governance in Canada,* edited by N. Nevitte, 37–70. Toronto: University of Toronto Press.

Nevitte, N., and M. Kanji. 2004. New Cleavages, Value Diversity, and Democratic Governance. In *Canadian Politics,* edited by J. Bickerton and A.G. Gagnon, 79–97. Peterborough: Broadview Press.

Pharr, S.J., and R.D. Putnam. 2000. *Disaffected Democracies: What's Troubling the Trilateral Countries?* Princeton: Princeton University Press.

Phillips, P.W.B., and W.A. Kerr. 2000. Alternative Paradigms – The WTO versus the Biosafety Protocol for Trade in Genetically Modified Organisms. *Journal of World Trade* 34, no. 4: 63–75.

Quintillan, S.P. 1999. Free Trade, Public Health Protection and Consumer Information in the European and WTO Context. *Journal of World Trade* 33, no. 6: 147–97.

Royal Society of Canada. 2001. *Elements of Precaution: Recommendations for the Regulation of Food Biotechnology in Canada.* Ottawa.

Salter, L. 1988. *Mandated Science: Science and Scientists in the Making of Standards.* Boston: Kluwer Academic Publishers.

Shrader-Frechette, K.S. 1985. *Science Policy, Ethics, and Economic Methodology.* Dordrecht, Holland: D. Reidel.

Shapiro, M. 1997. The Problems of Independent Agencies in the United States and the European Union. *Journal of European Public Policy* 4: 276–91.

Skogstad, G. 2001. The WTO and Food Safety Regulatory Policy Innovation in the European Union. *Journal of Common Market Studies* 39, no. 3: 485–505.

– 2003. Who Governs? Who Should Govern? Political Authority and Legitimacy in Canada in the Twenty-First Century. *Canadian Journal of Political Science* 36, no. 5 (December): 955–73.

– 2005. Contested Political Authority, Risk Society and the Transatlantic Divide in Genetic Engineering Regulation. In *Reconstituting Political Authority: Complex Sovereignty and the Foundations of Global Governance,* edited by E. Grande and L.W. Pauly, 238–60. Toronto: University of Toronto Press.

Smith, D.E. 1995. *The Invisible Crown: The First Principle of Canadian Government.* Toronto: University of Toronto Press.

Turner, R.S. 2001. Of Milk and Mandarins: rbST, Mandated Science and the Canadian Regulatory Style. *Journal of Canadian Studies* 36, no. 3: 107–30.

Vogel, D. 2003. The Hare and the Tortoise Revisited: The New Politics of Consumer and Environmental Regulation in Europe. *British Journal of Political Science* 33: 557–80.

Walker, V.R. 1998. Keeping the WTO from Becoming the 'World Trans-Science Organization': Scientific Uncertainty, Science Policy, and Factfinding in

the Growth Hormones Dispute. *Cornell International Law Journal* 31: 251–320.
– 2003. The Myth of Science as a 'Neutral Arbiter' for Triggering Precautions. *Boston College International and Comparative Law Review* 26, no. 2: 197–228.
Winham, G.R. 2003. International Regime Conflict in Trade and Environment: The Biosafety Protocol and the WTO. *World Trade Review* 2, no. 2: 131–55.
World Trade Organization. 1998. *EC Measures Concerning Meat and Meat Products (Hormones): Report of the Appellate Body.* Geneva.

13 Bringing Balance, Disclosure, and Due Diligence into Science-Based Policymaking

ROSS McKITRICK

Introduction

The questions posed in this book refer to the science guiding public policy as 'public science.' Chris Essex and I, in our book *Taken by Storm* (Essex and McKitrick 2002) used the term 'official science' to mean much the same thing. These terms are new, but the situation is not – the underlying problem of incorporating esoteric advice into public policy is as old as society itself. Recall, for instance, the story from Genesis chapter 41 about Pharaoh and his ominous dreams. Seven dying cows appear and eat seven fat cows; then seven shrivelled heads of grain swallow up seven plump heads of grain. The assembled prophets and fortune-tellers could make nothing of these dreams, but Joseph, the Hebrew seer, warned Pharaoh that it signified a coming famine. Seven abundant years would be followed by seven years of drought, and (said Joseph) the king must therefore immediately impose a 20 per cent tax on agricultural output, with the proceeds used to store up food. Pharaoh followed this plan, perhaps encountering protests along the way from the taxpaying public who were not privy to the dreams; but the famine came as predicted and the country was spared ruin. Of course the hero in the story is Joseph, but Pharaoh also deserves credit for making good use of advice from a source whose reliability was not easy to assess.

Individuals in positions of responsibility often have to look at uncertain information and make decisions that will have institutional force and thereby impinge upon the lives of many others. In a democracy those decisions should, at some level, receive the consent of those affected. Since the scientific information supporting the decision may be very specialized or technical, the people affected will typically not be in a position to agree or disagree with it and thereby offer informed consent to the policy. Hence it is understandable that scientific input to a democratic decision-making process will at times cause tension between the desire for the best information – however complex – and the desire for informed, voluntary consent by those being governed.

The need to adjudicate uncertain information in support of institutional decisions arises in many contexts, but I would like to compare three situations that are prominent in our time: court trials, business prospectuses, and government expert panels. My argument is that of these three, the latter is the most important, in terms of consequences for people's lives, but also the least systematic. I will focus especially on the mechanisms for bringing information to bear on environmental policymaking, arguing that the other two mechanisms provide ready-made models for improving decision mechanisms in the public sector. I will develop this argument in section three before turning in the fourth section to the three questions posed by this volume.

Trials, Finance, and Science Panels

Adversarial Proceedings in Court

Consider, first, a court trial. A man is charged with murder and faces life in prison. He claims he is innocent. An eyewitness saw him enter the building just before the murder. His fingerprint was on the knife, and there is fibre evidence the prosecution says can tie him to the murder scene. The defence says the eyewitness is wrong; the accused was in fact at home; that he used the knife when he visited the victim – an old friend – for dinner the week before, and the fibre could be from anyone who happens to have a light blue cardigan. What will the court do? The jury (or judge) must make a decision in the face of uncertain information that will either consign a man to prison or set him free. And in the process they may establish precedent for handling certain types of evidence or procedural disputes that could affect hundreds of trials in the future.

We all know what the basic shape of the process will be. The court will ensure that both the prosecution and the defence are adequately represented. Any idea that only the prosecution should present its case would be dismissed as a horrifying throwback to an earlier, tyrannical age. There will be a preliminary stage in which the prosecution must reveal its evidence and allow the defence to prepare a response. Both sides will describe their expert witnesses and the court must approve their qualifications, as well as approve the evidence that will be brought out in court. When the trial gets underway, the prosecution will bring in its witnesses, who will go over the evidence point by point. Every expert and every witness will be cross-examined by the defence attorney, whose aim will be to expose any weakness in the case. There will be no pretence that the prosecutor or the defence is 'balanced' – each side is there to make its own argument as strongly as possible. Then the defence will present its case, subject in turn to cross-examination by the prosecution. Witnesses will stay on the stand for as long as it takes to complete each cross-examination, and every inconsistency or hearsay utterance can be probed until it is either straightened out or demolished. Evidence will be presented in such a way as to ensure that everybody in the room agrees on what it is. There may be disputes about what it means but everyone has to agree – including the judge – that the exhibits are admissible as evidence, and the jury must be given the opportunity to see exactly what each one is. Finally both sides will sum up, and the court will render a verdict. But even then the process is not over, because the losing side can launch an appeal if they find new evidence or if they dispute the fairness of the first trial.

Courts can and do make mistakes. But it must be admitted that the process ensures that the decision draws in both sides of the story. This is a key point. The attorney on each side is there to present the case only for his or her side. If it turned out the defence lawyer was also spending a significant part of his time assisting the prosecution he could be disbarred for unethical conduct. Ultimately what we care about is that the *process* is balanced, not the individual participants. *Indeed it is necessary for the individual lawyers to be one-sided for the process itself to be balanced.* One reason is that there is often a public stigma attached to one side or the other. Sometimes the prosecutor is considered a heel for pressing charges and going to trial. Sometimes the defence is seen as trying to get rotten criminals off on cheap technicalities. The court cannot let these perceptions interfere with its function. Both sides must have vigorous representation for the process to work, and if the court is not satisfied that each side is properly represented the trial may not proceed, or the decision may be subject to later overturning on appeal.

We will consider some contrasts with the policymaking process later, but for the moment a particularly noteworthy point is the *principle of balance*: for a process to be balanced does not require that each participant be neutral; it requires that the contrasting points of view be deliberately sought out and presented to each one's best advantage.

Financing a Public Corporation

Consider, second, an investment prospectus.[1] To be specific, suppose it is a proposed mining operation. The main backer is a large established mining company with publicly traded shares. It is preparing a public prospectus in advance of issuing new shares in a subsidiary company that will mine for gold in the forests of, say, Venezuela. They claim that their prospectors have established the likely presence of about one million recoverable ounces, making the deposit worth about $300 million. They would like the subsidiary shares to trade on a public stock exchange.

A prospectus must contain audited financial statements. Auditing is carried out by specialized and highly paid professionals, and, for large corporations, the audit is virtually a full-time occupation. A company issuing an exploration prospectus must provide a qualifying report on its geological properties by an independent geological professional. The geologist must be truly independent of the issuing firm and would not begin to conduct the analysis unless he or she had inspected the physical cores and verified independently all data used to summarize their contents. Standard due diligence involves travelling to the site and obtaining new core samples by drillers who are also independent of the issuing company. Both the auditor and the independent geologist must approve the relevant language in the prospectus and provide signed consent letters to the securities commission. The prospectus itself is reviewed by two sets of securities lawyers – one for the issuing corporation and one for the underwriter or broker acting as agent (note the implicit principle of balance). Then the prospectus is reviewed a third time by the securities commission. Any apparent errors or concerns of the securities commission must be dealt with, regardless of whether it is material to the results. The process is expensive and painstaking. After all this, the officers and directors of the corporation have to sign a form certifying that they have made 'full, true and plain' disclosure, which means not only certifying that everything in the prospectus is true to the best of their knowledge, but also that they have not omitted anything from the prospectus which is material. These steps are required no matter how small the dollar amounts involved in the stock issuance.

Despite the multiple layers of due diligence for prospectuses, frauds still occur. One of the most famous cases recently was the 1997 scandal involving Bre-X, in which millions of dollars were lost over a phony gold mine in Indonesia. In this case there were lapses in due diligence. The drill core was never made available for inspection. During its main boom, Bre-X never issued a prospectus. When it listed on the Toronto Stock Exchange, it filed an ore reserves study by a well-respected engineering firm which contained the caveat that the ore reserve calculation relied on company information and that no examination of drill core or verification was carried out. The fraud was only exposed when, at a late stage in arranging the buyout of the company by a larger consortium, due diligence was finally undertaken in the form of a third party drilling new core samples. When it turned out there was no gold the entire scheme came crashing down – as did Bre-X's chief geologist, who (allegedly) threw himself out of a helicopter.

At this point it will be useful to summarize two additional principles that are manifest in both court proceedings and in the rules governing issuance of a prospectus. First, the *principle of disclosure* is that a party asking others to invest their resources on the basis of some factual or technical information must make full, true, and plain disclosure concerning the information, including the exact origin of the data and the exact form of any computational analysis applied to it. Second, the *principle of due diligence* states that anyone conveying information with the expectation that readers will act upon it (including investing their own or others' money in response to it) has verified, and can personally attest to, the truthfulness of the information.

Thus we see three principles at work in other decision-making contexts: balance, disclosure, and due diligence. I now look at the use of scientific information in public policy formation, and ask whether these principles are upheld therein.

Science-Based Public Policy Formation

This topic requires more discussion than the others as there is no formally codified process to be described: the situation is to some extent a free-for-all. To narrow the discussion I will primarily be referring to environmental policy, which can range from local issues (e.g., whether cities should ban cosmetic pesticides on lawns) to global issues (e.g., whether nations should implement the Kyoto Protocol). In each case some complex science must be reviewed to provide a basis of information for a policy decision.

The policy process specific to environmental issues has an extra strike against it: the enormous apparatus of activist lobbying. Table 13.1

Table 13.1 Annual revenue for fiscal year shown of some top environmental groups in North America and Europe

Group	Revenue	End of fiscal year	Source
Nature Conservancy	$923,010,000	30 Jun 02	Annual Report
Natural Resources Defence Council	$46,442,001	2002	Annual Report
The Trust for Public Land	$126,797,000	31 Mar 02	Annual Report
Environmental Defence	$43,841,405	30 Sep 02	Annual Report
Sierra Foundation	$73,112,136	31 Dec 01	charitynavigator.org
Sierra Club	$75,441,137	31 Dec 01	charitynavigator.org
Greenpeace Worldwide	$179,181,280	31 Dec 01	Annual Report
Worldwide Fund for Nature (WWF)	$117,800,000	2002	Annual Report
David Suzuki Foundation	$3,760,314	2002	Annual Report
Friends of the Earth USA	$4,391,503	Jun 01	Annual Report
TOTAL	$1,593,776,776		

lists some of the best-known environmental and conservation groups in the world, along with their annual revenues for the most recent year available (as of summer 2003 when I got the list together). All figures are converted into U.S. dollars. Worldwide these groups raise about $1.6 *billion* each year. By far the most successful fundraiser is the Nature Conservancy, with revenues of over $900 million and assets of $2.7 billion. Annual expenses at the Nature Conservancy just on administration and fundraising as of June 2001 were $99.2 million U.S. dollars (www.charitynavigator.org). While annual revenues like this would not quite get Big Green onto the Fortune 500 list, it is still a surprising amount of money.

By way of comparison, think about monetary policy for a moment. It is a highly technical area of economics, involving complex institutions and specialized theory. There are many ongoing debates within the profession over methodology, theory, and practical policy. Decisions in this area can have large effects on millions of lives for generations. Policy debates are vigorous but are conducted at a cordial, professional level. There is very little public activism, and media treatment tends to be routine and low-key, focusing on such things as daily stock tables or reports of speeches by central bankers.

Environmental policy is also highly technical, involving many branches of science as well as law and government. And there are many ongoing professional debates over methodology, theory, and practical policy, where decisions can affect people's lives for years to come. But unlike monetary policy, there is massive public activism to contend with, debates are often acrimonious even in professional circles, and media treatment leans heavily towards an alarmist interpretation of events.

It is the activism in particular that makes environmental science and policy such a difficult area. There is nothing remotely like Environmental Defence, Greenpeace, or the Sierra Club hanging over monetary policy. People do not chain themselves to bank machines and agitate for higher reserve requirements; movie stars do not campaign for fixed exchange rates; there are no publicity stunts to raise awareness about the bond yield curve. 'Activism' takes the form of think-tanks releasing studies on, say, international monetary policy coordination or bank mergers. 'Activism' is professional, technically informed, and courteous, befitting the topic.

The puzzle is not why activists are so quiet on financial issues but why they are so noisy on environmental issues, which are not intrinsically more important than financial issues. Part of the explanation is that some people do believe environmental issues are intrinsically more important, supremely important perhaps, though public opinion polls do not reflect this. Another part of the explanation is that many people seem to feel it is OK to hold and voice strong opinions on environmental issues on the basis of good intentions but little technical understanding, whereas in other areas people would want to have both in hand before becoming publicly active.

It is in the context of this unique storm of public activism that environmental policies are deliberated. A problem for the policymaker is that deciding on a contentious issue requires greater understanding of the science than the politician typically possesses. So an administrative layer has to be established to convey the science into the policymaking process.

In *Taken by Storm* we call this administrative layer 'official science.' We pointed out that official science is not, itself, science, even though it is typically assumed that it speaks for science. Its membership is somewhat open. There are individuals who are appointed, such as advisers to cabinet ministers and leaders of UN panels like the Intergovernmental Panel on Climate Change (IPCC), but others insinuate themselves into the role, like editors of popular science journals like *Nature* and *Sci-*

ence, journalists for magazines like *National Geographic* and *Discovery*, chairs of committees at royal societies and national academies, and so forth.

Official science is a necessary function, but it can easily find itself trying to perform an impossible task. Nature does not easily yield up its mysteries. Albert Einstein is said to have put it as follows: In the realm of the seekers after truth there is no human authority. Whomever attempts to play the magistrate there founders on the laughter of the Gods. Often when reading the overconfident tone of government and IPCC publications on climate change, I wish the would-be magistrates who write these documents had heeded this gentle caution.

The reality that many scientific questions are unanswerable runs up against the political constraint that official science is called upon, not to convey information per se, but to create an impression of *certainty*. This may seem like an unfair judgment, but recall Truman's famous lament about needing a one-handed economist. Official science acts as an agent to principals (namely politicians) who must go out in public to defend a contentious decision, and they want to be able to point to solid science as their support. For such a purpose the official science does not have to be *right*, but it does have to appear certain. That the subject might be fundamentally uncertain is no obstacle – after all the long-range prediction of climate is not possible, but that does not stop world leaders from making decisions on the assumption that it is routinely and accurately accomplished.

Lest that last statement strike the reader as controversial, I should offer a supporting citation:

> In climate research and modeling, we should recognize that we are dealing with a coupled non-linear chaotic system, and therefore that the long-term prediction of future climate states is not possible. The most we can expect to achieve is the prediction of the probability distribution of the system's future possible states by the generation of ensembles of model solutions. (IPCC 2001b)

Not only are long term predictions of climate states impossible, but even short-term prediction is effectively impossible. In the winter of 2004 Environment Canada predicted that summer 2004 would be exceptionally hot and dry.[2] It turned out to be unusually cold and wet. This provoked little public reaction, because no one expects computer models to be able to forecast the weather months in advance. The problems of predicting the climate years or decades in advance are even more intractable, not less: as we detailed in *Taken by Storm* there is less of a theoretical basis for understanding the state and dynamics on the climate scale than there is for observable weather. Yet the public accepts long-term climate forecasts at face value, based on endorsement by official science. In this respect official science has done its job well on the climate file, though the amputation of one hand from the body of expert opinion was not without some bloodshed.

That Official Science (in particular the IPCC) is geared towards providing certainty rather than information can be illustrated by the conspicuous change to the widely quoted 'conclusion' of the 2001 IPCC *Third Assessment Report*. The draft version of the *Summary for Policy Makers*, released in April 2000, at the close of the scientific review process, read as follows:

> *From the body of evidence since IPCC (1996), we conclude that there has been a discernible human influence on global climate.* Studies are beginning to separate the contributions to observed climate change attributable to individual external influences, both anthropogenic and natural. This work suggests that anthropogenic greenhouse gases are a substantial contributor to the observed warming, especially over the past 30 years. However, the accuracy of these estimates continues to be limited by uncertainties in estimates of internal variability, natural and anthropogenic forcing, and the climate response to external forcing. (IPCC 2000)

This paragraph is suitably non-committal, and conveys the intrinsic uncertainty in the subject. For this reason, presumably, it was deemed to be unsuitable for publication during the subsequent 'government review' phase, wherein it was rewritten into the much more definitive form published nine months later:

> In the light of new evidence and taking into account the remaining uncertainties, most of the observed warming over the last 50 years is likely to have been due to the increase in greenhouse gas concentrations. (IPCC 2001a)

There was no great advance of science in those nine months to justify this amplification of certainty: some officials (never identified) simply stripped out the relevant uncertainties because, in effect, that was their job.

Another illustration of the drive for certainty in official science was brought to light by an article in *New Scientist* in September 2004 concerning the fact that newer, more complex climate models are not providing anticipated reductions in the range of future global warming scenarios.

> Some climate scientists find these new figures disturbing not just for what they suggest about the atmosphere's sensitivity to greenhouse gases, but also because they undermine existing predictions. Uncertainty about those predictions is stopping politicians from acting to halt global warming. So, they argue, even suggesting that the model results are less certain could be politically dangerous.
>
> But other climate scientists fear creating a spurious certainty about climate change. Since we don't know what the future holds, they say, we shouldn't claim to know. These people see the predictions of climate models as less like a weather forecast and more like a bookmaker setting odds for a high-stakes horse race. There are no 'dead certainties.' They say that humanity has to act prudently and hedge its bets about future climate change in the absence of certainty. We will, they argue, never be able to see through the clouds, and politicians will just have to accept that. (Pearce 2004)

The bind for regular scientists who might protest the intellectual compromises inherent in official science is that it is hard to speak out on behalf of a position of uncertainty. If someone with the ear of Pharaoh claims to know something that cannot be verified one way or the other (such as the state of the future climate), the critic's profession of ignorance hardly amounts to a compelling rebuttal, and it guarantees that politicians will be less inclined to seek his advice anyway. Early in the process of researching climate change, some regular scientists tried to warn the public about the peril created by this epistemological asymmetry. For instance, Craig Bohren, an atmospheric physicist at Penn State, had this to say in a 1994 speech:

> The government's response to clamoring from an electorate frightened by global warmers to do something about global warming is to recklessly toss money to the wind, where it is eagerly grasped by various opportunists and porch-climbers ... I have never understood Gresham's law in economics – bad money drives good money out of circulation – but I do understand this law applied to science. Incompetent, dishonest, opportunistic,

> porch-climbing scientists will provide certainty where none exists, thereby driving out of circulation those scientists who can only confess to honest ignorance and uncertainty. (cited in Essex and McKitrick 2002: 301)

In a policymaking setting, faced with the need for certainty to underpin a public decision, some form of official science panel is inevitable. Consider, for example, the debate over whether lawn chemicals are a threat to health in cities. A government wants advice on the science, and by some chain of rumour, acquaintance, and political jockeying, Professor Bland is selected to form a panel and write a report. The panel settles on a particular view. The Bland Report comes out, a thousand pages long, dense with footnotes, all boiling down to a conclusion, which as it happens was precisely the view that Professor Bland and the other panellists held before writing the report.

So far so good. The problems begin when some other experts start to object. They say they were not consulted, or that the Bland Panel overlooked important evidence. But by now the government has institutionalized the Bland Report, referring to it as the scientific basis for a contentious decision. The opponents are the 'sceptics,' the minority, the outsiders; industry hacks or environmental alarmists as the case may be. It does not matter how many of them there are, how big the errors are that they find in the Bland Report, or how good their own arguments are. They do not have the institutional standing to produce a report of their own. Even if they sign joint letters of protest they are individuals speaking against an institution. To the extent they are politically embarrassing, the government can easily deal with them by playing whack-a-mole.

So the expert critics get frustrated and drop out of the debate. While this solves the immediate political problem, the result is that their expertise gets lost just when it is most needed.

While the overall process seems to have little consistent structure, at least as compared to trials and prospectus audits, there are two elements of codification worth mentioning. First, official science often faces a formal requirement that the information on which the conclusions rest must be drawn from 'peer-reviewed' journals. While this appears to be a constraint, in practice it is not, since a limitless variety of views can be supported with reference to peer-reviewed publications. Peer review is nothing like a jury trial or a financial audit; it is typically little more than an informal advice to an editor. An editor selects reviewers to provide anonymous advice about whether a paper ought to be published. Obviously an editor can (if he or she chooses) select reviewers who are known to be favourably or unfavourably disposed towards the paper, thus turning the process into a rubber stamp for foregone conclusions. Depending on the journal, the paper may go through a tough screening that weeds out error, but there are lots of journals publishing lots of papers of very inconsistent quality, and if someone is determined he or she can usually get published in a 'peer-reviewed' journal somewhere.

The 'real' review is the later process of replication, challenge, critique, and debate that goes on among specialists in a field after a paper has been published. This is a slower process and can take years or decades to proceed. However if a study is particularly pertinent for a policy debate, there is no reason in principle a science panel could not accelerate the process by hiring consultants to replicate it.

Indeed peer review can be perverted these days in even more insidious ways. In March 2004 a document was distributed via Canadian climate change research networks inviting participation in a multi-university consortium organized out of the UK. Entitled 'Economic and Technological Dimensions of Stabilizing Atmospheric CO_2 Concentrations: An International and Comparative Analysis,' the plan proposed a tightly scheduled series of consultations with interested researchers on the economics of stabilizing the carbon dioxide concentration in the atmosphere. It was taken for granted in the proposal that this would be, first, feasible, and second, desirable, and the proposal was to discuss modelling strategies for assessing the costs. There had been copious work on the costs of CO_2 emission cuts already, but the organizers had in mind new studies that build in rather fanciful assumptions about ambitious technological innovations that somehow could be induced by fiat, presumably reducing the projected costs of the policy. What made the proposal disturbing is not so much its tendentiousness or the ill-posed set of questions motivating it, but the transparent plan to produce papers on a timetable to ensure they could be used by the IPCC in its next (Fourth) *Assessment Report*, due out in 2007, including purchasing a special issue of a 'leading journal' (FEEM 2005):

> Based upon preliminary consultations held in Potsdam, Utrecht and Milan, diverse modeling teams would undertake to generate results relating specifically to questions surrounding atmospheric stabilisation that would be of interest to, and on a timescale for, the IPCC Fourth Assessment. The final results would be written up as a series of papers from the

> different teams. The results would then be collated and published as a Special Issue of a leading journal (see Umbrella Programme), in time for the results to be incorporated in the IPCC AR4.

The authors proposed careful coordination with the IPCC at each step to ensure that the research work was 'relevant' to their report.

Had this proposal come from, say, Exxon, the outcry would have been deafening. Suppose a major fossil energy corporation was soliciting participation in a series of studies based on vaguely defined methodology with a predetermined set of conclusions in mind, and with sufficient budget in place to purchase a special issue of a leading journal so as to guarantee the results come out in a 'peer-reviewed' forum in time to be used by the IPCC, which would have been involved along the way steering the research agenda. There would have, rightly, been considerable protest against the distorting of the peer-review process and the disingenuity of the IPCC steering research into print so that it could then cite it in its assessment report. That the funding is coming from governments does not make the situation any less disturbing.

Thus a stated reliance on the standard of 'journal peer review' does not amount to much of a standard at all. The second codification is unique to the United States, where the Federal Data Quality Act (FDQA) has been in force for three years. This act refers to the use of scientific information in government documents disseminated to the public and referred to as support for government policy. It imposes requirements of 'objectivity' (whether the disseminated information is presented in an *accurate, clear, complete,* and *unbiased* manner and is as a matter of substance *accurate, reliable,* and *unbiased*) and 'utility' (referring to the *usefulness* of the information to the intended users). A study containing 'influential scientific or statistical information' must also meet a 'reproducibility' standard, setting forth transparency regarding data and methods of analysis, 'as a quality standard above and beyond some peer review quality standards.'

The FDQA is already having an impact on the kind of scientific work done in the U.S. government, since the act prohibits public money from being used to disseminate documents that fail to meet its standards. Perhaps the best-known examples of its impact concerned the U.S. National Assessment on Climate Change of 2000 and the U.S. *Climate Action Report* of 2002. The Competitive Enterprise Institute (CEI)[3] sued the Administration in 2002 to prevent dissemination of the National Assessment on the grounds that it fails to meet the act's standards. The administration settled the lawsuit by adding a disclaimer to the National Assessment website stating that the document was produced by an outside agency and was not subject to the requirements of the act, effectively prohibiting its use for policymaking within the government.[4] The CEI then petitioned the EPA to cease dissemination of the *Climate Action Report*. The EPA rejected the petition, not because they disputed that the document fails to meet the FDQA standards but on the grounds that the EPA does not actively disseminate the report in question (the matter is under appeal).

I will conclude this section by offering a personal opinion on whether the process of connecting climate science to climate policy meets the tests implied by the principles of balance, disclosure, and due diligence. Considering the costs of implementing the Kyoto Protocol in Canada and elsewhere (McKitrick and Wigle 2002; McKitrick 2003) as well as the enormous complexities of the underlying science(s), I hold that the standard should be set very high for the policymaking process, and these three principles do not represent an onerous or unusual burden. Yet I find the IPCC and Environment Canada fail to meet them.

That the IPCC advocates for only one point of view is no secret. In a February 2003 interview in *Le Monde*, Rajendra Pachauri, the chairman of the UN Intergovernmental Panel on Climate Change, made the following comment:

> The fact is that our climate is changing and the consequences are very serious. Global warming demands dramatic behavioural changes on the part of individuals and societies, and we know that these changes are difficult to accept and to put into practice ... The Flat-Earth Society has only a handful of members today, and they continue to meet every year, to assert that the earth is indeed a slice. It is the same with climate change – you may deny it, but it is a fact. (Godoy and Suri 2003)

He conflates 'climate change' (which is a permanent and natural state of the Earth's environment) with human-induced global warming, but this does not disguise his point that anyone disagreeing with his position on the science (and the policy agenda he assumes it necessitates) can be dismissed as a Flat-Earther. In saying this he is following an earlier IPCC tradition: in a widely reported Reuters article of April 2001 his predecessor Robert Watson publicly dismissed the suggestion that there is a division of expert opinion on climate change: 'I personally believe it's something like 98-2 or 99-1' (McLean 2001), referring to the proportion of scientists who he said support the IPCC position.

There is nothing wrong with the IPCC holding a particular view on climate change; the problem is that it is an institution viewed by governments as a balanced adjudicator of the science, rather than as an advocate for one particular point of view. In fact many scientists have published research that casts doubt on the IPCC conclusions, with more coming out all the time, and many have stated publicly that they disagree with the IPCC on professional grounds. What is frustrating to an observer and practitioner like me is that because the IPCC is a government institution and the critics – no matter how many or how competent – are individuals, there is a presumption that the institution must be right.

Environment Canada has also never sought to evaluate the arguments of scientists who dispute their position on global warming: indeed they do not seem to believe that there are any grounds for disputing their position. In their *Action Plan 2000* they summarized the science with the statement 'Our scientific understanding of climate change is sound and leaves no doubt that it is essential to take action now to reduce emissions' (Environment Canada 2000: ch. 7). Around the time this was published I attended a meeting on research funding with (among many others) a senior official at Environment Canada. In discussing climate change he lamented that he did not know of any experts who could present counterpoints to the dominant views he was hearing within his ministry. I promptly took out a piece of paper and wrote down the names of at least a dozen scientists at universities, and the topics they could specifically address. I suggested he find the money to put on a workshop where these experts could come in and discuss their research. He was initially quite positive about the suggestion, but after thinking about it he handed me the list back and said that while he would love to see it happen, he could not do it himself. His explanation was worded roughly like this: The minister has spoken and the prime minister has spoken. If I spend government money on an event that openly contradicts their views I would lose my job.

This was a career civil servant, with considerable seniority, and he was unwilling to serve as a conduit for a balancing perspective out of fear of being fired. So, of course, no such consultation ever took place. Not every ministry is as unbalanced as Environment Canada, but listening to the utterances of former Environment minister David Anderson it is clear he only ever got one type of information, with the certainty quotient running extremely high. In response to my written suggestion in 2001 that he meet some of his sceptics, he wrote back about having 'conclusive proof that the climate has changed and that this is the result of human activities' thereby dismissing the need to hear from others (quoted in Essex and McKitrick 2002: 56). More recently (February 2004) he spoke in Toronto, where he castigated the suggestion that there is any uncertainty on global warming: 'Sir David King, the chief scientific advisor to the Government of the United Kingdom puts it bluntly. "Climate change is real." Full stop. Not, "maybe." Not, "sort of." *Real*' (Anderson 2004). Here again he conflates climate change with anthropogenic global warming, but the context of the speech made it clear he means the latter. This aggressive claim of certainty would be remarkable for any area of science, but for a topic like climate change it is demented. In any case the point is that there is no balance to the treatment of the subject in Environment Canada or the IPCC, because no contrasting points of view are sought out.

As for the principles of disclosure and due diligence, I have been involved in an ongoing effort to replicate one of the central aspects of the IPCC's case for global warming, the so-called 'hockey stick' graph of Mann, Bradley, and Hughes (1998; 1999) purporting to show that the climate of the late twentieth century is unusually warm compared with the past thousand years.[5] The episode sheds considerable light on the failure of the IPCC, Environment Canada, and other agencies that promote the global warming scare to implement either principle.

When my co-author Stephen McIntyre first looked into replicating the Mann, Bradley, and Hughes result he immediately discovered that the data could not previously have been requisitioned by anyone, since it took some considerable digging by the originating author to come up with a usable version of the data. When we published a study (McIntyre and McKitrick 2003) enumerating many errors in that dataset Professor Mann promptly disowned the file we had been sent, instead pointing to a new FTP source, the contents of which turned out to differ from what had been described in the original *Nature* publication. When we pointed this out to *Nature* they ordered publication of a corrigendum (Mann, Bradley, and Hughes 2004). In subsequent correspondence we have uncovered further discrepancies in the data identification and the methodological description, but *Nature* has effectively given up trying to resolve them.

The influence of the Mann, Bradley, and Hughes hockey stick on the IPCC conclusion cannot be overstated. In IPCC (2001a), it appears as Figure 1b; in IPCC (2001b), it appears as Figures 2-20 and 2-21; in IPCC (2001c), it appears as Figure 5; and in IPCC (2001d), it appears as Figures 2-3 and 9-1B. The IPCC *Summary for Policymakers* used this figure as the basis of its prominent claim that it is likely 'that the 1990s has been the warmest decade and 1998 the warmest year of the millennium' (3) for the northern hemisphere. The hockey stick graph has also been reprinted countless times and used by governments around the world, including Environment Canada. Yet none of these bodies made any attempt to verify the results or replicate the study. After publishing our first paper on the subject we were contacted by a scientist who had tried to warn the IPCC about its use of the hockey stick graph without proper evaluation. In his email, which I must quote anonymously, he said:

> I was one of a myriad of 'reviewers' of the IPCC 2000, prior to its publication. One of the major concerns I expressed was the high level of credence given to the Mann et al. temperature history, without it having been seriously subjected to testing. I strongly recommended that this had some dangerous implication, should the reliance upon that research prove premature.

Another scientist commented in print some time later that his own, similar misgivings about the graph had prompted him to remove it from his paleoclimatology teaching notes (Muller 2003). Clearly the IPCC does not perceive its duty to include due diligence in any form that would be recognizable in a private sector setting.

The duty of disclosure is closely related in this instance. After Steve McIntyre and I published our preliminary work, we were criticized by Mann, Bradley, and Hughes for not using the 'right' data or the 'right' methods, though we had used what was then available and requests for further details had been refused. In the aftermath of our publication we were able to work out some more specifics of the methodology and data, but to this day some important details remain undisclosed. We were unable to convince *Nature* to force release of these details even though they had published the original article and relied on some undisclosed computations in their subsequent refusal to publish a critique we submitted. We unsuccessfully petitioned the U.S. National Science Foundation (NSF), which funded the research, to enforce its own policies on grant recipients by requiring Mann, Bradley, and Hughes to disclose their computer code and computations; the NSF argued these were the personal property of the authors. We spent an enormous amount of time in 2003–4 extracting some of the information behind the

Mann, Bradley, and Hughes 1998 paper, but eventually it became clear the institutions involved had no willingness to enforce their existing disclosure requirements on published authors. Much less have the IPCC or Environment Canada ever verified the information behind the conclusions they publish, in particular those based on the studies of Mann, Bradley, and Hughes. We know they have never sought disclosure of this information, let alone felt an obligation to provide such disclosure to those readers in government and the general public whom they feel are obliged to take their position seriously.

A Better Mechanism for Official Science

It should not be surprising at this stage of the argument that I would like to see official science placed under an obligation to meet the three principles I have outlined above: balance, disclosure, and due diligence. If these are deemed too onerous for the government to meet when proposing policies that will cost the taxpayers tens of billions of dollars, how then does the government justify requiring the courts and the private sector to meet them in situations where much smaller amounts of money are usually at stake? There is no justification for any such double standard. The principles could be implemented as follows.

The Principle of Balance

In *Taken by Storm* Chris Essex and I spelled out a detailed proposal for introducing balance into settings where scientific information is used to decide major policies. Suppose a municipality is trying to decide whether to ban lawn pesticides. Rather than forming one panel, a city should form two: one would be asked to produce the strongest possible case for the ban and the other the strongest possible case against. Then each team would be asked to write a rebuttal to the other's. The final report would consist of all four documents and would not contain an executive summary.

Does this sound strange? Two teams? Handpicked so they hold foregone conclusions? Sure. Let them be as biased as they like. Let them self-select their members and tilt together into their preferred position. Remember that for the process to be balanced does not require that the individuals be neutral, only that the contrasting points of view be well represented. In the end the two teams' reports will be set side by side. If they are evenly matched, so be it. That is the honest message of the science. And any process that fails to convey it is perpetrating a fraud on the public.

In the case of climate change, the day is far spent and it may seem like a pipe dream to hope for a balanced process, but then again the issue is going to be here for a long time and there is no reason to settle for the flawed institutions currently dominating the discussion. There are opposing views, and it is not obvious which is correct on any particular question. Governments ought to form two groups with equal funding and adequate membership in each. One group could be called Heads and the other group Tails. The job of the Heads group would be to produce a report making as strong a case as possible that human activity is causing a significant climate change that will have harmful consequences. The Tails group would have the job of making as strong a case as possible to the contrary.

Since we would have done away with the artificial labels of 'mainstream' and 'marginal,' a wider range of participants would likely come forward, especially on what today is maligned as the 'sceptical' side.

Each group would be asked to produce, say, a three-hundred-page report, as well as, later, a one-hundred-page rebuttal to the other group. The complete eight-hundred-page document would be released without a summary, but with an index. It would be submitted to the world's governments without either panel being asked to render a decision on which team's report is stronger.

Each government then would have to decide for itself. They could, if they like, consult internal and external experts for their opinions. But even if one government made the mistake of setting up a national official science group to render a verdict and write a summary, it would not bind on any other country.

Let me stress that I do not think this process would be appropriate for scientific controversies in general. I have in mind situations where costly public policy decisions depend on difficult scientific controversies. The point of the process just described is to make sure that both the prosecution and the defence, so to speak, are represented. The current institutions are more like courts in which the prosecutor also serves as the judge and is not required to have the defence represented at all.

The Principle of Disclosure

In addition to the above mechanism I would like to see rules established (akin to the FDQA in the United States) governing what kind of science can be invoked for public policy. For a study to be referred to in a policy-relevant assessment it is not enough that it have passed journal peer review. It must meet a standard of disclosure in which the data and computational methods used to derive the results are either freely available or, in the case of proprietary data, have been supplied on request to an independent third party, who has certified the reproducibility of the results. Many academic journals are toughening up their disclosure rules. The *American Economic Review*, for instance, late in 2003 adopted a rigid new policy that, as a precondition of publication, any paper being published in the journal must supply the data and computational code for archiving on the journal website. This came about as a result of a project in which two authors were unable to reproduce a majority of the results in a recent edition of the *Review*, mostly because of unavailability of the data or the obscurity of the methods. The rule I have in mind would stipulate that in cases where full, true, and plain disclosure of data and methods cannot be verified, a study cannot be referred to in assessment documents which will be cited in support of a policy decision.

The Principle of Due Diligence

In many policy settings there are a few studies which are of identifiably pivotal importance. In air pollution policy, a handful of very influential studies look at the correlation between pollution levels and health effects or mortality. In climate change policy there are a handful of papers (including the Mann, Bradley, and Hughes papers cited above) that strongly influence the conclusions of the IPCC, and in turn groups like Environment Canada. A requirement of due diligence would require that the expert staff at government ministries or specialized bodies like the IPCC must verify that they can reproduce these results starting from scratch in another computer package. This would follow on from disclosure but would also ensure that the analytical methodology was accurately stated and the process of redoing the results would bring to light any errors in methodology. I envision either specialized statistical consulting firms would emerge to do this replication work, or there would be a build-up of more technically sophisticated staff in government bureaucracies. It would also force researchers who want their work to be influential on policy to be more conscientious in publishing their data, methods, and code in easily accessible formats.

Public Science in Liberal Democracy: Three Questions

To conclude, I turn to the three questions posed, which I hope the foregoing has laid some groundwork for answering.

1. Can science retain independence and objectivity in the face of demands to meet commercial and public policy objectives?

At the risk of sounding trite I must object that *science* is a thing, not a person, and as such is trivially independent, objective, and impersonal. It always will be these things, as will the stars above and the potted plant in the corner, and this is nothing special one way or the other. *Scientists* however, as human beings, are neither independent nor entirely objective. I have been influenced in my (admittedly limited) thinking about the nature of scientific knowledge by Michael Polanyi's fascinating book *Personal Knowledge* (1958).

As to objectivity, knowing is a skillful act involving a personal commitment to what one believes to be true, coupled with an awareness of one's capacity to be wrong, and consequently is inherently subjective in the sense that all knowledge must be 'somebody's' knowledge. Yet, as Polanyi emphasizes, knowledge is not simply subjective, since objective truth is the external reality of which the knower is focally aware. Skills, including particular elements of an intellectual framework, must be indwelt at an uncritical and precognitive level for them to support knowledge of the real, and as such any serious scientific training must cultivate a deep personal commitment to a field and its assumptions. Hence, as to 'independence,' let us ask: independent of what? The scientist only acquires training by subordinating his or her judgment to a community of expert practitioners – a scientist attempting to be independent of the scholarly community would never acquire the knowledge nor indwell the tools of analysis sufficiently to get started in the field. Nor is a scientist independent of the interested public. The long effort required to acquire advanced scientific knowledge is motivated by a passionate interest in the subject, and this inevitably brings the academic scientist into repeated contact with people who share that interest, including members of the public. We would think of someone who tried to avoid all such interaction as being eccentric, if not senile.

Can scientists be independent and objective? No, nor would we want them to be. They are dependent on one another and on the communities in which they live, and their knowledge reflects the unavoidable personal commitment to the skills subordinated in the act of comprehending the objective reality hidden from the casual observer. We do ask that scientists be honest and suppress any preferences they may hold as to the outcome of analyses or experiments in favour of letting the data speak transparently. But users of the knowledge for the purpose of political or institutional decision making have a fiduciary duty to establish processes that do not rely for their validity on the voluntary honesty and balance of the participants, any more than we rely on the honesty and balance of the prosecutor in a trial – even though we might consider her honest and fair – yet still insist on independent counsel for the defence. Hence the process by which scientific information is conveyed into a policy process can best be made independent and objective by imposing mechanisms to ensure balance, disclosure, and due diligence, just as is routinely done in other decision-making contexts.

2. In what ways is scientific discourse privileged in the formation of public policy? Can there be a genuine public discourse if one party is privileged?

If the concern is that scientists get inordinate sway over policy formation, I would say scientific discourse has not been privileged. As an economist I lament the absence of economic reasoning in many policy settings where economics has direct bearing. I realize there are non-economists who lament the perceived extent of economics influence in policy formation, but this lament is overstated. Policy is driven by polling data, and technical arguments are at a disadvantage in this setting, since activists can use rhetoric and demagogy to persuade the public to support worthless and costly measures. When the counterargument requires careful construction and cannot be communicated in sound bites, then any technically dense communication is at a disadvantage.

But scientific discourse has been privileged in one sense: by being held to a much lower standard in terms of balance, disclosure, and due diligence. Academic research, even when being used to drive multibillion-dollar public investments, is done to standards that would never be acceptable in the business sector. This is not necessarily a problem for the academic purpose being served, since researchers have to have considerable leeway to make their mistakes in public in order to ensure scholarly communication remains open and important topics are probed thoroughly. The problem arises when governments assume journal peer review amounts to a standard of verification similar to what would be applied in a business setting or a trial procedure. This is a disastrous assumption.

3. How can scientific knowledge and scientific methodology be made compatible with the interdisciplinarity and integration required of public policy discourse and formation?

I have set out suggestions in the previous section that would address these issues, at least to some extent. But let me reverse the question: how can the interdisciplinary bodies involved in public policymaking (e.g., the civil service and the interested academic sector) be convinced of the need to grapple with the scientific knowledge and methodologies relevant to understanding the issues at hand? There is an onus on both sides. The users of the knowledge have to meet the producers of the knowledge halfway. It is not necessary for politicians to be scientists, or even for all their bureaucrats and advisers to be scientists. But they have to be prepared to do the hard work of learning some of the science, including the mathematical foundations.

I have, on several occasions, addressed audiences of influential bureaucrats on technical issues relating to climate change science and policy. While there are many intelligent and well-trained government staff members and elected officials who genuinely want to grapple with the technicalities, I have also encountered more than a few who are intellectually lazy, expecting to have complex mathematical ideas reduced to a thin, sweet milk then spoonfed into them. But many of the important scientific issues of the day just cannot be reduced this way without fundamentally misrepresenting them. There are people who are happy to present their message in snappy sound bites because their message is simplistic and shallow. But if we want sound policy we have to have a mechanism for communicating honest, complex, deep science into the policymaking process, without distorting or stripping down the content along the way. My suggestions, as laid out in the previous section, are aimed at doing just this.

NOTES

1 I am grateful to Stephen McIntyre for the description of prospectus requirements described in this section, as well as for details about the Bre-X case.

2 See comparison of forecast to actual temperature anomaly fields at http://

weatheroffice.ec.gc.ca/saisons/charts_e.html?season=jja&year=2004&type=t.
3 See www.cei.org.
4 See http://www.usgcrp.gov/usgcrp/nacc/.
5 Details are available in McIntyre and McKitrick (2005a,b), and on the website www.climateaudit.org.

REFERENCES

Anderson, D. 2004. Remarks to the Economics Club of Toronto. 20 February. http://www.ec.gc.ca/media_archive/minister/speeches/2004/040220_s_e.htm (accessed 27 April 2007).

Environment Canada. 2000. *Action Plan 2000*. Ottawa: Queen's Printer.

Essex, C., and R. McKitrick. 2002. *Taken by Storm: The Troubled Science, Policy and Politics of Global Warming*. Toronto: Key Porter Books.

Fondazione Eni Enrico Matei (FEEM). 2005. Economic and Technical Dimensions of Stabilizing Atmospheric CO_2 Concentrations. http://www.feem.it/feem/pub/networks/stabilizationgroup/default.htm (accessed 9 July 2007).

Godoy, J., and S. Suri. 2003. Hot Air over Global Warming. *Le Monde*, 19 February.

Intergovernmental Panel on Climate Change (IPCC). 2001a. *Summary for Policy Makers of the Contribution of Working Group I to the IPCC Third Assessment Report*. Cambridge: Cambridge University Press.

– 2001b. *Climate Change 2001: The Scientific Basis. Contribution of Working Group I to the Third Assessment Report.* Cambridge: Cambridge University Press.

– 2001c. *Technical Summary of the Contribution of Working Group I to the Third Assessment Report*. Cambridge: Cambridge University Press.

– 2001d. *Synthesis Report of the IPCC Third Assessment Report*. Cambridge: Cambridge University Press.

Mann, M.E., R.S. Bradley, and M.K. Hughes. 1998. Global-Scale Temperature Patterns and Climate Forcing over the Past Six Centuries. *Nature* 392: 779–87.

– 1999. Northern Hemisphere Temperatures during the Past Millennium: Inferences, Uncertainties, and Limitations. *Geophysical Research Letters* 26: 759–62.

– 2004. Corrigendum. *Nature*, 1 July.

McIntyre, S. and R. McKitrick. 2003. Corrections to the Mann et. al. (1998) Proxy Data Base and Northern Hemisphere Average Temperature Series. *Environment and Energy* 14, no. 6: 751–71.

– 2005a. The M&M Critique of the MBH98 Northern Hemisphere Climate Index: Update and Implications. *Energy and Environment* 16, no. 1: 69–100.

– 2005b. Hockey Sticks, Principal Components and Spurious Significance. *Geophysical Research Letters* 32, no. 3 (12 February): L03710 10.1029/2004GL021750.

McKitrick, R. 2003. Budget 2003 and the Kyoto Process. In *The 2003 Federal Budget: Conflicting Tensions*, edited by C. Beach and T. Wilson, 171–92. Kingston: John Deutsch Institute, Queen's University.

McKitrick, R., and R. Wigle. 2002. The Kyoto Protocol: Canada's Risky Rush to Judgment. C.D. Howe Commentary, October. http://www.uoguelph.ca/~rmckitri/research/papers.html (accessed 27 April 2007).

McLean, W. 2001. UN Expert: Climate Change Skeptics a Tiny Minority. Reuters, 5 April.

Muller, R. 2003. Medieval Global Warming. *MIT Technology Review*, 17 December.

National Assessment Synthesis Team. 2000. Climate Change Impacts on the United States: The Potential Consequences of Climate Variability and Change. Washington, DC: U.S. Global Change Research Program. http://www.usgcrp.gov/usgcrp/Library/nationalassessment/overview.htm (accessed 27 April 2007).

Pearce, F. 2004. Harbingers of Doom? *New Scientist*, 24 July, 43–6.

Polanyi, M. 1958. *Personal Knowledge*. Chicago: University of Chicago Press.

U.S. State Department. 2002. *Climate Action Report: The United States of America's Third National Communication under the United Nations Framework Convention on Climate Change*. Washington, DC: U.S. Government Printing Office. http://yosemite.epa.gov/oar/globalwarming.nsf/content/Resource CenterPublicationsUSClimateActionReport.html (accessed 27 April 2007).

14 Technoscience in an 'Illiberal' Democracy: The Internet and Genomics in Singapore

ZAHEER BABER

The issue of the relationship between science and democracy has been tackled by sociologists in a number of different ways. Given the fact that democracy is an essentially contested concept with various models and practices associated with it, this state of affairs is hardly surprising. Up until ascendancy of the social constructivists, the concept of science itself was essentially uncontested and perhaps even uncontestable. There was little disagreement about the essential ingredients of science and, a few dissenters notwithstanding, not much debate over the desirability of the constant and continuous advancement of science which was unambiguously equated with human progress.

For an earlier, pre-constructivist generation of science scholars such as Robert Merton (1942; 1973) and Bernard Barber (1952), the question of the relationship between science and democracy was largely a matter of specifying the set of social structures and cultural norms that either inhibited or facilitated the growth of science and technology. Thus, Merton's famous and much-criticized paper titled 'Science and Technology in a Democratic Order' (1942) specified the four institutional imperatives or norms – namely, universalism, communism, disinterestedness, and organized scepticism – that were supposed to be the crucial institutional supports for the advancement of science and technology. Against the background of claims about epistemologically distinctive Nazi and proletarian science in the Soviet Union, Merton was keen to defend the autonomy of science, relatively free from pressures from the state, church, or the economy. For him, universalism and organized scepticism were the guiding principles of both science and democracy and both institutions provided symbiotic resources for each other. In Merton's view, science as an institution became differentiated from other institutions and emerged as a relatively autonomous institution in the West. Liberal democracy, sharing similar attributes, could guarantee and protect the autonomy of science from a host of extra-scientific influences. At stake was nothing less than the epistemological purity of science, which could be preserved best under conditions of liberal democracy that allowed for a multiplicity of competing and balancing interests.

The social constructivist revolution heralded in the 1970 by a number of determined British and European sociologists successfully contested the very notion that science and technology were untainted by social and political interests (Mulkay 1979; 1985). Working under theoretical umbrellas and clusters, the social constructivists succeeded in opening up the 'black box' of science to social scrutiny. In so doing, science itself became a contested concept, much like democracy, and the construction and stabilization of scientific facts came to be regarded as process that was aided not simply by the brute facts of nature but also by the shifting alliances of social and political forces. In effect, as study after study sought to demonstrate, social and political factors played an insignificant role in the closure of many scientific controversies. The emergence of a full-blown constructivist perspective sought to provide a corrective to the traditional Mertonian view of science that had to be protected from possible distortions by social forces. The collision course between the relativist possibilities opened up by the constructivists and the implicit defence of the realist position by an earlier generation of scholars came to a head during the 'science wars' triggered off by Alan Sokal's now-infamous paper (Sokal 1996). Much of the bitter and at times mean-spirited aspects of the so-called 'science wars' have now largely subsided and given rise to a degree of rethinking of hardened positions taken up by proponents of both sides. A number of scholars, most notably the philosophers Philip Kitcher (2001) and Ian Hacking (1999) and the sociologist Thomas Gieryn (1999) have succeeded in creating a viable theoretical space for those not ready to commit themselves to mutually exclusive epistemological binaries such as relativism versus realism. The Mertonian view of the relationship of science and democracy had been modified by a number of scholars who have combined the best elements of constructivism with a structural perspective (Baber 1994; Klein and Kleinman 2002; Kleinman 2000; Kleinman and Vallas 2001).

One of the major consequences of the constructivist revolution has been the reconsideration of the relationship between science and democracy. Whereas the pre-constructivist sociologists sought to present democracy as the ideal social and political system in which neutral science and technology would take root and prosper, the constructivist position opened the possibility of thinking about the democratic possibilities inherent in different forms of technoscience. Thus if one accepts the argument that the construction of science and technology is not just driven by the material facts of nature and allows for some role for social forces and interests, then the final product, whether in the form of theories, explanations, knowledge, or artefacts, necessarily carries the imprint of social and power relations. If specific interest groups directly or indirectly influence the stabilization of scientific knowledge and artefacts, then the issue emerges as to whether science and technology can be democratized to best represent the interests of society. As the debates over genetically modified food, stem cell research, nanotechnology, climate change, and so forth indicate, the Comtean vision that a scientific dictatorship of sorts driven by presumably neutral facts, figures, and calculations would alone count for the future contours of society was thoroughly misplaced. Of course, as is well known, the Comtean project was anything but neutral and objective, representing as it clearly did a patently political and ideological agenda. Even though Karl Marx scornfully dismissed the Comtean project as a 'cookbook full of recipes for the future' (1967: 16), scientism and the cult of expertise in almost every sphere of life continue to be strong forces, as do movements against the potential total domination of social life by Max Weber's 'specialists without spirit, sensualists without heart' (Weber 1930: 182).

Not all constructivists have explicitly deployed their perspectives for prying open the black box of scientific knowledge with the goal of democratizing science. Indeed, some of them have been stridently criticized by Langdon Winner (1993) for playing lame apolitical games without much regard for the intersection of specific forms of scientific knowledge with the inequities of power relations. However, significant aspects of constructivism have been deployed by a number of scholars to study the politics of science and indigenous knowledge (Agrawal 2002; Gupta 1998; Gururani 2002; Philip 2003; Proctor 1991). On the other hand, scholars such as Vandana Shiva have also used constructivism to conjure up promises of a kinder, gentler, and infinitely more democratic science and technology. While cautious constructivists started out by insisting that the book of nature cannot be read directly and nature by itself is not the only ingredient of scientific knowledge, some of the more enthusiastic proponents of this idea ended up concluding that since scientific knowledge is socially constructed and influenced by specific social interests, it could easily be redirected and reconstructed along democratic, egalitarian lines, unhindered by any structural constraints.

Although few social constructivists have explicitly argued that nature plays no role in the constitution of scientific knowledge, some over-enthusiastic proponents of 'democratic science' trade on this assumption. What the results of decades of 'science and technology studies' have made abundantly clear is the fact that both social and natural factors are simultaneously implicated in the construction and deployment of scientific knowledge. As the sociologist Thomas Gieryn (1999) has pointed out, the debate over whether science is natural or social or realist versus relativist is really a non-issue, since it is both. Scientific knowledge is simultaneously natural and cultural. Gieryn (1999) and Kitcher (2001) deploy the analogy of maps and the practice of cartography to make their point. Maps are simultaneously representations of physical and political reality but are not coterminus with it. A unified mimetic map of everything on earth would be worse than useless. The unity-of-science perspective believes in the construction of an ideal atlas that would eventually represent nature as it really is. This ideal, impartial, and disinterested map or atlas is not possible. Any map represents a particular point of view and facilitates a particular set of action, be it golfing, hunting, hiking, or riding the subway. No ideal map would ever capture everything of interest to everybody. Maps do represent particular interests and simultaneously offer not neutral but strategic truths that orient particular courses of action. Maps that are misleading are worse than no maps. Like maps, the significance of scientific results is intrinsically intertwined with strategic concerns, and we cannot appeal to some overarching project whose value transcends all others (Kitcher 2001: 95). However, as Kitcher has argued, the fact that interests are necessarily implicated in the construction of maps and scientific theories does not entail the abandonment of the ideals of objectivity and 'significant truth.' Building on the findings of a diverse range of science and technology scholars, Kitcher contends:

> I take moral and social values to be intrinsic to the practice of the sciences. My argument for this view doesn't depend on abandoning the idea that the sciences yield truth about nature or on giving up the ideal of objectivity. (65–6)

Just because maps represent particular interests does not mean they do not offer truths. These 'truths' are not pure neutral abstractions, views from nowhere and everywhere. Rather, they represent 'significant truths' from the point of view of both those who construct and those who deploy the maps or scientific knowledge for specific ends. Similarly, the myth of purity of science that conceives of science as searching for abstract, ahistorical truths is problematic because, like maps, 'the sciences are surely directed at finding significant truths' (Kitcher 2001: 65). Despite the constructivist revolution, or perhaps because of it, Kitcher seeks to preserve its most valuable insights without giving up on the conception of significant truths or objectivity. As he puts it:

> There is no need to abandon the everyday conception that inquiry yields truth about independent objects. Nor should we suppose that ideals of objectivity are misguided and that, because of rampant under determination, scientific decisions are made, perforce, by invoking moral, social and political values. (85)

At the same time he criticizes 'the myth of purity' or the 'claim that gesturing at the absence of any particular intent is enough to isolate a branch of inquiry from moral, social or political critique' because 'the significance of scientific results is entangled with practical concerns, and we cannot appeal to some overarching project whose value transcends all others' (91–5). Steve Fuller's project of 'social epistemology' also raises these important issues that bring power and questions of political economy to the centre stage of the analyses of science and technology (Fuller 1991).

Kitcher and Gieryn have raised an important issue. By selectively accepting and deploying the valuable insights generated by the constructivist perspective, they seek to question 'scientism' that sets up science as a value that transcends all other values and social concerns. At the same time they do not wish to argue, as some hyper-constructivists do, that science is simply the consequence of social, political, and moral struggles. Taking the second position would entail the possibility of constructing a kinder, gentler, and more democratic science without having to deal with some intractable constraints posed by nature, natural objects, and, of course, social structures and agents. In other words, Kitcher does not want to reject the idea that scientific facts are made and that this making is a social, political, and moral process that contributes to the making of not just any facts but significant facts or truths.

However, while nobody disputes the fact that scientific facts are made and not found, it ought to be obvious to anyone with the exception of the most die-hard anti-realist that scientific facts are not made up. If they were made up, the task of democratizing science would be easier than imagined and possibly meaningless. As many scholars have pointed out, it is precisely the creative tension between structure and agency, or between the reality of constraints and the possibility of transformative actions that contest those constraints, that allows for the simultaneous regularity and contingency of social life (Bhaskar 1979; Giddens 1986). If figuring out democratic solutions in theory or on a piece of paper were the only structural constraint in the way of a just and equitable life for everyone, then the issue would have been resolved quite some time back. With the exception of misanthropists, it would be hard to imagine many who would not welcome democratic technoscience.

Nowhere is the tension between constraints and possibilities more evident than in the republic of Singapore. At first glance, an island only six hundred square kilometres with about four million inhabitants would seem an unlikely place to examine the relationship between science, technology, and democracy. Yet, Singapore and perhaps China constitute the sites where some of the most far-reaching scientific revolutions in genomics are set to unfold. While China has never claimed to be a liberal democracy, Singapore has had an elected government since 1965. Elections are held every four years and the British parliamentary system is firmly in place. Sceptics who point out that despite elections the People's Action Party has never been voted out of power since the inception of the republic also acknowledge the election of a handful of opposition members of parliament since 1984. The script is however more complicated. Singapore is a democracy because it indeed does hold elections every four years, and voter turnout is necessarily high as it is compulsory. However, many observers have labelled Singapore polity as an 'illiberal democracy' since freedom of expression is circumscribed (Bell et al. 1995; Chua 1996). The consequences of the state control of the print and electronic media may not be so apparent all the time but becomes salient during elections. Under existing laws, public discussion of issues deemed 'political' is severely curtailed, and more than five individuals cannot get together to discuss so-called 'sensitive' issues. A 'public entertainment licence' is required for any public speech, and the scripts of all plays to be performed, even in the theatre studies department of the National University of Singapore, have to be vetted in advance by government officials. A few years ago, a speaker's corner, along the Hyde Park model, was inaugurated, but despite the initial interest, it is deserted much of the time. Potential speakers had to register at the local police station and provide the title of their talk; 'sensitive' issues discussing ethnicity and religion were off limits. An obvious consequence of these highly restrictive laws, inherited during the colonial period, but applied with a vengeance that modern technologies of control such as computers make possible, is the situation of an electoral democracy but without the guarantee of the free exchange of views, critiques, and contestation that constitute the essential ingredients of liberal democracy. The self-conception of a 'communitarian democracy,' where individual demands are deemed secondary to the claims of the presumed community, is promoted by the state. Attempts to open up avenues for freedom of expression have frequently been criticized for espousing alien Western liberal views that seek to weaken putatively Asian communitarian values. Because the state presumes to speak for the community, independent civil society organizations were banned until the mid-1980s. The few civil society organizations that did emerge after that period have been careful not to cross the OB, or 'out of boundary,' markers. Apparently there was no need for independent collective expression of views that took issue with specific state policies because the state saw itself as representing the interests of all sections of the society.

Since the mid-1980s, two major technoscience initiatives by the state have had important ramifications for democracy in the island city-state as well as global implications. In an era when some of the major transformative technologies such as genomics, genetic engineering, and nanotechnology are engendering global unease and civil society activism, the developments in Singapore are not limited to the city-state.

Personal Computers and Visions of an Intelligent Island

In the elections in the 1980s the ruling party (PAP) that had always won a hundred per cent of all the parliamentary seats in the elections was in for a major blow to the body politic. A member of the vehemently anti-PAP party, the Workers Party's, Joshua B. Jeyaratnam, defeated a sitting MP. In the next elections, this shock was further intensified when two other PAP members lost their seats. The structural context of these defeats was the recession of the 1980s, driven partly by the emergence of a number of competitors for the manufacturing jobs that Singapore had managed to attract since the 1960s. Indonesia, Malaysia, and Thailand were now able to attract some of these jobs, and the cost of labour in Singapore was on the rise. The circuits of global capital were being restructured, and Singapore was no longer a favoured destination for low-end manufacturing. The loss of jobs as a consequence of these global transformations was partially responsible for the electoral setbacks for a government whose major source of legitimacy in the past had been the transformation of a collection of sleepy villages into a modern, industrialized society. The Singapore state, forever planning for the future, sought to capitalize on its superb infrastructure that its neighbours were deprived of and move into high-end manufacturing. Coincidentally, at the same time, the personal computer was introduced and the state was quick to adopt a policy of the computerization of the society. By the time the internet made its appearance, the state had seized upon the metaphor of Singapore as 'an intelligent island,' presumably different from its neighbours, which could settle for the non-intelligent, low-end manufacturing jobs.

The setting up of a National Computer Board, and later the Infocom Authority, to transform the country into an 'intelligent island' was highly effective, and soon Singapore was ranked among the top five most wired countries in the world. The social and political consequences for this enthusiastic embrace of the PC and the internet were many. In terms of democratization, even though computers enhanced the already super-enhanced surveillance capacity of the city-state, the very technology also enabled certain tentative democratic changes. Up until the 1980s any independent civil society organizations were banned. After the electoral shocks of the early 1980s, the government gingerly experimented with allowing three such organizations. The idea was to allow some controlled letting off of steam to prevent further erosion of political legitimacy. All three civil society organizations made it very clear that they would play by the rules and would not seek to embarrass the government. However, with the wiring of the city-state, a number of virtual civil society organizations emerged. Laws against more than five people discussing issues deemed to be political, although still on the books, were toothless as the internet chat lines exploded with contentious discussions that were simply unthinkable in the past. Censorship, pursued aggressively in the age of print, radio, and TV media, was also weakened in an era of web-based publications. *The Economist*, frequently a target of censorship, was available freely over the 'Net. Although the state had, through a skillful deployment of

Gramscian coercion and consent, wired up the island for purely economic reasons – e-commerce, the dot.com fever, financial hub of Asia, and so forth – the unintended consequence of this initiative was the opening up of spaces for discussion and a measure of social activism that literally fired up the political discourse beyond anyone's imagination. The most prominent of these clusters of civil society organizations include Sintercom, Talking.com, and ThinkCentre, among others. Although the political impact of these virtual activists should not be exaggerated, given the lack of any meaningful engagement with the state, they had a very real impact on opening up genuine political spaces. Previously, the state had sought to define politics as strictly the business of professional politicians. Such definitions of political reality had been partly accepted and were partly enforced by an all-powerful state that has intervened in every aspect of their citizens' lives ranging from dating, mating, recreation, and procreation to death rituals. The unintended consequences of a technoscience initiative adopted in response to the changing contours and circuits of global capital have triggered a series of political changes that are still underway (Ho, Baber, and Khondker 2002; Baber 2006).

The dot.com fever did not last long, but the high-end technological infrastructure enabled the rulers of the intelligent island to seek to capitalize on another emergent technology that its neighbours were simply technologically unprepared for. The unfolding genomics revolution presented Singapore with the possibility of positioning itself as an offshore location, deploying the model that had been so successful in the 1960s, when it was opened for transnational manufacturing.

Existing biotechnology institutes associated with the National University of Singapore were rapidly reconfigured in line with an ambitious master plan costing billions of dollars for transforming the country into a hub for biomedical manufacturing. The institutional node of this genomics initiative is the almost-ready futuristic Biopolis that will spatially connect the two major universities and researchers with the global biomedical manufacturing giants. Almost all the major manufacturers already have a strong presence in Singapore and occupy prime locations in the Biopolis. In line with the agenda of positioning itself as the prime location for the offshore production of genomic knowledge, biology is a compulsory subject for all secondary school students. Buzzwords such as 'knowledge society,' 'innovation,' and 'critical thinking,' which are of course evident in other societies, are impossible to avoid in Singapore.

More than any other fields of knowledge, genomics and nanotechnology have engendered the most visible social responses from civil society activists across the globe. The divide between Europe and North America over genetically modified food is the most graphic example of the concerns expressed over the possible long-term social and natural ramifications of these new technologies. However, while in most democratic societies there is some limited space and collective capacity – but a capacity that ought not to be exaggerated – civil society engagement in Singapore, the internet-based groups notwithstanding, continues to be severely limited. Although on paper a committee advising the government on social and ethical issues is in place, it is unlikely that the technocratic administrators of the country will listen to them seriously if the payoff is more investments by biomedical manufacturing giants and other corporate entities. The fact remains that the most ambitious and determined genomics and biomedical initiative is unfolding in an illiberal democracy where the room for meaningful public discussion and engagement with the state is very limited if not actually non-existent. Of course, the real influence of civil society actors and the so-called 'public' in non-illiberal democracies should not be exaggerated. However the mechanisms and modalities of such engagement are in place, and once in a while the public does indeed have a say, directly or indirectly. The point here is obviously not to contrast the lack of democratic engagement in Singapore with some pre-existing template that actually exists elsewhere.

While there is no dearth of discussions about the essence and practice of democracy, its key spirit has been well articulated by the economist Amartya Sen. For him, democratic systems can be distinguished from others on at least three counts: one, political freedoms; second, the support they grant people in their efforts to express their claims to political attention; and, finally, the constant opportunity the citizens and the state have to learn from each other (Sen 1999: 146–59). The essence of a democratic polity remains the opportunity for citizens at some level to have some input or at least the theoretical opportunity to have inputs, however minimal, on crucial issues, including major technoscientific initiatives. The opportunities for Singapore citizens on this score are severely limited as the state claims to speak for the entire community. This is not to say that the Singapore state has taken decisions that are detrimental to the interests of the community at large. A major factor in the continuing legitimacy for the existing government is the fact that over the years it has taken decisions that have led to dramatic economic improvements in the lives of its citizens. However, a number of dissenting voices have emerged that question the technocratic, narrow productivist logic inherent in every policy and decision.

As such, the major initiatives in genomics and biotechnology underway in Singapore are not subject to the usual visible public engagement. The plans for the Biopolis include a number of projects that are the subject of intense debates elsewhere. Thanks to globalization, not just of capital and manufacturing, but also of knowledge, information, and expertise, the results of research conducted in Singapore will find their way around the world. Alan Coleman, closely associated with cloning the Dolly sheep, is now working in Singapore, as are other top scientists from around the world. Civil society activism as well as economic and trade factors may influence a number of evaluations of technoscience in Canada such as genetically modified wheat or bovine growth hormone (rBGH). However, with globalization, civil society organizations limited by the boundaries of the nation-state are increasingly ineffective in meaningfully engaging with developments that are beyond its reach. There is no doubt that technoscientific innovations in Singapore, with little or no input from citizens' groups, will find their way to other societies in one form or another.

To return to Kitcher and Gieryn's analogy of maps and scientific knowledge, the genomics initiative currently underway in Singapore seeks to map out the significant truths of science from the perspective of corporate and economic interests. The turn to genomics occurred after the attempt to harness the internet for the dot.com fever was not as successful as anticipated. Although the wiring up of Singapore allowed for the emergence of some civil society organizations, they have no influence on the research agendas of the scientists housed in the futuristic Biopolis. The influence of corporate and economic interests on the texture and trajectory of technoscience is obviously not an issue that is peculiar to Singapore. There is no shortage of analyses of the enormous influence private corporations and other interest groups have on the presumably neutral scientific agendas (Baber 2001; Epstein 1996; Etzkowitz 2002; Kleinman 1994; 1995; 2000; 2003; Kloppenburg 1988; Proctor 1991). However, what is specific to the Singapore context is the relative lack of public debates and civil society activism that in other contexts serve to, at least in theory, contest the influence of corporate science. In Singapore the bottom line that informs the latest technoscience initiatives continues to be a one-dimensional, productivist economistic logic. However, the technoscientific maps that are currently under construction in Singapore will obviously be applied globally. While civil society actors might have some limited influence elsewhere, the science being mapped out in Singapore has no such social input. Singapore's largely successful attempt to move from industrial manufacturing to a leading offshore production site for scientific knowledge and research has global ramifications.

With the emergence of truly transformative technoscience, science, technology, and 'the public' are more than ever intimately intertwined. A short-sighted approach by civil society groups representing 'the public' has been the attempt to reject any such technoscience. Members of some scientific establishments have adopted a similar perspective by rejecting any involvement of 'the public' in deliberations on matters of science policy. The move to enhance the 'public understanding of science' has at times amounted to little more than making scientific and technological decisions and innovations acceptable to the public rather than involving them at least in some stages of decision making (Fuller 1997). While a number of developments in genomics stir apprehension, these transformative technologies also offer tremendous opportunities such as the detection of lung cancer at an early stage and the production of synthetic insulin, among others. The choice really is not between a completely rejectionist stance and a scientistic attitude that seeks to determine, 'scientifically' of course, what technoscientific initatives are best for society. Both approaches are bound to encounter resistance and perhaps fail at various levels. As technoscience moves into the marketplace and the imperative of genuine democratic involvement in its trajectory and application becomes clearer, such involvements need not be perceived as unwarranted meddling by scientifically ignorant groups who want a voice in the issues that genuinely affect them. As more funds are allocated for gigantic technoscientific initiatives, the issues of political economy and who gets what, and why, continue to be salient (Fuller 1997). How best to incorporate the views of the presumed beneficiaries of transformative technoscience in the process of the production and application of science is of course the key question that continues to defy easy answers and platitudes but cannot be ignored.

REFERENCES

Agrawal, Arun. 2002. Indigenous Knowledges and the Politics of Classification. *International Social Science Journal* 173: 325–36.

Baber, Zaheer. 1992. Sociology of Scientific Knowledge: Lost in the Reflexive Funhouse? *Theory and Society* 21:105–19.

– 1994. Beyond Hyper-Constructionist Fads and Foibles: Towards a Critical Sociology of Science, Technology and the Environment. *Critical Sociology* 20, no. 2: 125–38.
– 2001. Globalization and Scientific Research: The Emerging Triple Helix of State-Industry-University Relations in Japan and Singapore. *Bulletin of Science, Technology and Society* 21, no. 5: 401–8.
– ed. 2006. *CyberAsia: The Internet and Society in Asia.* Leiden: Brill Academic Publishers.
Barber, Bernard. 1952. *Science and the Social Order.* Glencoe, IL: Free Press.
Bell, Daniel, David Brown, Kanishka Jayasuriya, and David Martin Jones. 1995. *Towards Illiberal Democracy in Pacific Asia,* New York: St Martin's Press.
Bhaskar, Roy 1978. *A Realist Theory of Science.* London: Harvester Wheatsheaf.
– 1979. *The Possibility of Naturalism.* London: Humanities Press.
Chua, Beng Huat. 1996. *Communitarian Ideology and Democracy in Singapore.* London: Routledge.
Epstein, Steven. 1996. *Impure Science: AIDS, Activism and the Politics of Knowledge.* Berkeley: University of California Press.
Etzkowitz, Henry. 2002. *MIT and the Rise of Entrepreneurial Science.* New York: Routledge.
Fuller, Steve. 1991. *Social Epistemology.* Bloomington: Indiana University Press.
– 1997. *Science.* Minneapolis: University of Minnesota Press.
Giddens, Anthony. 1986. *The Constitution of Society.* Berkeley: University of California Press.
Gieryn, Thomas. 1999. *Cultural Boundaries of Science: Credibility on the Line.* Chicago: University of Chicago Press.
Gupta, Akhil. 1998. *Postcolonial Developments: Agriculture in the Making of Modern India.* Durham: Duke University Press.
Gururani, Shubhra. 2002. Constructions of Third World Women's Knowledge in the Development Discourse. *International Social Science Journal* 173: 277–89.
Hacking, Ian. 1999. *The Social Construction of What?* Cambridge: Harvard University Press.
Ho, K.C., Z. Baber, and H. Khondker. 2002. Sites of Resistance: Alternative Websites and State-Society Relations. *The British Journal of Sociology* 53, no. 1: 127–48.
Kitcher, Philip. 2001. *Science, Truth and Democracy.* Oxford: Oxford University Press.
Klein, Hans, and Daniel Lee Kleinman. 2002. The Social Construction of Technology: Structural Considerations. *Science, Technology and Human Values* 27, no. 1: 27–52.

Kleinman, Daniel Lee. 1994. Layers of Interests, Layers of Influence: Business and the Genesis of the National Science Foundation. *Science, Technology and Human Values* 19: 259–82.

– 1995. *Politics on the Endless Frontier: Postwar Research Policy in the United States.* Durham, NC: Duke University Press.

– ed. 2000. *Science, Technology and Democracy.* Albany: State University of New York Press.

– 2003. *Impure Cultures: University Biology and the World of Commerce.* Madison: University of Wisconsin Press.

Kleinman, Daniel Lee, and Steve Vallas. 2001. Science, Capitalism and the Rise of the Knowledge Worker: The Changing Structure of Knowledge Production in the United States. *Theory and Society* 30:451–92.

Kloppenburg, Jack. 1988. *First the Seed: The Political Economy of Plant Biotechnology, 1492–2000.* Cambridge: Cambridge University Press.

Marx, Karl. 1967. *Capital, Vol. I.* New York: International.

Merton, Robert K. 1942. Science and Technology in a Democratic Order. *Journal of Legal and Political Sociology* 1: 115–26.

– 1973. *Sociology of Science: Theoretical and Empirical Investigations.* Chicago: University of Chicago Press.

Mulkay, Michael J. 1979. *Science and the Sociology of Knowledge.* London: Allen and Unwin.

– 1985. *The Word and the World: Explorations in the Form of Sociological Analysis.* London: Allen and Unwin.

Nowotny, H., P. Scott, and M. Gibbons. 2001. *Re-Thinking Science: Knowledge and the Public in an Age of Uncertainty.* Cambridge: Polity Press.

Philip, Kavita. 2003. *Civilizing Natures: Race, Resources and Modernity in Colonial South India.* New Brunswick, NJ: Rutgers University Press.

Proctor, Robert. 1991. *Value-Free Science? Purity and Power in Modern Knowledge.* Cambridge: Harvard University Press.

Sen, Amartya. 1999. *Development as Freedom.* New York: Knopf.

Shiva, Vandana. 1989. *Staying Alive: Women, Ecology and Development.* London: Zed.

Sokal, Alan. 1996. Transgressing the Boundaries: Toward a Transformative Hermeneutics of Quantum Gravity.' *Social Text* 46/47: 217–52.

Weber, Max. 1930. *The Protestant Ethic and the Spirit of Capitalism.* New York: Charles Scribner's Sons.

Winner, Langdon. 1993. Social Constructivism: Opening the Black Box and Finding It Empty. *Science as Culture* 16: 427–52.

15 Retaining Scientific Excellence in Setting Research Priorities: Lessons from the Consultative Group on International Agricultural Research (CGIAR)

CHRISTOPHER D. GERRARD

The accelerated pace of globalization has stimulated dramatic changes in trade, finance, intellectual property, information and communications technology, health, environment, and security. The opportunities and threats brought by globalization spill over national borders, and addressing the challenges they pose often requires collective action at the global level. In the absence of a global government able to establish and enforce policy regimes and rules and to collect taxes and raise revenues, global programs are increasingly being used to organize global collective action – particularly to meet the growing demand for the provision of global public goods.

Global programs are partnerships whose members contribute and pool resources toward achieving agreed-upon objectives; the activities of the program are global in scope; and the partners establish a new organization with a governance structure and management unit to deliver the activities. The World Bank contributes from its own profits about US$120 million a year to global programs. Its traditional partners have been other international (typically UN) organizations, bilateral donors like the Canadian International Development Agency (CIDA), foundations like Ford and Rockefeller, and, more recently, the Gates Foundation in the health sector. With the establishment of the Global Fund to Fight AIDS, TB, and Malaria in 2001 – now the largest global program in terms of disbursements – the 125 global programs in which the Bank is involved are disbursing about US$3 billion a year.

The Consultative Group on International Agricultural Research (CGIAR), which provides the raw material for the thoughts expressed in this paper, was the first global program that the World Bank helped to establish. It was founded in 1971 to scale up years of effort by the

Rockefeller and Ford Foundations in supporting four international agricultural research centres in Colombia, Mexico, Nigeria, and the Philippines.[1] As characterized by former chairman David Hopper, its original mission was a strategic, science-based focus on increasing 'the pile of rice on the plates of food-short consumers' (author interview with David Hopper, 17 January 2002). It was to use the best science in the advanced countries to develop improved agricultural technologies for the benefit of food-deficit countries and populations. Today, it supports fifteen autonomous international agricultural research centres, fourteen of which are located in developing countries,[2] with 8,500 scientists and staff in more than one hundred countries. It has a Secretariat housed in the World Bank and a Science Council (formerly called the Technical Advisory Committee) housed in the Food and Agricultural Organization (FAO) of the United Nations. From an initial eighteen, its membership has expanded to sixty-three, comprising twenty-four developing and transition economies, twenty-two industrialized countries, thirteen international/regional organizations, and four private foundations.

The CGIAR is widely regarded as an exemplary instrument of international cooperation. The Third System Review in 1998, which was chaired by Maurice Strong, described it as the 'single most effective use of official development assistance, bar none' (IRPFSSA 1998: vii). Impact evaluations of its productivity-enhancing research have demonstrated not only extraordinarily high rates of return to germplasm enhancement research – ranging from 40 to 78 per cent (Maredia et al. 2000 and Alston et al. 2000)[3] – but also sizable impacts on reducing poverty by increasing employment, raising incomes, lowering food prices, and releasing land from cropping – a phenomenon that has come to be known as the Green Revolution. Those involved in international health research have often lamented that there has not existed a similar Consultative Group on International Health Research, which, if it had existed in the 1980s, might have initiated international research for AIDS vaccines and drugs for developing countries a decade sooner than actually occurred.

This chapter assesses the challenge of retaining scientific excellence in setting research priorities, drawing on lessons from the CGIAR. It does so by addressing three key questions. First, can science retain independence and objectivity? Second, is scientific discourse privileged? Third, can science be made more compatible with policy discourse? I draw conclusions on each of these queries based on my experience with the CGIAR.

Can Science Retain Independence and Objectivity?

The CGIAR, in essence, is a mechanism to raise and to pool about US$450 million of public funds from the OECD countries and the World Bank to support strategic agricultural research in its fifteen centres. Although the CGIAR represents less than 4 per cent of total global expenditures on agricultural research, the fifteen centres have traditionally occupied a particular niche in the global agricultural research system in between the basic agricultural research that is conducted in advanced-country universities and the applied and adaptive research that is conducted in the national agricultural research systems (NARS) of developing countries. The fifteen centres were established to conduct strategic research of a global or regional public goods nature, such as developing enhanced parent lines for, say, rice, wheat, maize, and sorghum that could be used by numerous countries in applied cross-breeding programs. The CGIAR centres also possess 600,000 accessions of live genetic material that have been collected mainly from developing countries, which are a truly global public good. Maintaining these gene banks and the biological diversity which they represent is vitally important for the future world food supply, since monocultures of improved crop varieties create an ideal climate for new pests and diseases to emerge.

The most important decision that the members of the Consultative Group make annually is the allocation of resources to the system's different research and training programs. This involves making choices among at least five different dimensions: among the five commodity centres (that focus on specific crops or livestock), the four eco-regional centres (that focus on farming systems in specific ecological regions), the four natural resources management centres (that focus on fisheries, forestry, and water), and the two other centres (that focus on policy and gene pool conservation); among the five major research and training thrusts – increasing 'agricultural' productivity, protecting the environment, saving biodiversity, improving policies, and strengthening NARS;[4] among the developing regions of the world – Asia, Latin America and the Caribbean, Sub-Saharan Africa, and West Asia and the Middle East; among overhead costs (including gene bank operations and maintenance), long-term research programs, and short-term programs; and, since 2003, among Challenge Programs and other programs – Challenge Programs being high-visibility, time-bound partnerships among centres and other research institutions that are intended to address complex issues of global or regional significance.[5]

The Consultative Group adopted six principles in the early 1970s to guide this resource allocation process. First, each donor has the freedom to choose which centres or research programs within centres to support. Second, the centres are the only legal entities in the system that are accountable to their own governing boards and that have the autonomy to make decisions and allocate the resources that they receive. Third, decisions of the Consultative Group are by consensus. Fourth, the Technical Advisory Committee (now called the Science Council) should provide independent technical advice regarding priority setting and resource allocation. Fifth, the Consultative Group itself does not have a formal legal status. Finally, international politics should not influence CGIAR decision making.

Roughly until 1992, the CGIAR had a good track record in allocating its resources in accordance with high-priority global public goods research areas as determined by the Technical Advisory Committee (TAC). The system worked because of the way in which the World Bank allocated its annual financial contribution (now US$50 million) in totally unrestricted funds.[6] Although individual donors allocated their financial contributions to whichever centre or research program they wished, the World Bank followed a 'donor-of-last-resort' model in allocating its own $50 million contribution to fill the gaps between the research priorities as articulated by the TAC and the financial contributions to those priorities by the other donors. The TAC played a powerful role – like a 'scientific priesthood' – not only by recommending allocations of resources among centres, programs, and activities, but also in monitoring budgets and conducting centre-level and system-level reviews. The World Bank – as the Chair of the Consultative Group and as the only donor which made its financial contribution to the system as a whole – became the glue that made the system coherent and larger than the sum of its fifteen research centres.

Things began to change in 1992, when four new natural resources management centres – on forestry, agroforestry, water, and fisheries – were established and the United States and Canada reduced their funding by $24 million and $6 million, respectively, between 1992 and 1994.[7] The World Bank responded to this funding crisis by increasing its financial contribution from US$40 to 50 million, by encouraging other donors (notably Japan) to increase their contributions, and by changing the allocation of its own financial contribution from the 'donor-of-last-resort' model to a matching grant model, in which it began to indiscriminately match funding from other donors (at the rate of 12 cents on the dollar), whether in support of the TAC's priorities or not.

As a result, the CGIAR's expenditures on productivity-enhancing agricultural research – the system's comparative advantage and core competence – declined by 6.5 per cent annually in real terms between 1992 and 2001, while expenditures on improving policies and on protecting the environment – of questionable global or regional public goods character – increased by 3.1 per cent annually. During the same time period, overall donor contributions grew at an average annual rate of 0.7 per cent in nominal terms, declined by 1.2 per cent per year in real terms, and became increasingly restricted to the research priorities of the donors – from 36 per cent of total contributions in 1992 to 57 per cent in 2001.

Several factors explain the changing research mix and the increasing restrictions. First, germplasm improvement research has been unpopular in the constituencies of some key donors because of negative perceptions of the Green Revolution. Second, the CGIAR has correctly responded to the genuine second-generation environmental pressures on soils and water created by the radical change in farming systems during the Green Revolution – where research continues to be needed. Third, the rise of environmentalism, the 1992 United Nations Conference on Environment and Development in Rio de Janeiro, and growing environmental advocacy in donor countries led to growing demands on the CGIAR to respond to environmental concerns. Fourth, the failure of governments of developing countries and their donor supporters (including the World Bank) to make the necessary investments in their own NARS led CGIAR donors to turn to the centres to fill the national and local public goods gaps closer to the farmer, which should ideally be filled by the national systems. Collaterally, the maintenance and management of the CGIAR's gene banks, its unique global public good, suffered from inadequate funding.

The cumulative result of these trends has been the declining influence of independent scientific advice provided by the TAC, a concurrent transformation of the CGIAR's authorizing environment from being science-driven to being donor-driven, and a shift in the system from producing global and regional public goods toward providing national and local services. The CGIAR experience demonstrates how rules create incentives for behaviour which lead to outcomes that are less or more desirable, and how the sum of the interests of individual stakeholders in a global program do not necessarily define a global public good.[8]

Thus the CGIAR largely failed to retain scientific independence and objectivity over the last ten years in the face of growing demands from individual donors (and their domestic constituencies, such as northern non-governmental organizations). Instead, the CGIAR has responded to, among other things, environmental concerns of questionable global public goods character. The CGIAR would not be the first organization established for one purpose that then shifted towards another. It cannot be said, however, that its shift of focus in the 1990s from enhancing agricultural productivity towards environmental issues has been due to the absence of a global environmental program. In 1991, the international community established the Global Environment Facility (GEF) in 1991 specifically to provide grant and concessional funding to assist developing countries in undertaking investments in relation to various international environmental conventions (such as the Convention on Biological Diversity, the UN Framework Convention on Climate Change, and the Convention to Combat Diversification). The GEF is now roughly the same size as CGIAR – disbursing around US$450 million of investment grants to developing countries in 2003.

Is Scientific Discourse Privileged?

This case study involves international public discourse about international public goods, rather than national public discourse about national public goods. The debate has traditionally focused on policies internal to the organization itself, such as funding and research priorities. Now it is increasingly about global public policy issues such as biotechnology, genetic resource management, intellectual property rights (IPRs), and public–private partnerships in agricultural research – the results of which have spillover effects on other parties. Still, unlike the World Trade Organization or the Montreal Protocol on Substances That Deplete the Ozone Layer, the CGIAR does not claim to be a global rule setter and enforcer, say, with regard to trade in plant genetic material. The CGIAR is not an international convention; it is still an informal organization in which the only legal entities are the fifteen centres.

While biotechnology, genetic resource management, IPRs, and public–private partnerships in agricultural research are important global issues, these are also highly specialized 'small p' political issues that are not discussed, for example, at the UN Security Council. In addition to the sixty-three CG members, centre directors, and board chairs, the annual general meetings of the CGIAR are attended by numerous stakeholders with an interest in the outcomes of the Group's policy discussions – by representatives of industrialized and developing country research institutions, the commercial private sector, and civil society organizations. But the global public rarely pays much attention.

If the donors are pushing the CGIAR downstream towards the farmer, these global challenges are pulling the CGIAR upstream towards the global level. Concerning biotechnology,[9] informed experts consider the CGIAR's estimated investment in biotechnology research of $25 million, spread over several centres, to be too small, leading it to obsolescence (Lesser 2002). They consider the centre-by-centre and commodity-by-commodity approach that was pertinent when the CGIAR was established to be inappropriate in light of the revolutionary changes in the research methods in genomics. Critical investment in equipment needed for long-term research is hampered by the short-term project-by-project financing the CGIAR receives from donors and by the fragmentation of research across several centres. Consensus on the most controversial issue of genetically modified organisms (GMOs) has proved difficult to achieve, and this lack of consensus within the CGIAR threatens the system.[10]

Concerning genetic resource management, the CGIAR possesses 600,000 accessions of genetic material collected mainly from developing countries and held in trust by the centres with FAO as the trustee.[11] Although these are a truly global public good, donors have been unwilling to fund the overhead costs of managing the gene banks, which are being supported by the ever-diminishing unrestricted funding provided by a handful of donors together with the World Bank. Yet biodiversity conservation was high on the agenda of the UN Conference on Environment and Development in Rio de Janeiro in 1992 and at the World Summit on Sustainable Development in Johannesburg in 2002.

Concerning intellectual property rights, a combination of the Convention of Conservation of Biological Diversity in 1992, the World Trade Organization's TRIPS agreement (Trade-Related Aspects of Intellectual Property Rights), and the 2001 International Treaty on Plant

Genetic Resources for Food and Agriculture (ITPGR) have dramatically changed the intellectual property environment for the CGIAR. A recent British commission on IPRs reviewed these and other changes and advised global programs such as the CGIAR to confront the full implications of IPRs (Commission on IPRs 2002). It made recommendations to deal with 'the fundamental asymmetry in relations between developed and developing countries' (129).[12] The Commission also warned that the compromise wording of the ITPGR and its scope for patents on genes to be isolated from the CGIAR's plant genetic material requires vigilance. Yet, the CGIAR's system-wide office on IPRs, located at and paid for by the International Service for National Agricultural Research (ISNAR), has been understaffed and inadequate to meet the system's need for IPR capacity. Its funding has dwindled as other CGIAR centres have developed their own capacity, and ISNAR's core programs were recently absorbed by the International Food Policy Research Institute (IFPRI) and relocated to Africa under IFPRI governance.

Concerning public–private partnerships, the private sector is an increasingly important actor in agricultural research. It funded 35 per cent of the $33 billion (in 1993 international dollars) spent on agricultural R&D worldwide in 1995–6 and 70 per cent of the research on genomics, while possessing 80 per cent of the intellectual property emanating from it. Yet in 1995–6, private sector expenditure on agricultural R&D in developing countries was less than 6 per cent of all private sector expenditures in agricultural R&D worldwide, which is also less than 6 per cent of the total expenditures (both public and private) on agricultural R&D in developing countries.

The public sector and the CGIAR continue to have a strong comparative advantage vis-à-vis the private sector in developing technologies for poor and marginal farmers in developing countries, due to limited markets for such technologies, the limited purchasing power of poor and marginal farmers, and long-term returns to such research (Lesser 2002; Pardey and Beintema 2001; Spielman 2002). A partnership between the International Rice Research Institute (IRRI), the Rockefeller Foundation, Syngenta, and others, which developed 'golden rice,' highlights both the role of public–private partnerships and their difficulty in taking a new technology from the laboratory to the consumer.

Effectively responding to these four global challenges requires a system-level approach. But the system has responded only slowly to these challenges due to differing views, for example, about biotechnology and GMOs among its North American and European members. The collective action problems in reaching agreement within the CGIAR on such issues have proved daunting. The six founding principles that were adopted when the CGIAR consisted of fewer centres and less diverse constituents seem no longer suited to today's politically driven authorizing environment, wider research agenda, and expanding membership.

Can Science Be Made Compatible with Policy Discourse?

My answer to the first question posed in this chapter demonstrated how operational rules such as the matching grant formula for the allocation of the Bank's financial contribution to the CGIAR have influenced research outcomes. My answer to the second question demonstrated how governance rules such as consensus decision making have made it difficult for the CGIAR to reach agreement on new operational policies with respect to biotechnology, genetic resource management, IPRs, and public–private partnerships. My answer to this third question (can science be made compatible with policy discourse?) raises the discussion to the constitutional level. The CGIAR faces a number of constitutional-level problems, including an assignment problem not unlike that among different levels of government in constitutional theory and practice.

The CGIAR system is being pulled in two opposite directions. On the one hand, the CGIAR centres are not conducting sufficiently coordinated research on the *highly decentralized nature* of natural resource management (NRM) research, which calls for effective partnerships with NARS to produce regional and national public goods in NRM. On the other hand, the system is *not sufficiently centralized* to deal with advances in the biological sciences and IPRs, which call for a more unified approach to research strategies and policies. Due to the skills, expertise, and scale economies involved, some functions should be centralized and others decentralized. The issue is how to bring these changes about. The CGIAR is an informal organization without a written charter and without an agreed-upon mechanism to reform the governance of the system.[13]

The World Bank convened a high-level donor meeting to establish the CGIAR in 1971. It also convened a ministerial-level meeting in February

1995, in the wake of the 1993-4 financial crisis, to re-establish the CGIAR on a firm financial footing. But, except for these meetings, the Consultative Group has been essentially a self-governing body in which the members who have to play by the rules are also the ones who make the rules. In my opinion, the self-interest inherent in such a process has clearly inhibited fundamental governance reforms in the CGIAR.

For example, the CGIAR's Third System Review in 1998 recommended that the CGIAR establish itself as an independent legal entity and adopt a corporate governance model in order to deal with IPR issues and public–private partnerships. Then the CGIAR Committee of Board Chairs and the Center Directors' Committee jointly recommended a decentralized Federation of Centers in 2000. While the two proposals differed in the degree of decentralization proposed, both acknowledged the need for a legal entity with a centralized board to enable system-level responses to IPR issues. Yet neither of these proposals was considered seriously by the Group.

A blocking coalition of members – it is not clear how large – clearly prefers the World Bank to continue to chair the CGIAR and to house the Secretariat, notwithstanding the conflicts of interest that arise out of this dual role. Donor members view the Bank's leadership role, its financial contributions, and its operational support as a seal of approval which gives them confidence to invest in the system. They are fearful that the Bank's and other funding might decrease if the CGIAR or its chairmanship were moved outside the Bank.

In reality, conflicts of interests among the roles which various Bank staff have played in the CGIAR, as well as inadequate Bank oversight from outside the vice-presidency of the Bank's involvement in the CGIAR, have compromised the Bank's capacity to exercise strategic leadership of the CGIAR and to press for reforms on a scale or at a speed that might be warranted. In particular, it has been problematic for the Environmentally and Socially Sustainable Development Network (ESSD) vice-president, who is also the CGIAR chair, to be both judge and advocate – to acknowledge the need for and to press for major reforms, while also making the case for continued funding to the Bank and other donors. If collective action organizations require political entrepreneurs, or champions, to bring about major changes, and if the Bank has traditionally been that champion, it has also been a compromised champion that, almost unavoidably, brings its own interests to the table.[14]

In fairness, the CGIAR has since 2000, under the leadership of Chairman Ian Johnson, instituted four significant reforms in its governance and management known as the Change Design and Management Process. These consist of establishing an Executive Council (ExCo) and a System Office, transforming the TAC into a Science Council, and initiating a programmatic approach to research in the form of Challenge Programs.

In my view, some of these reforms are creditable, others need revisiting, and, collectively, they do not go far enough. The chairman gets high marks for establishing the long-overdue Executive Council. But I am sceptical that the transformation of the TAC into a Science Council will achieve the objective of strengthening the role of independent scientific advice in the CGIAR, since the Science Council will focus primarily on science quality and play a small role, if any, in the important functions of priority setting and resource allocation. And the first four Challenge Programs have proceeded without first addressing the issues of system-level funding, priority-setting, science quality, and governance raised in previous evaluations of the CGIAR.

The CGIAR, like many of the 125 global programs in which the Bank is currently involved, represents an attempt at global governance and global policymaking without a global government. Its present constitutional-level challenges are both substantive and procedural – reconfiguring the governance of the CGIAR to promote greater efficiency, instituting tougher priority setting, and maintaining scientific excellence without sacrificing legitimacy and ownership. In my view, this now requires an intervention at a level above that of the Consultative Group itself, like the two high-level meetings that were held in 1971 and 1995. Such a meeting should seriously consider moving the CGIAR outside the Bank, establishing the CGIAR as an independent legal entity – among other things, to centralize its capacity to deal with the biotechnology revolution and IPRs – and appointing an eminent person (preferably a scientist) as chief executive officer of the CGIAR, as is currently the case for the Global Environment Facility. Consistent with its comparative advantage and core competencies, the strategic priorities of the CGIAR should respond more actively to changes in the global research context, giving more prominence to basic plant breeding and germplasm improvement, and focusing its natural resource management research tightly on productivity enhancement and sustainable use of natural resources for the benefit of developing countries.

Appendix 1 List of CGIAR Centres, 2003

Acronym	Full name	Location	Established	Joined CGIAR	Type of centre	Mandate
IRRI	International Rice Research Institute	Los Baños, Philippines	1960	1971	Commodity	Rice and rice-based ecosystems.
CIMMYT	Centro Internacional de Mejoramiento de Maïz y Trigo	Mexico City, Mexico	1966	1971	Commodity	Crops: wheat, maize, triticale.
CIAT	Centro Internacional de Agricultura Tropical	Cali, Colombia	1967	1971	Ecoregional	Crops: beans, cassava, tropical forages, rice. Agroecosystems: hillsides, forest margins, savannas, fragile African and Asian environments.
IITA	International Institute of Tropical Agricultura	Ibadan, Nigeria	1967	1971	Ecoregional	Crops: soybean, maize, cassava, cowpea, banana, plantain, yams. Sustainable production systems for the humid lowland tropics.
ICRISAT	International Crops Research Institute for the Semi-Arid Tropics	Patancheru, India	1972	1972	Ecoregional	Crops: sorghum, pearl millet, finger millet, chickpea, pigeon pea, and groundnut. Sustainable production systems for the semi-arid tropics.
CIP	Centro Internacional de la Papa	Lima, Peru	1971	1973	Commodity	Crops: Potato, sweet potato.
ILRI	International Livestock Research Institute	Nairobi, Kenya	1973[a]	1973	Commodity	Livestock diseases. Cattle sheep and goats. Feed and production systems.
IPGRI	International Plant Genetic Resources Institute	Rome, Italy	1974	1974	Other	Plant genetic resources of crops and forages. Collection and gene pool conservation.

Appendix 1 (*Concluded*)

Acronym	Full name	Location	Established	Joined CGIAR	Type of centre	Mandate
WARDA	West Africa Rice Development Association	Bouaké, Côte d'Ivoire	1970	1975	Commodity	Rice production in West Africa.
ICARDA	International Center for Agricultural Research in the Dry Areas	Aleppo, Syria	1977	1977	Ecoregional	Crops: wheat, barley, chickpea, lentil, pasture and forage legumes. Livestock: small ruminants.
IFPRI	International Food Policy Research Institute	Washington, DC, USA	1975	1980	Policy	Food policy. Socioeconomic research related to agricultural development.
ISNAR[b]	International Service for National Agricultural Research	The Hague, The Netherlands	1979	1980	Training	Strengthening national agricultural research systems.
ICRAF	International Centre for Research in Agroforestry	Nairobi, Kenya	1977	1991	NRM	Agroforestry. Multi-purpose trees.
ICLARM	International Center for Living Aquatic Resources Management	Penang, Malaysia	1977	1992	NRM	Sustainable aquatic resource management.
IWMI	International Water Management Institute	Columbo, Sri Lanka	1984	1991	NRM	Irrigation and water resources management.
CIFOR	Center for International Forestry Research	Bogor Barat, Indonesia	1992	1992	NRM	Sustainable forestry management.

[a] ILRI was actually formed in 1995 from a merger of ILRAD (the International Laboratory for Research on Animal Diseases) and ILCA (the International Livestock Center for Africa). ILRI was established and joined the CGIAR in 1973; ILCA in 1974.

[b] The Consultative Group recommended in 2003 that ISNAR's core programs be absorbed by IFPRI and relocated to Addis Ababa, Ethiopia, under IFPRI governance. This merger took place in 2004.

NOTES

The author is a lead economist in the Independent Evaluation Group (IEG)) of the World Bank. He is the joint author, along with Uma Lele, of the evaluation on the World Bank's involvement in global programs – *Addressing the Challenges of Globalization: An Independent Evaluation of the World Bank's Approach to Global Programs* (2004) – and one of the many co-authors, led by Uma Lele, of *The CGIAR at 31: An Independent Meta-Evaluation of the Consultative Group on International Agricultural Research* (2003). While the present paper draws heavily on the latter evaluation, the views expressed in this paper are those of the author and do not necessarily reflect the views of the World Bank, the IEG, or the other authors of these studies.

1 The International Center for Tropical Agriculture (CIAT) in Colombia, the International Maize and Wheat Improvement Center (CIMMYT) in Mexico, the International Institute of Tropical Agriculture (IITA) in Nigeria, and the International Rice Research Institute (IRRI) in the Philippines.
2 See Appendix 1 for the complete list of centres, their location and mandate, when they were established, and when they joined the CGIAR.
3 For a full discussion of the rates-of-return literature, estimation methodologies, biases, and other issues, see Gardner (2002).
4 This was the CGIAR's classification in 2001. Increasing 'agricultural' productivity includes germplasm enhancement and production systems research on fisheries, forestry, and livestock as well as crops. 'Germplasm' is the hereditary material of plants and animals that is capable of being transmitted from one generation to the next. DNA by itself is not germplasm; it is only germplasm when it has the capacity of being transmitted.
5 The CGIAR approved the first two programs on 'Water and Food' and 'Biofortified Crops for Improved Human Nutrition' at its annual general meeting in October 2002, a third program on 'Unlocking Genetic Diversity in Crops for Resource-Poor Areas' in 2003, and a fourth on 'Building Sustainable Livelihoods in Sub-Saharan Africa through Integrated Agricultural Research for Development' in 2004. These programs are intended to increase the scope for inter-centre collaboration, facilitate a wider range of partnerships, and tap new sources of funding from current and new donors.
6 As defined by the CGIAR, unrestricted funds are those which a centre may allocate to any of its programs or costs according to its institutional needs or priorities. Restricted funds are restricted either by attribution (to a particular research program or region) or by contract (to a project, subproject, or activity). Following the principle of donor sovereignty, a donor's allocation

of funds to one centre as opposed to another is not defined as restricted funding by the CGIAR.

7 In the United States, the Clinton administration transferred resources from agriculture to health. In Canada, the Chrétien government cut most foreign assistance programs in response to Canada's domestic fiscal crisis.

8 I am not arguing that the donor-of-last-resort method of allocating the World Bank's financial contribution worked perfectly. The donor-of-last-resort model also had some limitations, the primary ones being 'a stifling of Center initiatives in raising funds and an insulation of some Centers from economic realities. As a result, the Bank ended up providing more funding for some Centers over a longer period of time than might otherwise have been the case' (Anderson and Dalrymple 1999: 24).

9 'Biotechnology' has been defined by the Office of Technology Assessment (OTA 1989) of the U.S. Congress as 'any technique that uses living organisms or substances from those organisms, to make or modify a product to improve plants or animals, or to develop microorganisms for specific uses' (3). It encompasses both 'traditional biotechnology,' which includes well-established technologies used in commercially useful operations such as biological control of pests, conventional breeding of plants, animal vaccine production, and cell and tissue culture techniques, as well as 'modern biotechnology,' which includes recently available tools for expediting selection and breeding ranging from the use of recombinant DNA, monoclonal antibodies, molecular markers, and transgenic techniques, to genetically engineered live organisms used to modify a variety of characteristics in host plants and animals such as productivity enhancement, growth cycles, and resistance to a variety of environmental or genetic stresses.

10 The erosion of the non-political nature of the CGIAR is evident in its handling of the politically charged GMO debate. The CGIAR's Executive Council (ExCo) decided that the Science Council, which is normally expected to speak on major scientific issues of global importance, should not handle GMO issues, but that these should be left to a special standing panel on genetic resources established in 1994 to monitor and analyse developments in genetic resources policy, internationally and within the system. The panel is to recommend CGIAR action as necessary, including the engagement of the CGIAR in the negotiations of the FAO Commission on Genetic Resources for Food and Agriculture that led to the adoption of the International Treaty on Plant Genetic Resources for Food and Agriculture in 2001. ExCo has observed that the use of GMOs is not acceptable to certain CGIAR members.

11 The management of the genetic resources collections is governed by a stan-

dard agreement between FAO and each centre that holds material. The System-wide Genetic Resources Programme (SGRP) is charged with assisting the centres in fulfilling their obligations under the FAO agreement. The SGRP is administered through IPGRI as the convening centre.

12 These include (1) international negotiation of IPR policies and their implementation to ensure global IP systems contribute to the development of developing countries by stimulating innovation and increasing their access to technology and its products at competitive prices; (2) learning from the international experience, including understanding the rapidly evolving nature of international public–private partnerships; (3) assessing the implications of intellectual property rights for the Convention on Biological Diversity; (4) strengthening the capacity of international and national institutions involved in intellectual property rights issues; and (5) supporting the international architecture.

13 The CGIAR formally approved its first written charter at its annual general meeting in October 2004, in response to such a recommendation in IEG's 2003 evaluation. This includes a section on amending the charter in the future.

14 In response to the recommendation in IEG's 2003 evaluation to separate the Bank's oversight of the CGIAR from its management of the CGIAR, the Bank has now designated its Development Economics vice-president and chief economist to exercise the oversight function.

REFERENCES

Alston, J.M., C. Chan-Kang, M.C. Marra, P.G. Pardey, and T.J. Wyatt. 2000. *A Meta-Analysis of Rates of Return to Agricultural R&D: Ex Pede Herculem?* Research Report 113. Washington, DC: IFPRI.

Anderson, J.R., and D.G. Dalrymple. 1999. *The World Bank, the Grant Program, and the CGIAR: A Retrospective Review.* OED Working Paper Series No. 1. Washington DC: The World Bank.

Borlaug, N. 2003. Science and Hysteria. *Wall Street Journal*, 22 January.

CGIAR Committee of Board Chairs and CGIAR Center Directors' Committee. 2000. *Towards a Federation of Centers: The Report of the CBC/CDC Retreat.* The Hague, 2–3 September.

CGIAR Secretariat. Various years. *Annual Report.*

CGIAR Secretariat. 2000. *Background Paper: CBC/CDC Retreat: The Hague, Netherlands, 2–3 September 2000* and *Report of the CBC/CDC Retreat* (September 2000). Washington, DC.

– Implementation of Recommendations from the Third System Review of the CGIAR. Draft. Washington, DC.

Commission on Intellectual Property Rights. 2002. *Integrating Intellectual Property Rights and Development Policy. Report of the Commission on Intellectual Property Rights.* London, September.

Food and Agriculture Organization of the United Nations. 2001. International Convention for the Protection of New Varieties of Plants. Rome.

Gardner, B. 2002. *Global Public Goods from the CGIAR: Impact Assessment.* OED Working Paper. Washington, DC: The World Bank.

Herdt, R. 2000. Thoughts on the CGIAR. Paper presented at the Dresden MTM. Washington, DC: CGIAR Secretariat. (Revised 8/17/00).

International Food Policy Research Center (IFPRI). 2002. *Green Revolution: Curse or Blessing?* Washington, DC: IFPRI.

International Research Partnership for Food Security and Sustainable Agriculture (IRPFSSA). 1998. *Third System Review of the CGIAR.* Washington, DC: CGIAR System Review Secretriat.

Lesser, W. 2002. *Reviews of Biotechnology, Genetic Resource and Intellectual Property Rights Programs.* OED Working Paper. Washington, DC: The World Bank.

Maredia, M.K., D. Byerlee, and J.R. Anderson. 2000. Ex Post Evaluation of Economic Impacts of Agricultural Research Programs: A Tour of Good Practice. Presented at a workshop of the SPIA (Standing Panel on Impact Assessment) of the CGIAR's Technical Advisory Committee, Rome.

Office of Technology Assessment (OTA). 1989. *New Developments in Biotechnology: Patenting Life.* Special Report, OTA-BA-370. Washington, DC: U.S. Government Printing Office.

Operations Evaluation Department. 2003. *The CGIAR at 31: An Independent Meta-Evaluation of the Consultative Group on International Agricultural Research.* Washington, DC: The World Bank. The Operations Evaluation Department (OED) of the World Bank changed its name to the Independent Evaluation Group (IEG) in December 2006.

– 2004. *Addressing the Challenges of Globalization: An Independent Evaluation of the World Bank's Approach to Global Programs.* Washington, DC: The World Bank.

Pardey, P.G., and N.M. Beintema. 2001. *Slow Magic: Agricultural R&D a Century after Mendel.* Food Policy Statement No. 36. Washington, DC: IFPRI.

Spielman, D.J. 2002. *International Agricultural Research and the Role of the Private Sector.* OED Working Paper. Washington, DC: The World Bank.

U.S. National Academy of Sciences. 2000. *Knowledge and Diplomacy: Science Advice in the United Nations System.* Washington, DC: The National Academies Press.

16 Toward Centres for Responsible Innovation in the Commercialized University

DAVID H. GUSTON

In his *Science, Truth, and Democracy,* philosopher Philip Kitcher (2001) proposes 'well-ordered science' as the ideal to which the organization of the research enterprise should aspire. Distinct from government by 'vulgar democracy,' in well-ordered science a highly informed public, coupled with a public-spirited research community, sets overall research priorities. As a philosopher, Kitcher avoids a detailed comparison of this ideal to the reality of making science policy in the contemporary United States. But he does imply that reality falls short of the ideal on both counts of informed public participation and unselfish scientific service. He therefore recognizes the needs, respectively, for 'sociological information required to build realistic models' of the construction of 'tutored collective preferences' (135), as well as for 'a political theory of science that will consider the various ways in which the interests of actors and social institutions' relate to outcomes in well-ordered science (133n8).

Concurring with Kitcher's assessment of the shortcomings of the public and the scientific community, I am nevertheless more optimistic than he about our current state of knowledge on the involvement of the public in making technically complex decisions, as well as about the capacity of extant scholarship to guide the connection among actors, institutions, and outcomes in science policy. I draw on this knowledge to elaborate a proposal for university-based centres for responsible innovation that might contribute to a well-ordered science in the context of the commercialized university – a particularly important microcosm of science that is unlikely to achieve a 'well-ordered' structure spontaneously. The first section of this paper addresses the three premises about the commercialized university that ground my call for centres for responsible innovation. The next section of the paper elaborates the tasks that such centres would perform. The final section returns to Kitcher and describes how the operation of such centres is grounded in recent scholarship in the social studies of science and technology. The final section also rebuts two potential programmatic objections to centres for responsible innovation.

The Challenge of the Commercialized Academy

In December 2000, the vice-president for academic affairs at my (then) university, an economist by training, convened an interdisciplinary group of twenty-odd faculty in the Scarlet Room – a lush conference room in Rutgers's central administrative building. This vice-president wondered why Rutgers – with its two law schools, its recently powerful biomedical sciences, its long tradition in environmental research, its nationally regarded philosophy department, and its up-and-coming planning and policy school – had not yet put science together with law, ethics, and policy into some coherent enterprise. He challenged the faculty in attendance to pursue what he called an initiative in law, ethics, and the sciences.

The Scarlet Room group met only once more, spurring some new collaborations such as one on 'consumer acceptance of biotechnology food' but inducing no coherent university-wide effort. At that second meeting, however, the vice-president requested we participants circulate among ourselves proposals for what such an initiative in law, ethics, and the sciences might do. I offered to the group the bare-bones sketch of something I called the 'centre for responsible innovation.'

I had suspected that the concerns that drew the Rutgers group together were only narrowly cast as 'ethics' or 'law.' Something more subtle, perhaps, but more profound than the ad hoc collection of patenting genes, privacy in the Internet era, or ethics training for graduate students suffused the Scarlet Room. We were acknowledging that the primary task of the university – the creation and dissemination of new knowledge – has normative dimensions that we were not engaging. A similar acknowledgment suffuses the collection of work in this volume as well.

Invigorating the university's primary task with the necessary normative supplements requires accepting the following three premises, none of which should be remarkably disagreeable. The first is that large research universities will continue to be in the business of knowledge-based innovation for a very long time – on the order of generations at least. Even if students evaporate to distant nodes on the World Wide Web and pressures on tenure erode the status of faculty, universities and the people affiliated with them will still create new knowledge and seek to express it in forms scientific, artistic, and professional. People will continue to come together in various real and virtual ways under the auspices of universities to advance the creation and application of new knowledge.

One must also consider the marketing of new knowledge. That the commercial aspect of the university will not disappear is the second premise. The commercialized academy has, in various ways, been around for a very long time. Brooks and Randazzese (1998), for example, emphasize the role of the land-grant institutions in the nineteenth century and the rise of engineering education in the early twentieth century in creating linkages between industry and the academy. MIT pioneered entrepreneurial relations with industry in the first half of the twentieth century (Etzkowitz 2002), and in the second half of the century, 'academic capitalism' in many nations followed the globalization of research in particular and of the economy in general (Slaughter and Leslie 1997; Crow and Tucker 2001).

Over the last generation, universities in the United States have become increasingly involved in the commercialization of the new knowledge they produce with the encouragement of the Bayh-Dole Act of 1980. The Association of University Technology Managers provides an overview of the growth of technology transfer activities in the tenth annual edition of its survey of licensing and other university-based technology transfer practices (AUTM 2001). Participation in the survey has increased from 66 of the top 100 U.S. research universities to 94, and the total number of respondents increased from 130 to 190 institutions. These respondents report an increase from 6337 invention disclosures in fiscal year (FY) 1991 to 13,032 disclosures in FY 2000. New U.S. patent applications from these institutions increased from 1643 to 6375 over the same period. Licences and options executed increased from 1278 to 4362, and licensing income increased from $186 million to $1.26 billion (in current dollars) from FY 91 to FY 00. Respondent universities in the United States received $1.1 billion of the licensing income in FY 00, a sum equivalent to 50 per cent of research expenditures by industry in universities in that year and approximately 7 per cent of such expenditures by the federal government. As measured, albeit crudely, by employment and overall economic activity, Bayh-Dole technology transfer is said to have been responsible for approximately 260,000 jobs and $40 billion in economic activity in the United States in 2000.

Although some research (Mowery et al. 2001) has questioned the necessity and uniqueness of Bayh-Dole's contribution to the boom in intellectual property transactions by universities, such activity – regardless of origin – has created a tightening aggregation of interests around technology transfer. Etzkowitz and Ledesdorff (2000) label this new political economy the 'triple helix' of university–industry–government relations. This new political economy is fuelled not just through licences with royalties but also through licences with equity. AUTM – itself an example of the creation of professional and interest groups around the commercialized academy – reports that licences with equity shares to universities have increased from 142 in FY 95 to 372 in FY 00. The 454 start-ups formed in FY 00 – up from 223 in FY 95 – are most often (80 per cent) located in the same state as the reporting institution, thus contributing to a critical trend in local and regional economic development. The not-for-profit State Science and Technology Institute emerged in 1996 to improve collaborative programs in science and technology for regional economic development, connecting with science and technology commissions or offices in dozens of states and with the Science and Technology Council of the States, affiliated with the National Governors Association. At the federal level, even the National Science Foundation (NSF) has supported an intersectoral approach to research through its Engineering Research Centres (see Feller, Ailes, and Roessner 2002 for an evaluation) and its Science and Technology Centres (see Mervis 2002) inaugurated in the 1980s, its more recent Partnerships for Innovation program, and other collaborative endeavours.

Even if the new political economy of academic commercialization were not strong enough to presume its durability, one could argue – and I do – that it demonstrates through the narrow measures of the market that universities are behaving in a 'responsible' way – that is, in accord with the considered values of the wider community. The promise of research to contribute to economic expansion is rooted in the post–Second World War 'social contract for science' (Guston 2000), and fulfilling this promise is a critical element of scientific responsibility. But the economic contribution tells only a part of the story of responsible innovation. Thus, the third premise – what dawned on the Rutgers group, and the contributors to this volume as well – is that the enterprise of knowledge-based innovation has normative dimensions that science policy confronts only marginally or in ad hoc ways.

Such normative consequences begin with the costs, in addition to the benefits, of technology transfer and other university–industry interactions. On campuses, market values intrude on the scholarly enterprise by creating opportunities for conflicts of interest and commitment among faculty and students, changing norms of scholarly communication and materials transfer, and potentially devaluing learning for its own sake (even while the traditional values of education are also under siege by market thinking, with university presidents viewed as CEOs and students as consumers; see Slaughter 2001). Sheldon Krimsky (2004) reminds us that there are broader public consequences to commercialization in addition to these local ones, as universities and their faculties become more like interested participants in technical decisions and controversies and less like disinterested observers. The ideal of what Joel Primack and Frank Von Hippel (1974) called 'public science' is almost completely lost, as more than two-thirds of national R&D spending in the United States comes from private interests and a significant fraction of the remaining one-third of public money is potentially compromised by interests in commercialization or commingling with private funds. This is not to say that private funding or commingling is necessarily bad, but it is to suggest that financial interests at least complicate the perception of the credibility of technical claims should controversies arise.

In addition to behaving responsibly with respect to economically relevant research through the creation of offices of technology transfer, universities had begun a decade earlier to behave responsibly with respect to human and animal research subjects through the creation of Institutional Review Boards (IRBs). Both of these activities clearly need improvement, particularly at their intersection, as demonstrated by the tragic death of Jesse Gelsinger during a genetic therapy trial at the University of Pennsylvania. In this case, the director of the research housing the research held financial interests in a firm, founded by one researcher, which helped finance the institute (Sacks 2000).

Technology transfer and IRBs show that universities can adopt a responsible position when the opportunity arrives. Yet the protection of research subjects and the contribution of university-based research to the economy do not exhaust the scope of the ethical, legal, and social implications of innovation. The societal value of knowledge creation cannot be measured merely by licensing income, and the ethical duties of research are not entirely discharged by obtaining informed consent. Derek Bok realized this more than two decades ago when he delved into the 'social responsibilities of the modern university.' Bok (1982) asked whether, in a world remade by knowledge-based innovation (and still held hostage to the Cold War), academic freedom should 'extend to university laboratories that could produce discoveries of awesome power and destructive force?' (7). One need not put the question in terms so stark. There are many grades of distinction between academic freedom and intellectual servitude, just as there are many critical societal impacts to discern short of 'awesome power and destructive force.' The point, familiar to students of science and technology, is simply that knowledge-based innovation helps create many patterns of society that people must respond to, usually without having had any choice in those patterns – whether through design, accident, or neglect (Winner 1977).

Alas, as individuals, as knowledge communities, and as a society, we neither know enough about how these patterns are created, nor know enough about how to know enough about them. There is thus a role for knowledge creation and dissemination around the normative dimensions of knowledge creation itself, and universities therefore have a role in this reflexive enterprise. We must find subtler ways to manage the ethical, legal, and social implications of research, content not merely with profit and protection but with the (admittedly contested) prospect of helping people pursue more uplifting lives in more just societies.

The reflexive pursuit by universities of the ethical, legal, and social implications of the knowledge-based innovations they principally pursue is particularly critical given the loaded dice that Bok (2004) refers to: If it is hard to say 'no' to the incremental commercialization of universities because each new bit is not entirely unprecedented and because the negative consequences of each new increment cannot be as clearly discerned as the positive one, then at least universities can say 'yes' to the reflexive study of the interaction of innovation and society.

The Roles of the Centre for Responsible Innovation

To house this self-scrutiny of the ethical, legal, and social implications of knowledge creation, I propose that universities create centres for responsible innovation (CRIs). CRIs would accomplish this reflexive pursuit in parallel to the universities' traditional missions in teaching, research, and service.

The teaching role for centres for responsible innovation would primarily be as brokers of interdisciplinary teaching and learning opportunities and other educational resources. One obvious area would be supporting departments in their federally mandated and/or independent efforts to integrate ethics and responsibility issues into their curricula. The National Institutes of Health (NIH), for example, require that institutions applying for biomedical training grants provide coursework in scientific responsibility to trainees. The Office of Research Integrity in the Public Health Service announced a policy to require similar training in the responsible conduct of research for all grant applicants (ORI 2000), although Congress suspended the policy pending what it considers more appropriate procedures in promulgating it (Brainard 2001). The Institute of Medicine (2002) has issued a call not only for education in the responsible conduct of research but also for 'institutional self-assessments' and 'ethical impact' assessments of institutional conflicts of interest – exactly the type of activities to which a CRI could contribute. In other disciplines, the Accreditation Board for Engineering and Technology has required ethics training in its accreditation criteria for undergraduate engineering degrees (ABET 2000). The centres would assist relevant courses in finding faculty or professional expertise and course material.

Centres would also propose, plan, promote, and/or implement novel educational programs. Depending on the status quo ante at the particular university, such programs might include helping to create or support undergraduate coursework, minors, or majors in science, technology, and society or graduate programs in science and technology studies. They could also include graduate programs that 'cross-trained' natural scientists and engineers in business, policy, law, and the humanities, as well as vice versa. At Rutgers, for example, one program combines a master of science in biomedical sciences with a master of business administration, and MIT has a Technology and Policy Program that adds training in policy to the education of graduate students in the sciences and engineering. CRIs would also offer summer seminars for faculty and staff, as the Rutgers's Centre for the Critical Analysis of Contemporary Culture did several years ago when it hosted a 'course for colleagues' on the so-called science wars that attracted faculty from across the university.

The primary research contributions of such centres would be engaging in externally funded, cooperative research on the ethical, legal, and social implications of other research on campus. On a number of different research university campuses, they would thus constitute a kind of distributed capacity for technology assessment for which some have called (e.g., Sarewitz 1996). Ethical, legal, and social implications research – known by the acronym ELSI – has expanded over the last decade through 3–5 per cent set-asides for such large federal programs as the human genome, information technology, and nanotechnology initiatives. Although successful in funding a significant amount of research – the human genome ELSI program in NIH and the Department of Energy has funded more than $150 million of work since FY 90 – ELSI seems to have been more successful in placating or diverting calls for other kinds of innovation in ethics and responsible research than in meeting expectations of informing the science policy process or encouraging broader and more sophisticated political considerations of such issues as genetic privacy and discrimination (Kitcher 2001). In a detailed study of the human genome ELSI program, Lauren McCain concludes that 'the ELSI experience so far does not support an early contention ... that public science projects can sufficiently monitor and address their own social impacts. ELSI-type programs are unlikely to help shape public research agendas' (2002: 132).

Such expectations for ELSI research, however, may be too high, as ELSI work concentrates on the innovations after they emerge from the laboratory and because it has had few if any institutional links back to setting research priorities and other science policy tasks. Instead, research conducted or encouraged by CRIs could resemble 'real-time technology assessment,' which combines historical research, public opinion and communication, traditional technology assessment, and interactions between the lay public and active researchers (Guston and Sarewitz 2002). This ensemble of research attempts to encourage responsible knowledge-based innovation by understanding historical precedents for similar innovations, by providing links for communication and education among researchers, potential consumers, and the lay public, and by allowing the possibility of intervention in the innovation process by public values before the innovation is reified as a market product. Particularly interesting would be such interdisciplinary collaborations coordinated by CRI on knowledge-based innovations with potentially important impacts on the local communities around the university.

CRIs would also be involved in more traditional scholarly pursuits, such as organizing symposia of university and outside faculty for high-profile intellectual and public events and pursuing an aggressive publishing agenda from them. It would provide advice, assistance, and seed funds for faculty and students commencing externally funded projects that might benefit from ideas in responsible innovation. Specifically, they might assist natural scientists and engineers applying for NSF grants to articulate responses to NSF's criterion 2 – the broader social merit of the proposed research. In a sense, criterion 2 may be misguided without such collaboration because it assumes that researchers are qualified and capable of rendering coherent judgments about the social implications of their work, an assumption that is incomplete in the least. NSF does not have a good track record in using criterion 2 to make funding decisions (Mervis 2001), and there is some sense, first, that part of this problem is that applicants do not know how to frame the responses to criterion 2 issues and, second, that political and administrative pressure to use criterion 2 issues will not subside regardless.

The primary service role for a centre for responsible innovation would be reaching out beyond the university to local, state, and national decision makers, and to the public generally. This role would, of course, require close collaboration with university communication and government affairs offices. State legislators and their staff often believe that universities, particularly public ones, should be more helpful as informational sources (Jones, Guston, and Branscomb 1996). Such a centre would offer informational programs for public officials – grounded in research at the university itself – and would collaborate with a university's Washington office on such programs for national decision makers as well. It would also collaborate with Institutional Review Boards, Offices of Technology Transfer, and other elements of university administration on local projects of mutual interest. The centre would pursue a strategy for dissemination that includes non-academic outlets, including op-eds, cable television, and public to reach the broadest possible audience.

Importantly, interaction with the public mediated by CRIs should not – must not – be exclusively one-way. Centres for responsible innovation must also be conduits for the influx to the university of public concerns and lay perspectives. These voices need to be heard not only by university officials but by rank-and-file faculty and researchers as well. There are a variety of models for facilitating such expert–lay interaction, including: the consensus conferences (citizens' juries) pioneered by the Danish Board of Technology and recently imported to the United States and other countries (Guston 1999a; Hörnig 1999; Joss and Durant 1995); scenario development workshops in which experts and lay participants reflect on coherent descriptions of envisioned technological futures (Andersen and Jaeger 1999; Sclove 1999); focus groups (Dürrenberger et al. 1999); and participatory research and design such as that conducted by community-based research centres and 'science shops' (SCIPAS 2001).

As this list of potential functions suggests, the vision for the centre for responsible innovation is not of an organization that does utterly unique things, although some and perhaps many of its tasks will be novel in particular institutions. Rather, its novelty is in the mission of institutionalizing responsible innovation at research universities and the ensemble of activities in pursuit of this mission. If we remain committed to the idea of the synergy between research and teaching, then this commitment should extend to the university's endeavours to behave responsibly. In this view, failure to institutionalize such an ensemble of activities is not just a failure to pursue responsibility more actively, but it is failure to pursue efficiently the mission of the university.

Through centres for responsible innovation, universities would construct ongoing projects and relationships to help assure members of their own communities, leaders from the public and private sector, and the public at large that the university is engaged in responsible innovation – despite continuing commercial ties. CRIs would require some commitment of faculty and staff resources, but they may very well raise a significant share of their own funds from sources including ELSI programs in the genome, information technology, and nanotechnology initiatives, the Program in Science and Society at NSF, local and regional foundations concerned with the immediate implications of university research, and local corporate givers pursuing an enlightened approach to innovation. The first mover toward a centre for responsible innovation would capture headlines and, better, imaginations, as such an institution would be, to my knowledge, unique not just in the country but perhaps the world. With a centre for responsible innovation, a university would be able to brand itself in an increasingly competitive environment as at the vanguard of both intellectual creativity and social responsibility.

Discussion

Having considered the warrant for and functions of centres for responsible innovation, we can return to the philosopher Kitcher's concerns about the science policy system more generally: given that CRIs seek to involve the lay public in potentially critical roles in knowledge-based innovation and that they are in some sense unprecedented institutions, can we have any confidence that they will gain at least a toe-hold in the normative wilderness of research universities? Are centres for responsible innovation as described sufficiently well grounded in thought and experience to have even a hope of success?

Over the past decade, familiarity with and expertise in mechanisms for the participation of the public in various technical decisions – particularly environmental policy (e.g., Fischer 2000; Hisschemoller et al. 2001) – has increased tremendously. In science and technology policy circles, perhaps the most notable achievement has been the spread of Danish-style consensus conferences, or citizens' panels, from the smaller nations of northern Europe to larger nations including the United States and Japan, and to some developing world nations as well. Governmental bodies, universities, and private groups have implemented such citizens' panels, as well as other participatory mechanisms, including the scenario workshops, focus groups, and community-based research centres (or 'science shops') mentioned above. With the increased conduct of participatory activities has come an increased attention in the science policy literature. Joss (1999) and Chopyak and Levesque (2002), in particular, provide useful summaries of the variety of mechanisms in use, their apparent strengths and weaknesses, and evaluations of their experiences. The literature reveals that such participatory mechanisms have demonstrated, at least under closely supervised and mediated situations, that members of the lay public can interact helpfully with experts to conduct assessments of technical controversies and policies, develop plans for technological futures, and design and conduct certain technical inquiries alongside traditional researchers. From these experiences, it is plausible, even likely, that a well-prepared CRI could succeed in engaging laypersons and communities around universities in substantive activities with university researchers.

Over the last decade, scholars have also made considerable progress in understanding how the institutions of science policy operate, between politics and science, to produce societal outcomes. Instead of focusing on the 'estates' of science, the professions, administration, and politics that characterized the 'spectrum from truth to power' (Price 1965), they examine the institutions that manage transactions across these boundaries, which they conceive of as less static than Price did. This 'important strand of current scholarship about science and society' (Sonnert and Holton 2002) has yielded a description of 'boundary organizations' (Guston 2000; 2001). Examples include: offices of technology transfer (Guston 1999b), the Health Effects Institute (Keating 2001), agricultural extension (Cash 2001), the International Research Institute for Climate Prediction (Agrawala, Broad, and Guston 2001), and the Subsidiary Body for Scientific and Technological Advice to the UN Framework Convention on Climate Change (Miller 2001). Critical to the success of boundary organizations is their ability to manage the cooperation between scientists and non-scientists in the pursuit of ends of mutual interest, what Guston (2000) calls 'collaborative assurance' (144). Centres for responsible innovation can engage in collaborative assurance, for example, by providing services to natural science and engineering units (by assisting with ethics curricula, grant applications, and information dissemination to the public and decision makers) while at the same time providing services to society (by helping attune scientists and engineers to societal needs). Although no person 'may serve two masters,' boundary organizations like CRIs can serve both science and society by helping each to achieve the teaching, research, and service goals both want to occur responsibly.

Creating new institutions, even those about which one might have some confidence based in theory and practice, entails some risk. One potential risk of creating centres for responsible innovation is that CRIs will end up institutionalizing criticism of science. The honest response to such a concern is that CRIs will indeed institutionalize science criticism, but criticism in the sense of constructive engagement with the quality and contribution of the scientific enterprise. Such criticism, more along the lines of art or food criticism (Chubin 1994), is exactly what such a centre should be after, and significantly what the commercialized university needs. The acceptability within the ranks of scientists of ELSI programs suggests, to the first approximation, that informed criticism can find an institutional home close to the research it engages (although it may have little influence). Moreover, there are other, albeit modest, proofs of concept of aspects of this enterprise. For example, nanotechnology researchers at Arizona State University have collaborated with the Consortium for Science, Policy, and Outcomes to propose a real-time technology assessment agenda and to cross-train ASU nanoscience and engineering graduate students in science policy under an Integrative Graduate Education and Research Traineeship (IGERT) grant from NSF.

A second possible programmatic criticism is that CRIs would contribute to what might be called the 'ELSI-fication of the social sciences.' That is, the important role of the social sciences in examining the implications of the natural sciences may be taken as the only, or the primary, role for the social sciences. John Steelman, one of the architects of the modern research establishment in the Truman administration and himself an economist, endorsed the view that 'competent social scientists should work hand in hand with the natural scientists, so that [societal] problems may be solved as they arise, and so that many of them may not arise in the first instance' (Guston and Sarewitz 2002: 95). Much more recently, the sociologist William Julius Wilson (2002) has argued that the social sciences must respond to the 'impetus to address policy-relevant issues ... that grow out of the struggles of nation states to adapt to the impact of rapid technological and economic changes on individuals, families, communities, institutions, and the society at large' (1).

Without predicting the likelihood of demands for greater relevance from social science research, ELSI-type work comprises a significant but not overwhelming share of the federal social science research dollar. Since 1991, NIH has spent just over $130 million in ELSI funds, including an estimated $21.5 million for FY 02. DOE has spent about $25 million over the same period on its genome-related ELSI research, including an estimated $2.8 million in FY 02.[1] Total NSF spending on the social and economic sciences is more than $70 million per year (AAAS 2002: 144), and of this sum, spending on the Program in Science and Society – NSF's closest cognate program to ELSI – is in the range of $4 million per year. (NIH, the Defense Department, and the Department of Education each fund more mission-related social science than NSF as well.)

Given that genome-related ELSI funding is nearly an order of magnitude greater than NSF's, a more likely result than the ELSI-fication of the federal social science portfolio is the overwhelming general program of research in the ethical, legal, and societal aspects of innovation by particular innovations that the federal government has already identified and cultivated for support. With significant ELSI-type funds coming only after the federal government has committed to a multi-year, multi-billion-dollar investment, the results of such research may be enlightening but they have a vanishingly small chance at influencing the policy environment or the trajectory of the innovation. The risk of ELSI-fication seems not as great as the risk that only specific technologies will be scrutinized, and only then after they have already been imparted with great momentum. The need for decentralized, university-based centres for responsible innovation therefore becomes even more apparent as a kind of early warning.

A third objection might be that there are few incentives to go ahead with such a challenging agenda, as well as many obstacles or opposition to overcome. While such a pragmatic objection should be of no great consequence for universities to do what is right, it is nevertheless reasonable to ask before embarking on the host of institutional changes that CRIs would necessitate. Such opposition would likely include traditionalist perspectives in academic science that, not having grasped the changes in the governance of science wrought by the earlier examples of responsible innovation like human subjects protection and technology transfer, still demand an unfettered scientific autonomy that is as inappropriate as it is passé.

But the point is not to force cooperation with the CRI agenda but rather to enhance the opportunities and competitiveness of researchers and universities that adopt it voluntarily. The incentives include helping university scientists and engineers do more with their research than simply producing knowledge, and helping them do more with their students than simply reproducing themselves. They include not just positioning the university, but protecting it, as a locus of responsibility and reflection in a competitive and litigious society.

Since the first drafting of this chapter, I have become the director of a new Centre for Nanotechnology in Society at Arizona State University (CNS-ASU) which embodies most if not all of the attributes described here of a centre for responsible innovation. Although CNS-ASU has encountered some hesitancy, we have encountered still more enthusiasm about its work among the nano-scale science and engineering researchers with whom we collaborate closely. They have sought us out as research collaborators, as mentors and instructors for their students, and as models for societal implications in other technical areas; one has even attributed greater success in landing a large grant proposal to perspectives that we have encouraged. This is not to say that creating CNS-ASU as a centre for responsible innovation has been easy or complete, but it is to say that proof of concept exists, as does the market for these activities among academic scientists themselves.

Conclusion

In this chapter I have argued that universities should make a demonstrative commitment to responsible innovation in the face of the challenges wrought by the commercialization of the academy. Such a commitment is warranted because universities will continue to be in the business of knowledge-based innovation for the foreseeable future; because the commercialization of knowledge-based innovations has been, is, and will continue to be integral to some aspects of the university as well; and because doing no harm and contributing to the economy do not exhaust the normative demands on the research enterprise. University-based centres for responsible innovation can help satisfy these additional normative demands, even while assisting members of the university community to achieve the substantive goals they have for teaching, research, and service. Grounded in recent scholarship in public participation and the design of science policy 'boundary organizations,' they can draw university research and the values of communities and needs of decision makers closer in fruitful ways. CRIs can help create microcosms of 'well-ordered science,' one university at a time.

NOTE

1 Data provided by Howard Silver and Angela Sharpe of the Consortium of Social Science Associations.

REFERENCES

Accreditation Board for Engineering and Technology (ABET). 2000. *Criteria for Accrediting Engineering Programs*. Baltimore: ABET. http://www.abet.org/images/Criteria/eac_criteria_b.pdf (accessed 5 July 2007).

Agrawala, S., K. Broad, and D.H. Guston. 2001. Integrating Climate Forecasts and Societal Decision Making: Challenges to an Emergent Boundary Organization. *Science, Technology, and Human Values* 26, no. 4: 454–77.

American Association for the Advancement of Science (AAAS). 2002. *AAAS Report XXVII: Research and Development, FY 2003*. Washington, DC: AAAS.

Andersen, I.E., and B. Jaeger. 1999. Scenario Workshops and Consensus Conferences: Towards More Democratic Decision-Making. *Science and Public Policy* 26, no. 5: 331–40.

Association of University Technology Managers (AUTM). 2001. *AUTM Licensing Survey, FY 2000*. Northbrook, IL: AUTM.

Bok, D. 1982. *Beyond the Ivory Tower: Social Responsibilities of the Modern University*. Cambridge: Harvard University Press.

– 2004. The Benefits and Costs of Commercialization of the Academy. In *Buying In or Selling Out? The Commercialization of the American Research University*, edited by D.G. Stein, 32–47. New Brunswick, NJ: Rutgers University Press.

Brainard, J. 2001. Ethics-Training Rule Is Suspended. *The Chronicle of Higher Education* (2 March): A27.

Brooks, H., and L.P. Randazzese. 1998. University-Industry Relations: The Next Four Years and Beyond. In *Investing in Innovation*, edited by L.B. Branscomb and J.H. Keller, 361–99. Cambridge: MIT Press.

Cash, D. 2001. In Order to Aid in Diffusing Useful and Practical Information: Agricultural Extension and Boundary Organizations. *Science, Technology, and Human Values* 26, no. 4: 431–53.

Chopyak, J., and P. Levesque. 2002. Public Participation in Science and Technology Decision Making: Trends for the Future. *Technology in Society* 25: 155–66.

Chubin, D. 1994. How Large an R&D Enterprise? In *The Fragile Contract: University Science and the Federal Government*, edited by D. Guston and K. Keniston, 118–44. Cambridge: MIT Press.

Crow, M.M., and C. Tucker. 2001. The American Research University as America's *de facto* Technology Policy. *Science and Public Policy* 28, no. 1: 1–10.

Dürenberger, G., H. Kastenholz, and J. Behringe. 1999. Integrated Assessment Focus Groups: Bridging the Gap between Science and Policy. *Science and Public Policy* 26, no. 5: 341–9.

Etzkowitz, H. 2002. *MIT and the Rise of Entrepreneurial Science.* New York: Routledge.

Etzkowitz, H., and L. Ledesdorff. 2000. The Dynamics of Innovation: From National Systems and 'Mode 2' to a Triple Helix of University-Industry-Government Relations. *Research Policy* 29: 109–23.

Feller, I., C.P. Ailes, and J.D. Roessner. 2002. Impacts of Research Universities on Technological Innovation in Industry: Evidence from Engineering Research Centers. *Research Policy* 31: 457–74.

Fischer, F. 2000. *Citizens, Experts, and the Environment: The Politics of Local Knowledge.* Durham, NC: Duke University Press.

Guston, D.H. 1999a. Evaluating the First U.S. Consensus Conference: The Impact of the Citizens' Panel on Telecommunications and the Future of Democracy. *Science, Technology, and Human Values* 24, no. 4: 451–82.

– 1999b. Stabilizing the Boundary between U.S. Politics and Science: The Role of the Office of Technology Transfer as a Boundary Organization. *Social Studies of Science* 29, no. 1: 87–112.

– 2000. *Between Politics and Science* New York: Cambridge University Press.

– 2001. Special Issue: Boundary Organizations in Environmental Policy and Science. *Science, Technology, and Human Values* 26, no. 4: 399–408.

Guston, D.H., and D. Sarewitz. 2002. Real-Time Technology Assessment. *Technology in Society* 24: 93–109.

Hisschemoller, M., R. Hoppe, W.H. Dunn, and J.R. Ravetz. 2001. *Knowledge,*

Power, and Participation in Environmental Policy and Risk Assessment. New Brunwick, NJ: Transaction Press.

Hörnig, S. 1999. Citizens' Panels as a Form of Deliberative Technology Assessment. *Science and Public Policy* 26, no. 5: 351–9.

Institute of Medicine. 2002. *Integrity in Scientific Research: Creating an Environment That Encourages Responsible Conduct.* Washington, DC: National Academy Press.

Jones, M., D.H. Guston, and L.M. Branscomb. 1996. *Informed Legislatures: Coping with Science in a Democracy.* Lanham, MD: University Press of America/CSIA.

Joss, S. 1999. Special Issue: Public Participation in Science and Technology. *Science and Public Policy* 26, no. 5: 290–3.

Joss, S., and J. Durant. 1995. *Participation in Science: The Role of Consensus Conferences in Europe.* London: The Science Museum.

Keating, T. 2001. Lessons from the Recent History of the Health Effects Institute. *Science, Technology, and Human Values* 26, no. 4: 409–30.

Kitcher, P. 2001. *Science, Truth, and Democracy.* New York: Oxford University Press.

Krimsky, S. 2004. Reforming Research Ethics in an Age of Multivested Science. In *Buying In or Selling Out? The Commercialization of the American Research University,* edited by D. G. Stein, 133–52. New Brunswick, NJ: Rutgers University Press.

McCain, L. 2002. Informing Technology Policy Decisions: The U.S. Human Genome Project's Ethical, Legal, and Societal Implications Programs as a Critical Case. *Technology in Society* 24: 111–32.

Mervis, J. 2001. NSF Scores Low on Using Its Own Criteria. *Science,* 30 March.

– 2002. Science with an Agenda: NSF Expands Centers Program. *Science,* 26 July.

Miller, C. 2001. Hybrid Management: Boundary Organizations, Science Policy, and Environmental Governance in the Climate Regime. *Science, Technology, and Human Values* 26, no. 4: 478–500.

Mowery, D.C., R.R. Nelson, B.N. Sampat, and A.A. Ziedonis. 2001. The Growth of Patenting and Licensing by U.S. Universities: An Assessment of the Effects of the Bayh-Dole Act of 1980. *Research Policy* 30: 99–119.

Office of Research Integrity (ORI). 2000. PHS Policy on Instruction in the Responsible Conduct of Research. http://ori.dhhs.gov/html/programs/rcrcontents.asp.

Price, D.K. 1965. *The Scientific Estate.* Cambridge: Belknap Press of Harvard University.

Primack, J., and F. Von Hippel. 1974. *Advice and Dissent: Scientists in the Political Arena.* New York: New American Library.

Sarewitz, D. 1996. *Frontiers of Illusion: Science, Technology, and the Politics of Progress.* Philadelphia: Temple University Press.
Sacks, S.M. 2000. The Case of the University of Pennsylvania Health System. *SciPolicy: The Journal of Science and Health Policy* (Fall): 122–53.
Sclove, R.E. 1999. The Democratic Politics of Technology: The Missing Half, The Loka Institute. http://www.loka.org/idt/intro.htm.
Slaughter, S. 2001. Professional Values and the Allure of the Market. *Academe* (September/October): 22–6.
Slaughter, S., and L.L Leslie. 1997. *Academic Capitalism: Politics, Policies, and the Entrepreneurial University.* Baltimore: The Johns Hopkins University Press.
Sonnert, G., with the assistance of G. Holton. 2002. *Ivory Bridges: Connecting Science and Society.* Cambridge: MIT Press.
Study and Conference on Improving Public Access to Science (SCIPAS). 2001. Living Knowledge: The International Science Shop Network, Project Technical Reports Compendium, vols. I.A, I.B, and II. Amherst, MA: Loka Institute. http://www.livingknowledge.org.
Wilson, W.J. 2002. Expanding the Domain of Policy-Relevant Scholarship in the Social Sciences. *PS: Political Science and Politics* 35, no. 1: 1–4.
Winner, L. 1977. *Autonomous Technology: Technics-Out-of-Control as a Theme in Political Thought.* Cambridge: MIT Press.

17 Citizens and Biotechnology

RAHUL K. DHANDA

Two of the most enriching concepts to ever capture the imagination of scholars are science and democracy. Both have received tremendous attention and, arguably, success as the dominant forms of thought in the world today. Science has justified the massive expenditure of government funds worldwide, and democracy has justified a range of interventions, including trade embargoes, international sanctions, and even war. Interestingly, both, in the rhetoric of their practitioners, are considered constantly refined ideals, but ideals nonetheless.

To comment on either requires a tremendous measure of expertise, but to comment thoughtfully on both is an even more daunting task. Despite this dual challenge, it is my intention to complicate the calculus even further – by adding the additional variables of private industry and bioethics to the analysis. Having participated in society as a scientist, a corporate employee, and as a commentator on bioethics, I have played in too many realms to be an expert in any. But from my experience it is clear that all deserve their place in democracy. While I could present a scientific, bioethical, or a corporate interpretation of the relationship between democracy, science, and technology, I prefer to speak with what should be the most powerful voice of all – as a citizen. It is not enough as a citizen to be concerned exclusively with justice, profit, or the pursuit of knowledge – one must explore and place them all in a context that characterizes their interrelated benefits and problems. While there are a wide range of types of democratic citizens in the world, I can only speak as a participant in the democracy in the United States of America. Narrowing my scope one step further, I must also disclaim that my observations will focus on a particular kind of science – that of biotechnology.

To participate in science, business, and democracy demands a reconciliation of notions that, on their face, are not easily reconciled. In a culture that values a wide variety of freedoms, corporations represent a particularly confounding factor. When the freedom to operate leads to a conflict between corporate profit and the myriad concerns that surround the scientific enterprise, like those of safety and ethics, many are sanguine about the choices that may be made. Yet, in a participatory government, both citizens and their empowered officials should strive to work with scientists, universities, and corporations alike to obviate as many of the anticipated negative externalities as possible. Protecting public safety, in short, is not an overreach of the government, but to do so the government undoubtedly requires mechanisms that keep pace with emerging technologies.

To fully understand the relationship between the government, science, and industry, it is useful to consider the lowest common denominator that links them together – money. Time is also a factor, and briefly surveying both in the context of research is a necessary first step in determining which challenges require the attention of the public. The public is involved with the scientific enterprise at this stage by providing the grants that stimulate early-stage science. Taxes and other sources of revenue provide the government with the funds that it makes available to scientists. As research proceeds, the technology will, in the best of cases, become available to the public after years of experimentation and vetting validate the technology. The benefit to the public is the fruits of the research, while the benefit to scientists is the capital to pursue expensive research interests.

If both the public interest and autonomy of the scientific enterprise are to be observed, then attention must be paid to the early stages of research and the later dissemination to society – the twin aspects of public–scientist interactions. Although related, they are very different. The former, early-stage research, represents the opportunity for private institutions (corporations) and the public to directly influence the direction of research by controlling the transfer and conditions of funding, while the latter, dissemination, represents the moment when the public will be exposed to the technological application. Throughout this process, science itself is caught between satisfying the social good and the tastes of the market.

With such strong forces vying for the fruits of the scientific endeavour, it is not surprising that many question whether the corresponding research can be pursued free of influence. I think it is very clear that it both does and does not. This glib response is admittedly simplistic, but it is difficult to see how there can be anything but a paradox. Consider the example of the human genome project. Upon its earliest conception, many thought it unworthy of substantial funding, and it took some years before it was approved as a multi-billion-dollar, government-sponsored project (Davies 2002). During its execution, it achieved an incredibly high profile, and both companies and scientists promoted it as one of the most important scientific undertakings in the twentieth century. When it was completed, many were left wondering when there would be a return on the investment. Active human genes numbered only in the 30,000 range, far less than the anticipated 100,000, and applying the information is looking to take longer than the sequencing project itself. At different points in the process, scientists, citizens, and corporate employees wavered in their opinions of the project. If one were to question whether the public was served, whether corporations pushed the project, and whether scientists maintained their autonomy, the answer would be yes and no all the way down the line.

To elaborate, many scientists participated in this endeavour, and many more did not. For those who did not, they fell into two categories: some felt that this area of research was not of interest; and some did not win the grants for which they applied. In both cases, one of choice and one of disappointment, the evidence that scientists retain independence is challenged. Consider the first category. To perform their research, these scientists did not respond to any of many requests for proposals offered by the sponsoring agencies, the Department of Energy or the National Institutes of Health. Instead, they assessed their research interests and applied their grant-writing resources towards other efforts. In the other group, scientists hoped to do this research, but they were left wanting. Unfortunately, for those that do not want to or are not chosen to participate in NIH sponsored studies, resources are limited. Certainly, they can appeal to other organizations, but, for career advancement and sustained research funding, they will eventually have to tap into government funding. If their agenda is not closely aligned with that of government sponsors, then the public funding route is closed to them unless they bend to meet it. Choice, in this case, is not the same as autonomy, since there are few choices available.

If government funds do not materialize for these scientists, seeking funds elsewhere is limited to NGOs and corporations. In either case, the research agenda is likely to be restricted to fairly narrow confines. Interestingly, in the biosciences, there exists considerable diversity of research foci and government funding agencies. The NIH alone contains twenty centres that have provided billions of U.S. dollars to a wide range of research efforts over the past few years. Thus, the realistic research activities of scientists are limited, but the availability of research options is still impressively large.

Research agendas often boil down to money, which is not a logical deduction from democracy, as governmental systems are unlikely to emphasize what Marx called 'the means of production.' More likely, the economic system, in this case capitalism, will influence the allocation of funds. In the simplest terms, money is not distributed equally – it is earned based on the merit of research projects, investigator competence, conformity with research priorities, and affiliation with successful peer groups. Sociologist Robert Merton termed this the Matthew effect, where 'eminent scientists get disproportionately great credit for their contributions to science while relatively unknown scientists tend to get disproportionately little credit for comparable contributions' (Merton 1968). By itself, this pattern is neither good nor bad for the public, although many would presume that funds ought to be distributed evenly to those attempting to increase knowledge through methodologically acceptable efforts. That research is subject to the tastes of funding sources, even if those sources derive from tax dollars contributed by the public, challenges purist pursuits. Further, availability of corporate funding adds another layer. The belief that science ought to drive the agenda and not those affected by it creates a public–private dichotomy too often neglected by the institutional interests of industry, the public, and science. Science, we will see, is perhaps more private than it ought to be, regardless of the preferences of industry and the predilections of the public.

Private Funding, Public Knowledge

Because few, if any, researchers enjoy the wealth and freedom to pursue their own research, funding sources define both the boundaries and the direction of science. When scientists search for capital from private industry, the research is likely to be more confined in its scope than with public funds. Corporate research requires very precise focus, outcomes, development, and timelines. Although there are exceptions, generally working with a corporation entails acting as a closely controlled contractor. In contrast to the NIH funding setting, where research with public funds focuses on a project developed by a scientist for the scientist's purposes, private money places science in the service of capital. In both cases, the findings are likely to result in novel inventions, and possibly even patents. In the private sector, these inventions would become the property of the funding agent, while in the NIH model the Bayh-Dole Act places ownership in the hands of the centre where the research was performed. The invention in this case is available to out-license, with any resulting income likely to be split between the institution, the inventor's lab, and the inventor. Interestingly, in both cases, a corporation is likely to 'own' a valuable technology at some point in its life cycle.

Whether the Bayh-Dole Act and this licensing pattern increase the likelihood of science making its way into the private domain is unlikely to inspire debate. It almost certainly has. Indeed, very few patentable, biotechnical advances are being kept from private industry. While this may seem at odds with the public value of the research, and the public funding behind it, it is a necessary step. The argument that public funds ought to lead towards public 'ownership' of technology is problematic if posed against the necessity to commercialize technology to deliver public 'value.' Stated another way, the public can only derive value from its publicly funded research if that research is eventually situated within a corporation.

Before we continue with this provocative thought, it is important to understand how the terms 'public,' 'state,' 'government,' and 'private' are defined in this analysis. While scholars often argue about the definition of 'democracy,' less debate tends to surround these other words, yet they are critical to the democratic condition. These are interdependent terms. More precisely, 'public' is not usefully defined unless done so in contrast to 'private.' In this essay, these two words are derived from John Dewey's *The Public and Its Problems* (1927). In this text, Dewey draws distinctions based on action. Actions that beget consequences that affect more parties than those involved in a transaction are deemed 'public,' while those where the consequences are limited to the principals are 'private.' Further, when consequences are anticipated to reach beyond the transacting principals (to varying degrees), an entity must emerge to mediate those actions on behalf of the public – that entity is the 'state.'

In the simplest sense, pinpointing the institutions that are private in the corporate–scientific interface requires identifying those expectedly influenced by the transaction. More precisely, if the scientist and those within the corporation are the only ones affected by the interaction, then the transaction is private. If these terms are satisfied, few arguments can be made in favour of frustrating the transaction.[1]

Where parties are capable of a private transaction, the transaction itself must be questioned to determine whether the event is a public one. Corporate-sponsored activity in the area of medical research, for example, now involves a series of other actors. Consider genetic information and drug development. A company that specializes in genetic technologies is likely to target diseases that have a genetic component. After finding the gene (or more likely, genes) associated with the disease, that company may decide to develop a diagnostic to test for the disease as well as to use the information to identify drug targets. To identify drug targets, researchers would have to investigate the proteins created and influenced by the 'disease' gene and then try to develop compounds that either mimic the 'non-disease' variety of that gene or interrupt the functioning of the disease gene.

While this general description of discovery and innovation may seem irrelevant to the democratic impulse, the details suggest the contrary. Consider how genes are discovered. The common conception is that advanced sequencing techniques lead to the identification of functional components of the genome, but that is only one part of the effort. Sequencing efforts along the lines of the human genome project determine one version of the string of bases (ACGT) that compose DNA. Particular patterns of bases will suggest where a new gene starts, while further assays can determine whether that portion of the genome is really a gene or a red herring. Still, this method does not provide enough information to offer insight into the gene's function. For that, much more research has to be performed. To fully characterize a gene, studies must be performed on large populations of individuals with the disease in question. Often, this is where academic scientists are recruited to participate in research efforts. If a scientist and a company are interested in the mechanism of genetic disease associated with asthma, for example, both may collaborate to achieve the scientific and corporate end. The researcher may have access to a large patient population with asthma and the company may have data on genes hypothesized to be related to asthma. The two together can perform an association study. In these experiments, individuals with asthma (and 'control' individuals without asthma) would volunteer their blood, or possibly even lung tissue, for genetic analysis. Comparing DNA among individuals with and without disease is likely to pinpoint sequences that are common to asthma sufferers. After this analysis to identify gene candidates, lung tissue is even more valuable because scientists can look at those genes that are 'turned on' in the lungs for one population versus the other.

This crude sketch of the scientific process is very revealing. The private transaction between company and scientist has directly involved research participants. While the actors in this drama expand, the argument can still be made that the affair is a private one. The human research subjects are likely to have signed a contract known as a consent form that empowers the scientist to perform this study as well as to use the biological materials for commercially related research. Working through the example in more detail, or analysing an actual case, would reveal similar interactions and contracts involving the scientist's university, staff, and others that are affected by the transaction.

After wading through this laborious example, let us now focus on the science as well as the research participants. The goal of this research course is to discover the genes involved with disease as well as to utilize this new knowledge to discover an appropriate therapy. For these genetic association studies, individuals donate their DNA, but is that a private act? In terms of the U.S. Constitution, the Fourth Amendment would claim that it indeed is. However, this information, like many other forms of health information, carries knowledge of other persons with it. Traditionally, bioethicists would argue that my DNA and the information in my DNA will also offer insights into my kin. One mutation in my genome might suggest the same in my brothers, since we each have the same genetic donors (our parents). While a compelling notion, it is not entirely convincing. DNA is shared by all human beings, whether 'owned' by a company or not. The human genome is not a human's genome – it is humanity's genome. Discoveries related to the 'building blocks of life,' as DNA has been called, have tremendous reach beyond the laboratory, and the potential abuses of that knowledge, although arguably limited, certainly extend the transaction beyond those involved in research. This very public concern over private research is one to which this essay will return, but its value now is in illustrating how the life sciences as a set of related disciplines is unlikely to accomplish its tasks in private.

Two more abstract notions must be examined. The first is the assertion that the act of knowledge generation is or can be private. To claim that the creation of knowledge, or asserted truths, is a private endeavour is specious at best. Even within the confines of private corporations, the information will be revealed and will have a social effect

(Coase 1960). Patents prove this out; these devices are contracts promising the exclusivity of gain from knowledge to the inventor or another designate, but with the caveat that the underlying knowledge be disclosed. Seemingly, the privacy of this scientific transaction is fading with each layer.

The second notion is that individual participation in such research is a self-interested action. In the bioethics literature, as well as that of medical history (Jonsen 1998), one motivation stands out among the many offered by human research subjects when asked why they participate in research – they commonly assert a sense of civic duty (Rabinow 1999). Most research projects in the life sciences are decades away from practical application, and those involved with the process, whether investigators or human subjects, are usually well aware of the temporal distance between experiment and therapy. Further, these individuals are also commonly aware of the disappointing reality that most of these experiments are unlikely to result in a cure. Yet these volunteers participate, and, in participating, they make a commitment to society. Indeed, they participate because they expect this transaction to be a public one.

Despite vehement support for the value of science and an almost apologetic sympathy for the institution, Dewey recognized its ever-increasing public nature. Indeed, his meditations on democracy reveal a powerful lesson that remains to be fully learned in our society: 'To make physical science a rival of human interests is bad enough, for it forms a diversion of energy which can ill be afforded. But the evil does not stop there. The ultimate harm is that the understanding by man of his own affairs and his ability to direct them are sapped at their root when knowledge of nature is disconnected from its human function' (1927: 176). Dewey's argument is based on the direct effects of technological advance on human activity during the industrial revolution, when exploiting technology resulted in the even greater exploitation of individual labourers. His fear is that an enchantment with science might lead towards an uncritical faith in its pursuits. Further, these pursuits might be considered a virtue in themselves, rather than a contextual goal, thereby yielding an unnecessary instrumentality. He argued that 'man, a child in understanding himself, has placed in his hands physical tools of incalculable power. He plays with them like a child, and whether they work harm or good is largely a matter of accident. The instrumentality becomes a master and works fatally as if possessed by a will of its own – not because it has a will but because man has not' (175).

While contending that man's will is not powerful enough to challenge the instrumentality of science is extreme, we would be wise to anticipate and heed this warning. In effect, some entity must act to stave off the potentiality, and often this requires the resources of the state. The power and influence of applied scientific knowledge requires control. Whether that control is over the research, the technologists, or the resulting products, regulation plays an important role. Dewey (1927) recognized that certain phenomena, like those of public health or professional standards, demand public intervention. 'If I make an appointment with a dentist or doctor, the transaction is primarily between us. It is my health which is affected and his pocket book, skill and reputation. But exercise of the professions has consequences so wide-spread that the examination and licensing of persons who practice becomes a public matter' (51).

The scientific enterprise thus may be deemed public for any of a number of reasons: public funding, the impact of knowledge generation, or the expectations of research participants. When this is the case, Dewey and others have argued that the state must act on behalf of the public to mediate these transactions in anticipation of their social effects. Indeed, the government has already emerged as a participant in these transactions, as evidenced through licensure of doctors and through the concept of patents explained above. Similarly, the same forces involved in certifying physicians are at play in the case of protecting intellectual property – deference to the public drives both. Information is a very cheap commodity and trading in it would be abundant without protections. The ease by which ideas can travel creates a real disincentive to generate them. Thus, governments intervene to encourage innovation by offering exclusivity to those who develop an idea first. In the case of health care, the cures and diagnostics that are needed to maintain public health are likely to emerge more rapidly if those who create them can benefit. Patents are but one way in which the government has decided to intervene in the scientific enterprise because of the impact that technology has on the public.

Another important assumption embodied in the patent mechanism is that private industry is a critical component of the technology development and dissemination process. This is perhaps best illustrated by the peculiar intent behind the Bayh-Dole Act. The intent of this legislation is to shift assignable rights to inventions funded by government dollars. Before its adoption, NIH-funded research outcomes would be the property of the U.S. government. The Bayh-Dole Act, passed in the early 1980s, transferred the rights to invention to the universities hosting the research. Universities now own the inventions, but they are not expected to keep them. Universities have neither the ability nor the mechanisms to develop products, which are usually the form in which technologies are introduced to the public.

One might wonder why the U.S. government believes that capitalization of its investments ought to be done by private industry. This is in part due to design and in part by accident. The government is not in the practice of aiding corporate development efforts, but it does participate in basic research, known as discovery. Before a drug is developed into a therapy it will have been identified as a target through basic research techniques, and very likely that work occurred in connection with an NIH-sponsored lab, if not entirely so. No investigator would be able to acquire the funds to develop the therapy into a usable form from the government, as the government rarely sponsors anything more than basic research. Further, sources other than corporations are sparse given the cost of development (estimated to range from $300 million to $800 million [DiMasi, Hansen, and Grabowski 2003]) and the risk of failure. Those development costs are controlled, indirectly, by governmental regulatory bodies, in particular the Food and Drug Administration. The FDA, on behalf of the public, has set up safety standards and efficacy requirements that must be met for any new drug to be approved for sale.

Having identified some of the conditions by which biological sciences progress from idea to the public, it is now possible to investigate the implications of such activities. First, the circuitous process by which scientific ideas travel from general concept to public adoption requires some consideration. When an invention is ready for commercialization, private firms are usually engaged. In most cases, a university will license the technology to a company, which will then assume ownership and develop a product on its own or through the investigator's lab. The company may be an established entity, or it may be a new start-up that involves the 'scientific founder' in the process. If it is the former, the company's funds are not usually at issue; however, if it is a new start-up, which is increasingly common, the company will need to raise the necessary capital to prepare the technology for development. The company may seek venture capital, either from the market or through NIH in the form of Small Businesses Innovation Research (SBIR) or Small Business Technology Transfer (SBTT) grants. These grants are geared towards fostering the research of innovative ideas that have both public and commercial appeal. While the grants are subject to the same guidelines as any NIH funding (primarily requiring the protection of human research subjects and other ethical restrictions), they impose no commercial obligation towards the NIH. In effect, they are a type of welfare for the biotechnology industry. This money is akin to a seed round of funding through venture capital, which means that no more than a few million dollars will be granted over the first few years. Upon graduating from NIH support, the company is likely to be indistinguishable from any other and to enter the usual venture capital cycle. If it is lucky, its technology and management will drive the company to success and its products will be available publicly.

If this pattern is the norm, it raises some concerns about the public interest. If science has a public impact, which is clearly evidenced by the importance of public health to the populace and the licensure of health experts that care for it, is it adequately representing the public? Scientists set the research agenda far more effectively than any other body, given that their network is in control of the government funding agencies. Biotechnology research in particular raises three explicit public concerns that require investigation: are research agendas aligned with the public interest? does research reflect a respect for ethics considerations? and, related to the second question, will the public have access to the technologies that are commercially developed, although initiated with public funds?

Unfortunately, these are loaded questions, to which intuitive responses do not bode well for the public. More precisely, answering them from the 'ought' perspective, rather than describing how the issues that frame these questions are handled in practice, is, on first inspection, disturbing. At a glance, addressing the questions in the following way would not require great explanation, as anyone familiar with them would understand the rationale behind each answer.

Q. Should science be co-opted for corporate profit?
A. No.
Q. Does the corporate agenda drive the research agenda?
A. No.
Q. Are research agendas aligned with the interest of the public?
A. Mostly.
Q. Does research reflect a respect for ethics considerations?
A. Yes, if it is publicly funded.
Q. Will the public have access to the technologies that are commercially developed?
A. Yes.

Even the most reassuring responses above are unlikely to assuage concerns related to biological research. More disturbingly, the less contestable answers may be the ones that require the most revision. Taking the first query, one might ask what is so inherently bad about science working in the corporate interest? As discussed earlier, the economic system within which drug development fits is heavily dependent on incentives for the drug developer. The costs of bringing therapies to the public are exorbitant, and the resulting uncertain and often fleeting returns focus corporate interest on high-priced and high-demand candidate drugs. The difficulty for many is that this could translate into either cardiovascular treatments or erectile dysfunction therapies. The market determines what will or will not be developed because it is the market that will sustain or defeat any effort. In democracy, the market is composed of the public, which will choose what it will or will not consume. While it is tempting to goad the U.S. populace with the 'you got what you asked for' response, it is both petty and incorrect. I would posit (revealing a generous bias) that few individuals would ask the government to fund research on Viagra over AIDS (to be fair, Pfizer is dedicated to both). However, the disturbing truth is that markets, as a representation of the public, are not democratic, or they are perversely so at best. To vote with one's wallet is to weight votes. Those who can afford to buy products are the only ones who will have their voices heard. As such, the market-driven research agenda, a joint product of the public, the state, and the private, corporate sector is far from democratic. In this fundamental sense, democracy and science are at a crisis. Technology choices are made that are not necessarily compatible with the needs of the public, and those that are pursued represent a loss in efficiency as resources are applied to irrational projects when contrasted with public need.

The efficiency of firms and the framework of government funding and evaluation leave the public at the mercy of what Max Weber termed 'formal rationality.' The mechanics of bureaucracy have marginalized the values that once inspired the machine. The public, the state, and the private market have strived to create a model by which all can benefit, but the power of the model has dominated both the means and the ends. Basic research into a whole range of socially desirable areas is sponsored by the state, but the cost of transforming it into a commercial product is so large that only those technologies with huge expected returns are likely to command the attention of for-profit firms (arguably the only institutions capable of undertaking the translation of research into useful and used products). Despite intentions to the contrary, both FDA regulatory processes and the parallel patent system create de facto monopolies of most healthcare products. While these systems are intended to ease the burden of future development efforts, no one knows if the next development is going to be a cure for cancer or a hair growth product. As each cycle of innovation proceeds, there is concern that public and social goals, like public health, progressively fade and that others, like profit and cosmetics, take their place. The difficulty is that few would risk upsetting the system if the result might mean losing the cure for cancer.

If I am accurate in describing this pattern, then the question of whether science ought to be co-opted by industry is answered with a resounding 'no.' However I fear that the answer is an empty one. What use is independent science if its benefits never reach those who need it? Without corporate oversight and interests, one could argue that scientists would be freer to pursue the type of research that inevitably benefits humankind. The most generous reader would assume that the pure generation of knowledge is inherently beneficial to mankind, and, whichever direction scientists choose to take, they will eventually be applying the fruits of their labour to help society. Unfortunately, the idealization is one that is hard to accept. On an infinite timeline, this may be true, but so might the opposite.

While it would be most polemical to argue that this construct is the deliberate result of colluding politicians, scientists, and CEOs, that tack would be specious at best. Such effects are the result of democracy, but they reveal just one part of the range of actions available to the actors on this stage. Democracy itself is a staged governmental form – a dialectic in itself. It promotes change and encourages improvements over time. Dewey (1927) asserted that 'the unity of the democratic movement is found in an effort to remedy evils experienced in consequence of prior political institutions ... it proceeded step by step, and ... each step was taken without foreknowledge of any ultimate result, and for the most part, under the immediate influence of a number of differing impulses and slogans' (85).

Democracy is deliberate. More precisely, the leaders of a democratic state are deliberately chosen and removed by the public, ideally based on the ability of the politicians to perform on behalf of the public. As opposed to dynastic forms of political rule, there are no electoral accidents in democracy. If such is the case, one may wonder how it is that such a framework emerged to remove the public from the inner workings of the technoscientific process. In short, I believe, the appeal to efficient means of government, production, and investigation has led to a bureaucratic state that has placed too much in the hands of experts. Knowledge is incredibly specialized, and the government funding of training specialists has contributed to the isolation of expertise to disciplinary confines. Further, the many layers of American government have placed the citizenship far from their representatives.[2] These are just a few of the effects of a growing country, technological progress, and an increasingly sophisticated governmental structure.

Bridging the Public–Private Divide in Biotechnology

The Enlightenment notion that science inherently helps humanity is one that has been challenged effectively since Mannheim, and while the debate continues, an appeal to any discipline's virtues is unlikely to sway the discussion. Practical policies and institutions are likely to be the only way to bridge the public–private divide in biotechnological sciences.

To participate in and contribute to the debate, one must establish the metrics by which science in its current form can be considered beneficial to society. Externalities are at least one measure of the benefits of science. The scientific welfare offered by the state may aid the public through the jobs created by funding, as well as the educational effects. While many may contest the value of science, few would do so regarding the boons of education – it is important to remember that a considerable amount of research funding is actually used to subsidize PhD students. Also, in the best scenarios, the commercialization of the technologies will lead towards increased public health and a stimulated economy. Indeed, by business terms, government funding of research has resulted in a substantial return on investment to the public (Cutler 2004), especially through commercialization (albeit limited to those who can afford to buy these products).

Biotechnology, like nuclear research during the Cold War, is a contested arena, only in this case due to the ethical considerations that surround it. As a result, limitations have been placed on the research, outlining what is permissible and what is not. For instance, studies that involve human beings (regardless of who funds or who undertakes the research) must conform to certain standards that protect the human research subject. To protect the public, the government also does not fund research that it deems morally contestable. In the United States, for example, with research related to embryos, only procedures that will aid in the survival of an embryo is permitted. Effectively research on embryos is impossible with federal funds. These conditions have been extended to stem cell research. While some might counter that the current environment actually supports stem cell research – because stem cells are not embryos and the relevant stem cell lines were not created with federal funds – the reality is that ethical directives and processes are an important and binding constraint on U.S. government funding of some types of biotechnology research.

Policy directives and funding guidelines were once effective, prior to the rise of venture capital and private funding sources. However, as private capital has emerged, these ethical guidelines and policies became a suggestion to researchers rather than the rule. Today, while many believe that cloning is illegal and that stem cell research can only be performed on a limited number of cell lines, this is only true of federally funded efforts. Without commenting on the propriety of these innovations, we can see that the related debate is interesting. The NIH, under the direction of Harold Varmus, in the early 2000s issued statements of strong support for the technology. Representing the scientific community, the NIH, joined by over 80 Nobel laureates, appealed to the Bush administration to lift the ban on federal funds for stem cell research. At the same time, Leon Kass, the chair of the Bioethics Advisory Council to the President, advised that the research was too socially caustic to promote. A compromise was adopted, allowing research to proceed with public funds only where the research utilized already-established cell lines.

This example illustrates two points. First, it shows that governments will weigh ethical considerations in their decisions regarding research, which they should do.[3] Second, there is doubt that such considerations will have any effect in today's climate. Examining the recent history of embryo research in the United States illustrates this reality. Well before stem cell research reintroduced the spectre of the abortion debate into the medical research arena, in vitro fertilization had commanded much of the argument. In the late 1970s, Louise Joy Brown, the first 'test tube' baby, was born. The technology became the subject of intense debate. Parents who were unable to conceive lobbied for federal support, while many others felt that the research would likely result in damaging embryos or creating life that would not be given a normal lifespan (i.e., the technique would create more embryos than could practically be used). Ultimately, political decisions kept the technology from ever receiving federal funds, which worked to push the technology into the private sector. Without any governmental oversight in the United States, this technology proceeded to the clinic without any federal intervention – today, it remains a sector without any special medical regulation. Further, its movement outside of the regulated environment has allowed the technology to evolve with greater deference to technological innovation, the value systems of its customers, and the profit of practitioners than to the concerns of the public.

While many suggest the response to this problem should be increased legislation, the nature of technological progress is such that its pace is well beyond the Congress's ability to match its stride. Further, regulating science by law is so politically volatile that attempts are discouraged (drawing comparisons to Lysenkoism [Lewontin and Levins, 1976]). Until the last few decades, the public relied on federal funds both to inspire the research that would aid in health care and to limit the abuses that might arise in the process. Today, the mixture of private money, scientific autonomy, and ineffective government oversight has pushed research into any fringe that it wishes to find. While most efforts are responsible and effective, the public's place in this discourse is becoming more limited, and the trend is in need of reversal.

While there are many possible approaches that may provide solutions, the very concept that provokes the issue of conflict between science, democracy, and corporations offers a likely building ground for a solution. The flow of capital through these institutions has carried with it a directive of how to establish a framework for action. Until now, that framework has involved the government at the initial stages of funding and as a surveyor of safety and efficacy issues prior to a product's release. Corporations fill the gap in the middle, and the public fuels the government's support through taxes. Scientists choose the agenda based on their own interests and the available cash. But even Vannevar Bush (1945), the architect of postwar, government-sponsored science that puts scientific curiosity at the centre of the system, realized that there ought to be limits to the scope of publicly funded scientific research. In his words, he asserted that 'advances in science when put to practical use mean more jobs, more abundant crops, more leisure for recreation, for study, for learning how to live without deadening drudgery. Advances in science will bring higher standards of living, will lead to the prevention and cure of diseases, will promote conservation of our limited national resources, and will assure means of defence against aggression.' To address this apparent disconnect between objectives, he advocated scientific centres, run by scientists, that would allocate funds

Sociologist Alvin Gouldner questions whether any of these approaches is adequate to sustain democracy. In a trilogy of books on 'the dark side of the dialectic,' he asserts that two admonishments apply to the discussion of the separation of technical expertise, despite their interdependence, from other aspects of society. First, he posits that 'the operating personnel of the administrative, the political, and the ruling classes each develop specialized standards and skills for dealing with their own spheres, thereby making the latter less intelligible and less manageable by those in other spheres' (1982: 230). He asserts that such a process removes the ruling class, in his case the economic elite, from certain activities within society. If one were to apply this to democracy, then it is the public that cannot engage or be engaged by the multiple social spheres surrounding it. Gouldner's caution is further characterized by a distrust of control over the products of one sphere by the actors in another. 'The technical staff is governed by and subordinated to an officialdom which sets the goals, but which knows little about the technical processes used to realize them. The technical staff is alienated from the ends, the officials from the means' (255). Here a paradox arises. Because of the 'technocratic' influence on democracy, nontechnical individuals are given control of technology to serve a nontechnical public. Yet, the public requires its due, despite a lack of expertise and ability to govern the specifics of technological development.

Gouldner, Dewey, and others are adept at defining the symptoms, although very few offer effective solutions to the U.S. version of democracy. Dewey, conspicuously, refuses a prescription. Recognizing the vibrancy of the democratic state, Dewey makes no presumptions that his citizenship ought to cast greater influence than another's. It is the responsibility of the populace to work out the best path forward, despite the bureaucratic challenges anticipated by Weber and Gouldner. Even so, they offer the germ of an approach in their diagnosis. If, as they suggest, values are eclipsed by efficiency, then embedding the values (here, read as goals) of the public throughout the process of scientific and product development offers an important start.

Looking Ahead

Different people are likely to suggest different solutions to the challenges outlined in this paper, and my own attempt is likely to be construed as somewhat perverse, at first. My answer, in the biotechnology arena, is increased welfare spending in research, industry, and society. Hardly novel in itself, perhaps my allocation will be so. As I have described them, there are three key problems: the public has limited influence on the type of research pursued; privatization has removed ethical requirements in the pursuit of research; and privatization of research is necessary given the costs, but the burden of subsidizing these costs are placed on the public, despite public funds initiating research.

Considering one institution at a time, each type of welfare is directed at a specific end. Corporate subsidies offer a partial solution to all three issues. Corporations choose their research projects because of the potential return on the investment involved in the development. Responding to the market, they develop therapies based on customers' willingness to pay. Fortunately, this is not as dismal as many would think, as a number of biotechnology companies have made their fortunes through a commitment to orphan diseases (the trade-off of fewer customers is mitigated by higher prices), but on the whole, affluent customers are targeted because they will pay more for what they want. With respect to the second concern, if SBIR grants are allowed into more corporations, then some assurance can be given that any research utilizing tools or equipment bought with grant money will have to obey federal ethics standards, as outlined by the NIH. Finally, the subsidy will relax the cost incurred during regulatory approval efforts, diminishing the need to appeal to an affluent populace to recover the sunk costs involved with development.

In anticipation of one logical extension that might be offered, let me suggest that moving all research into the academic or public setting is an unlikely alternative. There will always remain products outside the realm of consumer 'need' that fit within the 'want' category. More importantly, there is some question about the capacity of the public sector to sustain the requisite capacity to innovate. For example, no university or government system could create the framework necessary to coordinate the vast number of agents involved with the drug development and approval process. In this regard, the capability and efficiency of market systems is unparalleled, and the cost of creating its mirror in the public domain is likely to be inefficient.

Research spending should also receive an increase in funding, or at least a more controlled disbursement based on the correlation between the project goals and the desires of the public. While the latter may or may not be noble, the deliberate nature of democracy requires a more active public. The argument that elected officials represent the public wears thin once we recognize that officials are often elected because individual voters may agree with the individual on a number of issues while disagreeing on others. This paradox of democratic vision almost calls for more referendums, but the logistics of such efforts are likely to be formidable. A more responsive, and perhaps more educated, officialdom may be the only effective means to that end.

Finally, to align the incentives of the public with the scientific research agenda, a single-payer healthcare system may be the best option. So long as corporate motives are tied to the whims of the affluent, then the development and accessibility of health care will rank as a low priority. Because the current system caters to this sector, a healthcare effort that commits to the welfare of the overall populace is necessary to realign the research, public, and corporate agenda. The pressures on industry to move in multiple directions based on the potential markets would fit and focus more on the largest market, the entire population, while the other opportunities would still exist. Bargaining power for determining drug costs would be squarely on the shoulders of elected representatives with public health and public need predominating. Further, if the immediate healthcare requirements of the populace are being satisfied, the development of cosmetic therapies is certainly possible and may even command a greater premium to support them.

Democracy, science, and industry form an inevitable and precariously balanced triangle. This is not necessarily by design but by necessity. Each institution represents individuals who are fundamental actors in the system, and each looks to provide some form of social benefit. While other factors are at play, this reality is often easily lost in discussions of the three. Resolving this tension is unlikely to prove simple; nor is it likely to occur rapidly when effort is applied. Yet it is the duty of every citizen to strive for a better democracy, one that reconciles private with public interests. Whether the efforts are similar to those suggested above or of a completely different variety, the only way that an empowered electorate is certain to fail to improve its own condition is to not try.

NOTES

This author owes a great debt to Eddie Glaude of Princeton University, who spent considerable time discussing the themes of this article throughout its development, especially the discussions of 'public' and 'private' distinctions.

Additionally, Jaime Staples at Harvard University graciously provided invaluable editorial suggestions. Finally, this article was developed with the support of the Jamestown Project at Yale Law School, although the essay's contents are solely the responsibility of the author.

1 Certainly, arguments can be made with respect to economic limits of trade, but if this were the case, then the transaction could be considered outside of these boundaries.
2 Town hall meetings from the eighteenth century are no longer the norm, and, while a citizen can look into the eyes of a politician on television, there is no emotional connection looking back. Inherently, the causes are not necessarily bad. Indeed, much good has come from the ability to speak to the public through one venue at one time, as with television. The internet continues to provide information to those still forming an opinion regarding elections and referendums, although *caveat emptor* to surfers.
3 This is not a comment on the morality of stem cell research; rather, an admission that the state should protect the rights of the public, and if those rights are frustrated due to ethical considerations, then such considerations should be addressed.

REFERENCES

Bush, V. 1945. *Science: The Endless Frontier.* Washington, DC: United States Government Printing Office. http://www.nsf.gov/od/lpa/nsf50/vbush1945.htm (accessed 26 April 2007).

Coase, R. 1960. The Problem of Social Cost. *Journal of Law and Economics* 3, no. 1: 1–44.

Cutler, D. 2004. *Your Money or Your Life: Strong Medicine for America's Healthcare System.* New York: Oxford University Press.

Davies, K. 2002. *Cracking the Genome: Inside the Race to Unlock Human DNA.* Baltimore: The Johns Hopkins University Press.

Dewey, J. 1927. *The Public and its Problems.* Athens, OH: Swallow Press.

DiMasi, J.A., R.W. Hansen, and H.G. Grabowski. 2003. The Price of Innovation: New Estimates of Drug Development Costs. *Journal of Health Economics* 22: 151–85.

Gouldner, A. 1982. *The Dialectic of Ideology and Technology.* New York: Oxford University Press.

Jonsen, Albert. 1998. *The Birth of Bioethics.* London: Oxford University Press.

Lewontin, Richard, and Richard Levins. 1976. The Problem of Lysenkoism. In *The Radicalization of Science*, edited by Hilary Rose and Steven Rose, 32–64. London: Macmillan Press.
Merton, R. 1968. The Matthew Effect in Science. *Science* 159, no. 810: 56–63.
Rabinow. P. 1999. *French DNA: Trouble in Purgatory.* Chicago: University of Chicago Press.

Contributors

Zaheer Baber is Professor of Sociology at the University of Toronto. He received his BSc from Aligarh University, his MA and MPhil from the University of Delhi, and his PhD from the University of Toronto. His areas of specialization are science, technology, and society; the sociology of knowledge; comparative and historical sociology; as well as social theory. His publications include *The Science of Empire: Scientific Knowledge* and *Civilization and Colonial Rule in India* (1996). He is also contributing editor of the *Bulletin of Science, Technology and Society*; associate editor of the *Asian Journal of Social Science;* and member of the Book Review Editorial Board of *Critical Sociology.*

Peter Cook is currently the chief executive of CO2CRC, a major research consortium of universities, industry, and government institutions. The centre, formed in 2003, seeks to decrease greenhouse gas emissions through the development and use of new technologies, particularly geosequestration. It is one of the largest programs of its type in the world. Dr Cook is a co-ordinating lead author for the 2005 IPCC special volume on carbon dioxide capture and storage technologies. He has held senior government, academic, and research positions in the UK, Australia, and France. Dr Cook is the author or co-author of more than 130 publications and has given many invited international lectures and keynote addresses on resource and environmental sustainability issues. He has been awarded the CBE by Britain for his services to science, the Public Service Medal by Australia, the APPEA Lewis G. Weeks Gold Medal, the Coke Medal by the Geological Society of London, the Alexander von Buch Medal by the Geological Society of Germany, and the French Order of Merit. Dr Cook is a fellow of the Australian Academy of Technological Sciences and Engineering.

Leon Harold Craig is Professor Emeritus of Political Science, having taught at the University of Alberta in Edmonton for thirty-three years. He received his BA from the University of Washington and his PhD from the University of Illinois. His main field of interest is political theory, along with traditional political philosophy, with a particular focus on Plato, Shakespeare, Hobbes, Rousseau, and Nietzsche. Throughout his career he taught courses in philosophy of science and social science. Dr Craig's most recent major publications are a book-length study of Plato's *Republic* entitled *The War Lover* (1994) and a book-length study of Shakespeare's political philosophy, focusing mainly on *Macbeth* and *King Lear*, entitled *Of Philosophers and Kings* (2001). He is currently finishing a critical re-examination of Hobbes's *Leviathan*, with chapters relating it to Melville's *Moby-Dick* and Conrad's *Heart of Darkness*.

Rahul K. Dhanda is a founding member and senior fellow of the Jamestown Project at Yale Law School. Mr Dhanda began his life sciences career at Harvard Medical School, developing technologies for rapid DNA-based cancer screening. After leaving the academic laboratory, he moved to private industry, where he has worked for numerous companies. Currently, he is a part of Boston Scientific's urology division's marketing management team. In past positions, he was founder and chair of the Bioethics Advisory Group of the Massachusetts Biotechnology Council, as well as a research fellow at Harvard University's Kennedy School of Government. Mr Dhanda was also the director of the bioethics program at Interleukin Genetics and a member of the company's business development division. He has worked as an independent consultant to many life science firms, is the author of several publications, holds numerous patents and applications, and has recently written a book that addresses the crossroads between industry and bioethics, entitled *Guiding Icarus: Merging Bioethics with Corporate Interests* (2002). He has a BA in history from Wesleyan University and an MBA from MIT's Sloan School of Management.

Robert Frodeman is chair of the Department of Philosophy and Religious Studies at the University of North Texas and is director of the Humanities Policy Project and the New Directions Initiative. He specializes in environmental philosophy, science policy, and interdisciplinary research methodology. He has held positions at the University of Texas and the University of Tennessee and has consulted for the U.S. Geological Survey during the last ten years. In 2001–2 he was the Hennebach Professor of the Humanities at the Colorado School of Mines. He is editor of *Earth Matters* (2000) and *Rethinking Nature* (2004), and author of *Geo-Logic: Breaking Ground between Philosophy and the Earth Sciences* (2003) and of numerous articles. He is particularly interested in exploring the nature of scientific knowledge and the contribution that social scientific and humanistic perspectives play in public policy debates.

Chris Gerrard, a Canadian national, is head economist in the Independent Evaluation Group of the World Bank in Washington, DC. He has a BSc in economics and mathematics from the University of Saskatchewan, an MPhil in economics from Oxford University, and a PhD in agricultural economics from the University of Minnesota. From 2001 to 2004, he was deputy team leader and co-author of a major IEG evaluation on the World Bank's involvement in global programs, which included the evaluation of the Consultative Group on International Agricultural Research. Prior to joining the World Bank in 1994, he was a professor of economics at the University of Saskatchewan, as well as head of the economics department for four years. His main research interests have been macroeconomic and sectoral adjustment in developing countries, agricultural policy and institutional reform, decentralization, natural resource management, and the theory and practice of collective action. His principal regional focus has been sub-Saharan Africa, and more recently central and eastern Europe and the former Soviet Union.

David H. Guston is Professor of Political Science and Associate Director of the Consortium for Science, Policy, and Outcomes at Arizona State University. He is also the North American editor of the international peer-reviewed journal *Science and Public Policy*. In 2002, his book *Between Politics and Science* (2000) won the Don K. Price Award from the American Political Science Association for best book in science and technology policy. Born in Oakland, NJ, he attended Yale University, where he earned his AB in 1987. He received his PhD in political science from the Massachusetts Institute of Technology in 1993. He has also worked at the U.S. congressional Office of Technology Assessment and the National Academy of Sciences, and is a fellow of the American Association for the Advancement of Science. His current research interests include science and technology policy, the role of experts in the policy process, and the societal implications of nanotechnology.

Sarah Hartley is both an adjunct professor in the Department of Political Science at Simon Fraser University and adviser to Genome British Columbia on the ethical, social, legal, economic, and environmental aspects of genomics research. She obtained her PhD in political science and environmental studies at the University of Toronto in 2005. Her doctoral thesis compared government responses to the environmental risks of agricultural biotechnology in Canada and the UK. Her interests include science, values, ethics, and public policy.

Deborah R. Hutchinson, a research geologist with the U.S. Geological Survey, Woods Hole Science Center, Massachusetts, currently pursues a broad range of gas hydrate studies using theoretical, geophysical, and drilling experiments in locations as diverse as U.S. waters and Lake Baikal, Siberia. Other professional interests include Quaternary and Cenozoic studies of large inland lakes, the crustal structure of passive continental margins, and the tectonic evolution of continental rifts. She holds an MSc in geology from the University of Toronto and a PhD in geological oceanography from the University of Rhode Island, and has spent her career with the U.S. Geological Survey in Woods Hole, Massachusetts. She recently rotated out of the science leadership and management position of Chief Scientist at the USGS Woods Hole Science Center.

Ian C. Jarvie, in the Department of Philosophy, Faculty of Arts at York University, Toronto, received both his BSc Econ and his PhD (in scientific method) from the University of London (LSE), where he was a student of Sir Karl Popper. His specialty within the philosophy of science is philosophy of the social sciences. His other specialties are aesthetics and the sociology of the mass media, especially film. He is managing editor of the scholarly quarterly *Philosophy of the Social Sciences*. He has taught at York University for almost forty years, where he is Distinguished Research Professor. His most recent book is *The Republic of Science: The Roots of Popper's Social View of Science, 1935–1945* (2001). He is a Fellow of the Royal Society of Canada.

Alan McHughen, FACN, is a biotechnology specialist at the University of California Riverside, and is a public sector educator, scientist, and consumer advocate. After earning his doctorate at Oxford University and lecturing at Yale University, he spent twenty years at the University of Saskatchewan before joining the University of California, Riverside.

A molecular geneticist with an interest in crop improvement, he has helped develop U.S. and Canadian regulations covering the environmental release of plants with novel traits. Having developed internationally approved commercial crop varieties using both conventional breeding and genetic engineering techniques, he has first-hand experience with issues from both sides of the regulatory process, covering both recombinant DNA and conventional breeding technologies. His award-winning book *Pandora's Picnic Basket: The Potential and Hazards of Genetically Modified Foods* (2001) uses understandable, consumer-friendly language to explode the myths and explore the genuine risks of genetic modification technology.

Ross McKitrick, an associate professor in the Department of Economics at the University of Guelph, holds a PhD in economics from the University of British Columbia. He was appointed assistant professor in the Department of Economics at the University of Guelph in 1996 and associate professor in 2000. In the fall of 2002 he was appointed as a senior fellow of the Fraser Institute. His area of specialization is environmental economics and policy analysis. He has published numerous scholarly articles in academic journals as well as commentaries in newspapers and other public forums. His research on climate change has been published in leading economics and science journals and was the subject of a front-page article in the *Wall Street Journal* in February 2005. His 2002 book *Taken by Storm: The Troubled Science, Policy and Politics of Global Warming,* co-authored with Christopher Essex, won the $10,000 Donner Prize (best book on Canadian public policy) and was a finalist for the 2003 Canadian Science Writers' Association Book Prize. Professor McKitrick has made invited academic presentations in Canada, the United States, and Europe, as well as professional briefings to two Canadian parliamentary committees and to government staff at the U.S. Congress and Senate.

Gordon McOuat teaches sciences studies and is the director of the History of Science and Technology Programme and an associate professor in the Contemporary Studies Programme at the University of King's College and Dalhousie University. He has been a research fellow at Cambridge University and senior research fellow at the Dibner Institute for the History of Science and Technology, MIT. He is also vice-president of the Canadian Society for the History and Philosophy of Science (CHSPS). His work covers the history, philosophy, and politics of classification systems and natural kinds, and the philosophy of science. Recent publications include *The Politics of Natural Kinds* (2003) and *The Logical Systematist: Archives of Natural History* (2003).

Carl Mitcham currently serves as Professor of Liberal Arts and International Studies at the Colorado School of Mines, where, in addition, he directs the Hennebach Program in the Humanities. He is also a faculty associate of the Center for Science and Technology Policy Research at the University of Colorado, Boulder, and adjunct professor at the European Graduate School in Saas Fee, Switzerland. His publications include *Thinking through Technology* (1994), *Visions of STS: Counterpoints in Science, Technology, and Society Studies* (2001), and the four-volume *Encyclopedia of Science, Technology, and Ethics* (2005), and *Science, Technology, and Ethics: An Introduction* (forthcoming).

Peter W.B. Phillips is Professor of Political Studies and a professional associate in Agricultural Economics and Management and Marketing at the University of Saskatchewan. He is also professor at large with the Institute for Advanced Studies, University of Western Australia. He received his MSc Econ and PhD at the London School of Economics. In the past he was a senior policy adviser in government in Canada and developed and fulfilled a five-year NSERC/SSHRC chair in Managing Knowledge-Based Agri-food Development at the University of Saskatchewan. His research concentrates on issues related to transformative technologies. Dr Phillips was director of the University of Saskatchewan's College of Biotechnology (2002–5) and a member of the Canadian Biotechnology Advisory Committee (1999–2007). He is currently a senior research associate with the Estey Centre for Law and Economics in International Trade and lead or co-investigator on seven major interdisciplinary research programs related to governing knowledge-based growth. He is lead editor or author of six recent books, including *Governing Transformative Technological Innovation: Who's in Charge?* (2007).

Jene M. Porter is Professor Emeritus in the Department of Political Studies, University of Saskatchewan. Professor Porter's areas of interest include the history of political philosophy, modern political philosophy, and the philosophy of social sciences. He received his PhD from Duke University, and after a posting at Drury College he moved to the University of Saskatchewan, where his area of interest was political philosophy. He is author of numerous articles and his books include *Political Philosophy: The Search for Humanity and Order* (with John Hallowell, 1997) and *Classics in Political Philosophy* (ed., 1989).

Eric S. Sachs is the director of scientific affairs for Monsanto Company, St Louis, Missouri. Dr Sachs joined Monsanto in 1978 and is widely known for his strong commitment to biotechnology stewardship, communications, education, and industry responsibility. He is the founder and first chairperson of the Agricultural Biotechnology Stewardship Technical Committee, an important coalition of biotechnology companies committed to addressing scientific issues central to responsible stewardship of agricultural systems. In his previous position as business director and team leader, Dr Sachs's responsibilities included the development, commercialization, and global expansion of Monsanto's YieldGard Bt corn technology. He regularly collaborates with university and government scientists, advocates science-based policy to regulatory agencies, and interacts with consumer media on a range of biotechnology issues and topics. As a leader and communicator, he successfully uses his knowledge of science and biotechnology, experience, and passion to demystify the science of biotechnology, to communicate the safety and benefits of agricultural biotechnology, and to address questions from the public related to this important technology.

Steven Shapin is Franklin L. Ford Professor of the History of Science at Harvard University. Dr Shapin received his BA from Reed College and his MA and PhD in history and sociology of science from the University of Pennsylvania. He was lecturer and reader at the Science Studies Unit, University of Edinburgh, and Professor of Sociology at the University of California, San Diego. He is the author of *A Social History of Truth: Civility and Science in Seventeenth-Century England* (1994), *The Scientific Revolution* (1996), and co-author of *Leviathan and the Air-Pump: Hobbes, Boyle, and the Experimental Life* (1985). He has written numerous papers on the history and sociology of science. His interests have included the history of science in seventeenth- to nineteenth-century Britain, the sociology of knowledge, and, specifically, the sociology of scientific knowledge, work and skills in relation to scientific practice, and aspects of the scientific role and scientific authority in the twentieth and twenty-first centuries. He is finishing work on a book provisionally titled *Science as a Vocation: Scientific Authority and Personal Virtue in Late Modernity*.

Grace Skogstad is Professor of Political Science at the University of Toronto. Her current work examines the impact of the World Trade Organization on countries' policy autonomy and its implications for democracy. She is also studying trade disputes arising from the World Trade Organization's SPS Agreement and transatlantic divergence in the food safety and biotechnology regulatory frameworks of the European Union, Canada, and the United States. Her books include *The Politics of Canadian Agricultural Policy-Making* (1987), *Agricultural Trade: Domestic Pressures and International Tensions* (with Andrew Cooper, 1990), and *Canadian Federalism: Performance, Effectiveness, and Legitimacy* (with Herman Bakvis, 2002, 2007). Professor Skogstad earned her BA and MA from the University of Alberta and her PhD from the University of British Columbia.

Larry Stewart is currently Professor of History in the College of Arts and Science at the University of Saskatchewan. His credits include being former editor of the *Canadian Journal of History/Annales canadiennes d'histoire*. He has written *The Rise of Public Science: Rhetoric, Technology and Natural Philosophy in Newtonian Britain, 1660–1750* (1992), *Practical Matter: Newton's Science in the Service of Industry and Empire 1687–1851* (with Margaret C. Jacob, 2005), and articles in leading journals. His special area of interest is the history of science in the seventeenth and eighteenth centuries.

Richard S. Williams, Jr, a senior research geologist with the U.S. Geological Survey, Woods Hole Science Center, Massachusetts, specializes in using airborne and satellite remote sensing technology to study changes in glaciers from a global perspective. Of special interest is Iceland, where he and Oddur Sigurðsson have been studying changes in glaciers in response to climate warming. He has BS and MS degrees from the University of Michigan and a PhD degree from Pennsylvania State University, all in geology. In addition to his work for the USGS, Dr Williams is an adjunct senior scientist, Woods Hole Research Center, and vice chairman emeritus, Committee for Research and Exploration, National Geographic Society. He has authored more than three hundred journal articles, book chapters, books, abstracts, and maps. Dr Williams is co-editor of the eleven-volume USGS Professional Paper 1386 series, *Satellite Image Atlas of Glaciers of the World*. He is co-leader of a project to map the dynamic cryospheric coast of Antarctica with sequential satellite images, *Coastal-Change and Glaciological Maps of Antarctica* (USGS

Map Series I-2600). Dr Williams has two glaciers in Antarctica named for him. He is co-author of the third edition of *Physical Geography: The Global Environment* (2004) and co-author (with Oddur Sigurðsson) of *Icelandic Ice Mountains*, an annotated, illustrated English translation of Sveinn Pálsson's 1795 'Glacier Treatise' (2004).